Agroforestry Systems

Agroforestry Systems

The Role of Trees in Ecosystem Services—A Special Issue in Collaboration with the 4th World Congress on Agroforestry

Special Issue Editors

Scott X. Chang

Yi Cheng

MDPI • Basel • Beijing • Wuhan • Barcelona • Belgrade

Special Issue Editors

Scott X. Chang
University of Alberta
Canada

Yi Cheng
Xianlin University District
China

Editorial Office
MDPI
St. Alban-Anlage 66
4052 Basel, Switzerland

This is a reprint of articles from the Special Issue published online in the open access journal *Forests* (ISSN 1999-4907) in 2019 (available at: https://www.mdpi.com/journal/forests/special_issues/agroforestry_ecosystem).

For citation purposes, cite each article independently as indicated on the article page online and as indicated below:

LastName, A.A.; LastName, B.B.; LastName, C.C. Article Title. *Journal Name* **Year**, *Article Number*, Page Range.

ISBN 978-3-03928-164-0 (Pbk)
ISBN 978-3-03928-165-7 (PDF)

Cover image courtesy of Scott X. Chang.

Contents

About the Special Issue Editors

Scott X. Chang (Professor of Forest Soils and Nutrient Dynamics) received his BS.c. from Zhejiang Agricultural University, MSc from the Chinese Academy of Sciences, and Ph.D. from the University of British Columbia. He has held academic positions in New Zealand and the United States prior to taking up his current position at the University of Alberta. His main research interests are in forest soils, soil nutrient cycling and plant nutrition. He served as Chair of the Soil Fertility and Plant Nutrition Commission of the International Union of Soil Science; President of the Association of Chinese Soil & Plant Scientists in North America; Chair for the Forest, Range and Wildland Soils Division of the Soil Science Society of America; and Chair of the Alberta Soil Science Workshop. He has served as an associate/guest editor or editorial board member for *Biology and Fertility of Soils*; *Pedosphere*; *Soil Science Society of America Journal*; *Canadian Journal of Soil Science*; *Journal of Environmental Quality*; *Environmental Science and Pollution Research*; *Forests*; and *Forest Ecology and Management*. He is a fellow of the Soil Science Society of America, the American Society of Agronomy, and the Canadian Society of Soil Science.

Yi Cheng is an Associate Professor at Nanjing Normal University. He is a Subject Editor of the *Journal of Agro-Environment Science*, published by the Chinese Society of Agro-Ecological Environment Protection, and is an Editorial Board Member of the *Journal of Isotopes* published by the Chinese Isotope Society. He has received funding for three projects from the National Natural Science Foundation of China, two projects from the National Key Research and Development Program of China, and one project from EuroChem Agro GmbH in Germany. He has published more than 30 peer-reviewed papers in international journals, including first author publications in *Earth-Science Reviews*, *Soil Biology and Biochemistry*, *Geoderma*, *Biology and Fertility of Soils*, and *Plant and Soil*. He has also served the scientific community by reviewing for over 30 international journals, such as *Plant and Soil*, *Geoderma*, *Soil & Tillage Research*, *Biology and Fertility of Soils*, *Agriculture, Ecosystems and Environment*, and *Journal of Environmental Quality*. His main research interests include (1) soil nitrogen cycling and its ecological and environmental effect; (2) soil use and restoration; and (3) the control of non-point source pollution and N_2O emission.

Preface to "Agroforestry Systems"

Agroforestry has widely been considered a more sustainable agricultural system than conventional agriculture and forestry, conserving biodiversity and enhancing the provision of ecosystem services without compromising production function. Agroforestry systems can provide multiple ecosystem services, such as carbon sequestration, biodiversity conservation, maintenance of soil fertility, prevention of soil erosion, improvement of water infiltration, reduction in wind speed, pest control, and provisioning of pollination. Therefore, it is of great interest to understand the kinds of ecosystem services an agroforestry system provides in specific locations or under certain ecological conditions. Studies to understand whether or not the services provided by agroforestry systems are enhanced by management practices, and to identify factors that affect the provision of ecosystem services in agroforestry systems, are needed. In addition, in certain circumstances agroforestry systems can have negative impacts on ecosystem services. Thus, identifying the best management practices that allow the simultaneous provisioning of multiple ecosystem services is still a challenge for agroforestry. In this special issue, a total of 10 papers have been selected to provide insight into the field of ecosystem services and environmental benefits from agroforestry practices, including biomass production, soil fertility enrichment, carbon sequestration, and pollination.

Three of the papers in this special issue deal with biomass production and soil nutrient cycling as influenced by agroforestry management practices. The paper by Joslin et al. [1] examines growth and nutrient budgets over two rotations in a slash-and-mulch improved-fallow agroforestry system. They found that the improved-fallow slash-and-mulch system increased the rate of biomass accumulation by 41%–64% in secondary succession compared to primary succession. Their results further revealed that phosphorus (P) + potassium (K) fertilization increased biomass accumulation by 9%–24% in secondary succession. Their study suggests that nutrient accumulation through biomass production was adequate to replace nutrients exported via crop root and timber harvest. Manimel Wadu et al. [2] explicitly investigated how long-term land use between cropland and forestland in shelterbelt systems affect soil P status. Their research showed that the difference in P status between cropland and forestland in shelterbelt systems was dependent on the site location, while test soil P was mainly governed by a soluble inorganic P that was associated with Fe and Al in both cropland and forestland. Their research provides insights into the links between test soil P, P fractions and P sorption properties in different land uses in shelterbelt systems. Their research highlights that P management needs to be land-use-type-specific for shelterbelt systems. The paper by Dan et al. [3] focused on the responses of gross rates of soil N transformation to temperature changes in a subtropical acidic coniferous forest soil. They revealed how temperature change affects soil N availability and NO_3^- retention capacity through changing soil gross N transformation rates using an ^{15}N-tracing technique. These studies greatly improved our understanding of how agroforestry systems regulate soil fertility and biomass production.

Another three papers included in this special issue addressed economic and the ecological characteristics, food supply and water regulation of agroforestry systems. The paper by Park et al. [4] assessed the diverse roles of home gardens and their sustainable management for livelihood improvement in West Java, Indonesia. They observed that larger landholdings had a significantly present higher net value than smaller landholdings when home gardens were dominated by fruit tree species. They further found that species richness, species diversity, and carbon stock did not differ among different types and sizes of home gardens in West Java. They highlight that a multi-layered and diverse species composition is considerably beneficial for the sustainable management of home gardens in terms of income generation and against urbanization and commercialization in West Java, Indonesia. Wagner et al. [5] studied ecosystem services provided by different shade tree species as perceived by farmers, and discussed possible factors influencing these perceptions. Their results indicated that food, fodder, and fuelwood emerged as the most important ecosystem services, and factors such as elevation, gender, and farmer group affiliation had negligible influence on farmers' perceptions. They found that the perception of the density of tree species as providing these three ecosystem services was significantly higher for farmers prioritizing these services as compared to farmers that did not consider all three ecosystem services in their top three priorities. Their results highlight the importance of understanding the factors underlying farmers' management decisions before recommending shade tree species. The third paper in this group, by Wu et al. [6], used stable isotope methods (water δ^2H and δ^{18}O and leaf δ^{13}C) to investigate plant water-absorbing patterns and water-use efficiency in a monocultural rubber plantation, and in an agroforestry system of rubber trees (H. brasiliensis) and sharp-leaf galangal (A. oxyphylla). They reported the occurrence of plant hydrologic niche segregation, whereas the water-use efficiency of rubber trees in this agroforestry system suggested that competition for water was weak. They suggested that the weak belowground competition between rubber trees and sharp-leaf galangal may benefit their long-term association in the intercropping system.

It is noteworthy that three review papers and one modeling related paper are included in this special issue. The review paper by Mayrinck et al. [7] summarized the currently available research-based knowledge regarding shelterbelt agroforestry systems to improve the understanding of the climate change mitigation potential of planted shelterbelts throughout Canada and the world. Their research showed that as trees in shelterbelts continue to grow, they are able to reduce and offset a significant portion of the carbon released from agricultural practices, while providing several other social and environmental benefits. In the modeling paper, Russell et al. [8] used CENTURY, a process-based model of plant–soil nutrient cycling, in an experimental mode, to evaluate the effects of different factors on carbon stocks in soil and biomass in monocultures (annuals or trees) and agroforestry systems. They indicated that increasing temperature and soil sand content had relatively small effects on biomass carbon accumulation. They concluded that the inclusion of trees with traits that promote carbon sequestration such as lignin content, along with the use of best management practices, resulted in the greatest carbon storage among the simulated agricultural systems. Another review paper by Marais et al. [9] summarized the current state of knowledge on natural capital accounting and analyzed how such an approach may be effectively applied to demonstrate the farm-scale value of agroforestry assets. Their review indicated that, with the further development of conceptual models to support existing tools and frameworks, a natural capital approach could be usefully applied to improve decision-making in agroforestry at the farm scale. Finally, Bentrup et al. [10] synthesized information on how temperate agroforestry systems influence insect pollinators and their pollination services, with a particular focus on the role of trees and shrubs. They proposed that agroforestry practices, with the appropriate use of species and management, can better support pollinator conservation than homogenous fields and farms without agroforestry.

As guest editors of this special issue, we thank the authors for their valuable contributions, and the reviewers for the constructive comments that improved the quality of the papers published in this special issue. The timely support of the editorial office helped to make this special issue a success.

Scott X. Chang and Yi Cheng
Guest Editors

References

1. Joslin, A.H.; Vasconcelos, S.S.; de Assis Oliviera, F.; Kato, O.R.; Morris, L.; Markewitz, D. A slash-and-mulch improved-fallow agroforestry system: Growth and nutrient budgets over two rotations. *Forests* **2019**, *10*, 1125.
2. Manimel Wadu, M.C.W.; Ma, F.X.; Chang, S.X. Phosphorus availabilities differ between cropland and forestland in shelterbelt systems. *Forests* **2019**, *10*, 1001.
3. Dan, X.Q.; Chen, Z.X.; Dai, S.Y.; He, X.X.; Cai, Z.C.; Zhang, J.B.; Müller, C. Effects of changing temperature on gross N transformation rates in acidic subtropical forest soils. *Forests* **2019**, *10*, 894.
4. Park, J.H.; Woo, S.Y.; Kwark, M.J.; Lee, J.K.; Leti, S.; Soni, T. Assessment of the diverse roles of home gardens and their sustainable management for livelihood improvement in west Java, Indonesia. *Forests* **2019**, *10*, 970.
5. Wagner, S.; Rigal, C.; Liebig, T.; Mremi, R.; Hemp, A.; Jones, M.; Price E.; Preziosi, R. Ecosystem services and importance of common tree species in coffee-agroforestry systems: local knowledge of small-scale farmers at Mt. Kilimanjaro, Tanzania. *Forests* **2019**, *10*, 963.
6. Wu, J.; Zeng, H.H.; Chen, C.F.; Liu, W.J.; Jiang, X.J. Intercropping the sharp-leaf galangal with the rubber tree exhibits weak belowground competition. *Forests* **2019**, *10*, 924.
7. Mayrinck, R.C.; Laroque, C.P.; Amichev, B.Y.; Van Rees, K. Above- and below-ground carbon sequestration in shelterbelt trees in Canada: A review. *Forests* **2019**, *10*, 922.
8. Russell, A.E.; Mohan Kumar, B. Modeling experiments for evaluating the effects of trees, increasing temperature, and soil texture on carbon stocks in agroforestry systems in Kerala, India. *Forests* **2019**, *10*, 803.
9. Marais, Z.E.; Baker, T.P.; O'Grady, A.P.; England, J.R.; Tinch D.; Hunt, M.A. A natural capital approach to agroforestry decision-making at the farm scale. *Forests* **2019**, *10*, 980.
10. Bentrup, G.; Hopwood, J.; Lee Adamson, N.; Vaughan, M. Temperate agroforestry systems and insect pollinators: A review. *Forests* **2019**, *10*, 981.

 *forests*

Article

Modeling Experiments for Evaluating the Effects of Trees, Increasing Temperature, and Soil Texture on Carbon Stocks in Agroforestry Systems in Kerala, India

Ann E. Russell [1],* and B. Mohan Kumar [2]

[1]   Department of Natural Resource Ecology & Management, Iowa State University, Ames, IA 50011, USA

[2]   College of Forestry, Kerala Agricultural University, KAU PO, Thrissur, 680656 Kerala, India; bmkumar.kau@gmail.com

*   Correspondence: arussell@iastate.edu; Tel.: +1-515-294-5612

Received: 16 August 2019; Accepted: 9 September 2019; Published: 14 September 2019

Abstract: *Research Highlights:* Agroforestry systems in the humid tropics have the potential for high rates of production and large accumulations of carbon in plant biomass and soils and, thus, may play an important role in the global C cycle. Multiple factors can influence C sequestration, making it difficult to discern the effect of a single factor. We used a modeling approach to evaluate the relative effects of individual factors on C stocks in three agricultural systems in Kerala, India. *Background and Objectives:* Factors such as plant growth form, management, climate warming, and soil texture can drive differences in C storage among cropping systems, but the relationships among these factors and their effects are complex. Our objective was to use CENTURY, a process-based model of plant–soil nutrient cycling, in an experimental mode to evaluate the effects of individual factors on C stocks in soil and biomass in monocultures (annuals or trees) and agroforestry systems. *Materials and Methods:* We parameterized the model for this region, then conducted simulations to investigate the effects on C stocks of four experimental scenarios: (1) change in growth form; (2) change in tree species; (3) increase in temperature above 20-year means; and (4) differences in soil texture. We compared the models with measured changes in soil C after eight years. *Results:* Simulated soil C stocks were influenced by all factors: growth form; lignin in tree tissues; increasing temperature; and soil texture. However, increasing temperature and soil sand content had relatively small effects on biomass C. *Conclusions:* Inclusion of trees with traits that promoted C sequestration such as lignin content, along with the use of best management practices, resulted in the greatest C storage among the simulated agricultural systems. Greater use and better management of trees with high C-storage potential can thus provide a low-cost means for mitigation of climate warming.

Keywords: agroforestry systems; carbon sequestration; climate change; cropping system; growth form; lignin

---

1. Introduction

Deforestation, land degradation, and other changes in land use resulted in estimated emissions of 0.9 ± 0.5 Pg C per year from 2005 to 2014, thus contributing to the 4.4 (± 0.1) Pg per year increase in atmospheric C [1]. Reforestation is a major strategy for reducing atmospheric CO_2 because trees take up CO_2 from the atmosphere and sequester it in the long-living tissues of woody biomass. Thus, maximizing forest cover and limiting agriculture's impact on atmospheric C play critical roles in limiting global warming to the 1.5 °C increase specified in the Paris Agreement [2]. Agriculture is also expected to play a role in climate change mitigation, but this strategy is limited to augmenting soil C stocks (0–40 cm depth) globally by 0.4% per year, which is indeed an aspirational goal [3,4].

This presents a dilemma in choosing between land use for monocultures of trees that store more C and monocultures of annual crops that store less C but provide major food staples. Agroforestry systems, in which agricultural crops are grown with trees (www.icraf.org), offer a mitigation solution that meets the needs for both food provisioning via annual crops and high C storage in trees. The C storage potential is especially high in the humid tropics where these systems can be very productive, and which also provide multiple products [5,6].

With >181 million ha of arable lands [7], India has considerable potential for soil C sequestration through agricultural mitigation activities [8]. Indian soils are estimated to contain soil C stocks of 20–25 Gt (0–1 m) [3]. Apart from soil C stocks, tree-based systems often contain significant pools of biomass carbon [9]. Agroforestry is thus one of the cardinal strategies for augmenting terrestrial C pools. India is also home to an array of agroforestry systems, e.g., tropical homegardens. Indeed, the South Asia region is often regarded as the "cradle of agroforestry" [10].

Although agroforestry systems abound in India, it is difficult to predict their potential for C storage, in comparison with other options that farmers may choose, such as monocultures of annual crops, palms, or trees. This is not surprising, given the uncertainties in the terrestrial C budget in general [2]. In agroforestry systems, many factors can influence C cycling and, thus, introduce variability in C stocks in biomass and soil. For example, site characteristics such as soil and land-use types, species involved, stand age, and management practices are major determinants of C stocks in soil and biomass in agroforestry systems [9]. Moreover, the relationships and feedbacks among the various factors are complex, rendering it difficult to discern the relative contribution of each factor to C storage. The use of process-based modeling can address this issue because it provides a holistic framework for evaluating a single factor at a time [11].

We used CENTURY, a general model of plant–soil nutrient cycling [12], integrated with field-based measurements, to evaluate the capacity of agroforestry systems to sequester C, in comparison with two other main types of agricultural systems in central Kerala, India. In this context, we addressed the following questions and their corresponding hypotheses:

1.1. How Does C Storage in Agroforestry Systems Compare with that in Monocultures of Annual Crops or Trees?

Certain combinations of plant species that are grown together in agroforestry systems demonstrate complementary patterns in their resource use and, thus, have higher productivity and C storage [9]. In general, the degree of complementarity and its effect on C storage depends on environmental conditions, in combination with the functional traits of the species present in the system [13]. In this study, we focused on the functional trait of growth form. Agroforestry systems in central Kerala often contain a wide diversity of growth forms, including grasses, annual crops, root crops, vines, palms, deciduous trees, and evergreen trees [14]. In our modeled comparisons of three agricultural systems, we considered two growth forms—annuals and trees. Simulated agroforestry systems contain both growth forms, whereas monocultures contain either an annual or tree crop.

We hypothesized that C stocks in both biomass and soil will increase with the proportion of trees in the agricultural system, if factors other than growth form are held constant. The rationale was that woodiness drives biomass C storage because perennial trees have a greater proportion of long-lived, non-harvested tissues that can accrue greater biomass than annual crops. Trees also tend to decrease soil bulk density and produce more organic matter (OM) that provides C inputs to soil, thereby ameliorating soil C stocks [15]. In addition, decomposition of soil organic matter (SOM) is generally slower in less disturbed, perennial systems than in annual cropping systems [16], which would also promote soil C storage. Nevertheless, we emphasize that agroforestry systems can, indeed, store large amounts of C relative to tree monocultures [9]. In both agroforestry systems and tree monocultures however, C storage depends on multiple factors, including traits of the planted species, their complementarity in resource use, and management practices such as fertilization, irrigation, and stocking density. Thus, in field-based comparisons of C storage among agricultural systems across sites, it is difficult to control for factors that may be confounded with growth form and, hence, it is challenging to evaluate the effect of woodiness alone.

1.2. To What Extent Does Variation in Traits among Tree Species Influence C Stocks?

If land is to be dedicated to trees as a mitigation strategy, it makes sense to maximize this approach and guide the design of agricultural systems by increasing our understanding of the tree traits that influence C storage. From a biogeochemical perspective, traits that differ among tree species can differentially influence production of OM and its turnover and decomposition, thus affecting the accumulation of biomass and soil C [17]. Carbon storage would likely vary directly with a trait such as growth potential, given that other traits were similar among species. We focused on a trait for which the outcome would be less obvious—lignin concentration of above- and belowground plant tissues.

We hypothesized that soil C stocks will increase with increasing tissue lignin content. The rationale is that decomposition of litter can decline with the lignin content of plant tissues [18], such that OM will accrue in soil. We also hypothesized that biomass C will be lower in species with higher lignin contents in tissues because plant growth can decline with increasing lignin contents in plant tissues [19,20].

1.3. Will Increasing Temperatures Influence C Stocks in Tree-Based Systems?

With climate warming being unequivocal [21], it is critical to evaluate the effects of increased temperature on C stocks in biomass and soil, given the uncertainties regarding the effects of higher temperatures on C dynamics, especially in lower latitudes [22,23]. While numerous aspects of climate change affect C cycling processes in agricultural systems, we focused on rising temperatures in this study.

We hypothesized that increasing temperatures will have a relatively small effect on biomass C stocks in a humid tropical setting. The rationale is that higher temperatures prevalent in lower latitudes are already optimal for plant growth [24], such that small increases in temperature will not affect plant productivity and turnover. However, if temperatures surpass the optimum, this increase may reduce plant growth and increase mortality, thereby reducing biomass C. We hypothesized that increasing temperatures will reduce soil C via the effect of increasing decomposition of SOM [25].

1.4. What is the Effect of Soil Texture on C Stocks?

We included this modeling scenario as a benchmark for gauging the effects of the other modeled factors, given that the effects of soil texture on C stocks are well documented [26–28]. Soil types vary substantially within the study area and across other potential areas for agroforestry, and soil texture and the fractions of sand, silt, and clay provide a quantifiable approach for modeling the effect of soil type. We hypothesized that C stocks in biomass and soil will decline with the increasing sand content of soil. The rationale is that soils with a higher sand content have a lower capacity for nutrient and water retention and, thus, support less plant growth, therefore lower biomass and soil C [28].

We addressed these questions by conducting four sets of modeling experiments in CENTURY, comparing simulated C stocks in biomass and soil in baseline runs with output after incremental changes in parameters to evaluate the long-term effects on agroforestry systems. The four experiments were designed to test the effects on biomass and soil C stocks of: (1) cropping systems that differed in the number and type of growth forms; (2) different levels of one tree trait—lignin content in tissues; (3) increasing temperature; and (4) soil texture. We evaluated the model predictions regarding the effects of growth forms on soil C stocks by comparing modeled output with measured changes in soil C over an 8-year interval in 48 agroforestry systems in central Kerala.

2. Materials and Methods

2.1. Study Sites and Experimental Design

This study was conducted in the District of Thrissur in the state of Kerala, situated in the southwest corner of India (between 10° and 10°47′ N, and 75°55′ and 76°54′ E). This region is ideal for comparing the effects on C stocks of agroforestry systems that differ in species composition for several reasons. First, crop and tree species richness and functional diversity are high, with 91 species in at least eight functional groups in the studied systems [14]. Moreover, Kerala farmers grow these crops in both monocultures and polycultures within the same farm, thus allowing for comparison of a paired monoculture with a polyculture while controlling for other factors such as soil type and climate [14]. Second, these systems are ancient, with continuous sedentary agriculture documented for >4000 years [29]. As such, this allows time for differences in species traits to have measurable effects on soil C stocks. Finally, this region is expected to experience an increase in mean annual temperature (MAT) [30], such that predictions about the effect of increasing MAT on C stocks will provide useful information for designing future agroforestry systems.

The climate is tropical, humid, with two monsoonal seasons. The summer monsoon season occurs from June to September, with the northeast or retreating monsoon occurring from October to November [31,32]. Mean annual rainfall in Thrissur District is 2750 mm (Kerala Agricultural University weather station, Vellanikkara, Thrissur District, state of Kerala, India, latitude 10°31′ N, longitude 76°17′ E, elevation 40 m, data unpublished), with rainfall increasing from the coast to the hills (part of the Western Ghats), and 75% of rain falling during the monsoonal months from June–September. Daily maximum temperatures of 36–39 °C occur during March and April, the pronounced dry season; the MAT is 27.8 °C (Kerala Agricultural University).

The parent material is pre-Cambrian, except for coastal areas with more recent sediments [31,32]. The predominant dry-land soil types range from Typic Kandiustults (clayey-skeletal Haplic Acrisols, Food and Agricultural Organization, FAO system) in coastal areas to Ustic Kandihumults (clayey-skeletal Humic Acrisols, FAO system) to Oxic Humitropepts (clayey Mollic Cambisols, FAO system) in the highlands [32] (H. Eswaran, personal communication). These soils do not contain carbonates, so soil C is equivalent to soil organic C.

In a previous study conducted in 1997 (Jan–May), we located 50 farms in which we established a set of paired plots (each 10 × 10 m) within a single farm. One of the paired plots contained a given plant functional group of crops (annual, palm or tree) grown in a monoculture and the other contained the same crop grown in an agroforestry system. The paired plots were on similar soil types, with the same cropping system and management for at least a decade. Local Agricultural Officers from the Kerala State Agriculture Department provided us with a list of farmers whose farms met these criteria, from which we randomly selected 50 farms in which farmers wished to participate in the study. The selected farmers represented a wide range of socioeconomic groups. Their landholding size ranged from 0.03 to 10 ha [14]. The most common crops grown in monocultures included: annual crops such as plantain or multipurpose banana (*Musa* × *paradisiaca* L. var. *Nendran*); palms, such as coconut (*Cocos nucifera* L.) or arecanut (*Areca catechu* L.); and trees, e.g., rubber (*Hevea brasilensis* Müll. Arg.). The agroforestry systems contained from two to >seven functional groups of crops, including: grasses, e.g., Napier grass (*Pennisetum purpureum* Schumach.); annual herbs; perennial herbs; root crops such as taro (*Colocasia esculenta* (L.) Schott); vines such as black pepper (*Piper nigrum* L.), palms; leguminous trees (*Erythrina indica* Lam., coral tree); deciduous trees (rubber); and evergreen trees (*Artocarpus heterophyllus* Lam., jackfruit). We revisited these farms for this study and re-sampled the soil eight years later (October–November 2005) if the cropping system had not changed since the previous sampling. At the 2005 sampling time, all of the annual systems previously sampled had been rotated to another crop, as part of the management plan. Thus, we have no data for the second sample time for annual cropping systems. Although the sampling months differed between the two sample years, we did not expect this to be a confounding factor because previous studies in the humid tropics indicated a lack of seasonal variability in soil C stocks [33].

2.2. Field and Laboratory Methods and Data Analysis

The objective of the field study was to obtain field-based data for characterizing the capacity of the model to quantify soil C stocks and their changes in agroforestry systems and monocultures of trees. Sampling the surface soil (0–10 cm) was sufficient to meet this objective, and further, this approach minimized the effects of destructive sampling on a family's farm. The objective of the modeling study was to characterize the potential for total C storage in agroforestry systems, so we modelled soil to a depth of 50 cm, in addition to modeling the surface soil.

In the field study, we re-sampled soil in plots that had been sampled eight years previously. It was not possible to compare changes in soil C in the annual monocultures, because these plots had been rotated to a different system during the interim between sample times. We re-sampled monocultures of palms to provide a basis of comparison for the other systems, although data were not available to parameterize the model for that system. Thus, 39 plots were sampled at both times.

For both sampling times, the methods were as described by Russell [14]. Briefly, we used a push-tube soil sampler to extract eight cores, 4.2 cm in diameter, in each plot at randomly selected points. Samples within a plot were bulked, with subsequent analyses conducted on the single bulked sample. We removed roots and rocks from samples, which we then air-dried and passed through a 2 mm mesh sieve. Sub-samples were dried at 105 °C for dry-weight conversion. Soil samples were finely ground for analyses of organic C and total N by dry combustion (DC) using a Thermo-Finnigan EA Flash (Series 1112, EA Elantech, Lakewood, NJ, USA). Duplicates were run on all samples. Bulk density was sampled one time by the soil core method [34].

To analyze differences in soil C among cropping systems and between sampling times, we first tested for homogeneity of variances and normality of distributions. The data did not fit the assumptions, but ln-transformed data did; thus, analyses were done with the transformed data. We used a generalized linear model procedure to conduct an analysis of variance (ANOVA) test with a time series analysis to test for differences in soil C between sample times and among cropping systems [35]. We also used ANOVA to test for differences among systems in changes of soil C over time. For post-hoc tests, we used Tukey–Kramer multiple comparison tests (experiment-wise error rate, $\alpha = 0.05$).

2.3. Modeling

The CENTURY model was used to simulate carbon and nutrient dynamics for many types of ecosystems, including tropical systems [25,36–39]. The CENTURY model contains a soil organic matter/decomposition submodel, a water budget model, a forest production submodel, and functions for simulating management activities and natural events.

We parameterized the model (version 4.5, forest mode) for climate in the Thrissur District using monthly maximum and minimum temperature and precipitation data (1983–2005) from the Vellanikkara weather station of Kerala Agricultural University Thrissur (Table S1). We used published data for crops, trees, and soils in these systems [14,40–42] to parameterize the model for the baseline run. We modeled a typical land-use-change scenario for the region, i.e., the "spin-up", that consisted of allowing the natural vegetation of semi-evergreen forest to grow for 1800 years, followed by slash-and-burning of the forest and establishment of a homegarden for 98 years (CENTURY "Schedule" file, Table S2). We then appended the experimental run to the spin-up land-use-change scenario. Table S2 contains the schedule file for the 100% annual cropping treatment in Experiment 1 on the effects of growth forms in cropping systems. For all experiments, the baseline tree in the model was "LUQD" for Luquillo deciduous, the typical tree modeled in CENTURY for tropical forests with distinct dry periods. This LUQD tree is deciduous during drought periods, as are rubber trees. In addition, the parameters for this tree type (listed in Table S3) best matched information from the literature on tree species in

Kerala agroforestry systems. For all systems that contained an annual crop, plantain banana was used ("BAN" in CENTURY, parameters listed in Table S4).

For each of the modeling experiments, we changed one set of parameters at a time to create a single treatment level, ran the model, and collected the output for aboveground biomass C and soil C (0–50 cm) at the end of the 100 year experimental run. The process was repeated for each treatment level and for the two soil depths (0–10 and 0–50 cm). Each experiment had five treatment levels. The baseline and experimental parameters are listed in Tables S2–S5. For modeling Experiment 1, regarding the effect of growth forms on cropping systems, the three systems included: (1) a monoculture of an annual crop, banana (plantain); (2) an agroforestry system (AFS) of annual crops and trees; and (3) a monoculture of trees. Typically, this annual crop receives OM additions and occasional irrigation. In contrast, these management factors are optional in the other two cropping systems. To allow for comparison among the three systems without confounding these management factors, we included treatment levels with (+) and without (0) OM addition and irrigation for the agroforestry and 100% tree systems. Thus, the five treatment levels were: Annuals +; AFS 0; AFS +; Trees 0; and Trees +. To compare our field-based soil C measurements with Experiment 1 output, model output for the 0–10 cm soil depth was also collected.

For Experiment 2, regarding the effect of tissue lignin content, the Trees 0 cropping system was used, i.e., a monoculture of trees with no OM additions or irrigation. The experimental range in the fraction of lignin levels varied by component, as follows: leaves, 0.05–0.25; fine roots, 0.18–0.36; fine branches, large wood and coarse roots, 0.25–0.45 (Table S3). These values were centered approximately around the baseline tree species, LUQD (leaves, 0.15; fine roots, 0.28; other components 0.35, Table S3). The experimental ranges were based on values for trees from the literature [20,43,44].

For Experiment 3, regarding the effect of increasing temperatures, simulations were run using the AFS + cropping system. Both minimum and maximum temperatures were increased for all 12 months (see variables "TMX2M(1–12)" and "TMN2M(1–12)' in Table S5). Starting with the current baseline temperatures, temperatures were increased incrementally across the five treatment levels to a maximum increase of 6 °C. Based on the United Kingdom Meteorological Office coupled climate model, Bhaskaran et al. [30] predicted temperature increases of 1–4 °C in the Indian subcontinent during cooler periods—the winter and monsoon seasons. The range for our experiment was intended to include higher temperature increases during hotter parts of the year.

For Experiment 4, regarding the effect of soil texture, AFS + was the simulated cropping system. The sand fraction parameter was decreased from 0.7 to 0.1 across the five experimental runs, and the clay fraction correspondingly increased from 0.1 to 0.6; the silt fraction varied less, from 0.20 to 0.30 (Table S6). The experimental textures thus ranged from sandy loam to loam to clay loam to clay soil. The ranges for sand and clay were centered near the baseline values: sand, 0.30; clay, 0.41, a clay/clay loam soil. The experimental range encompassed the soil textures found in this region [32].

3. Results

3.1. Model Output

3.1.1. Experiment 1: Effects of Growth Form(s)

In comparisons of the three systems, simulated C stocks in aboveground biomass were 430% greater in the AFS + system (annuals and trees) than in the Annual + monoculture, both of which received OM additions and irrigation (Figure 1a). The monoculture of Trees + had only 34% greater C stocks than the AFS + system. Soil C stocks were 90% higher in the AFS + system than in the Annual + system, but only 10% higher in the Tree + system compared with the AFS + system. Thus, total C increased by 270% in the transition from all annual crops to the agroforestry mixture of annuals and perennials, and by 25% in the transition to the monoculture of trees.

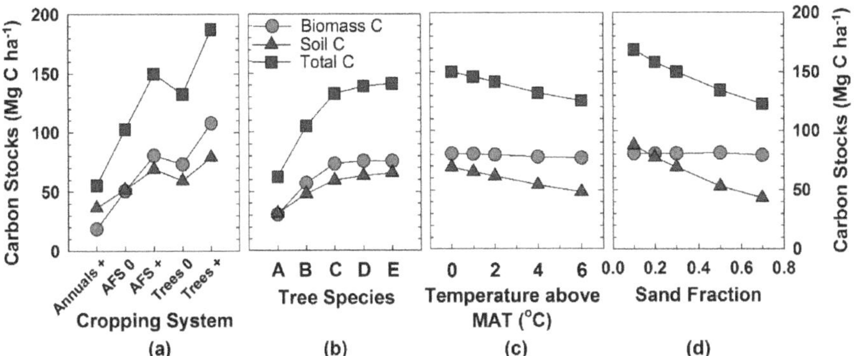

Figure 1. Simulated C stocks in four modeling experiments. (**a**) Experiment 1, regarding the effect of cropping system. The five treatments either received (+) or did not receive (0) organic matter (OM) additions and irrigation. (**b**) Experiment 2, regarding the effect of tree species. Lignin content in modeled tree species (A–E) increased from 0.25 to 0.45 in woody tissues, 0.05 to 0.25 in leaves, and 0.18 to 0.36 in fine roots. See Table S3 for tissue lignin fractions (variables WDLIG (1–5)). (**c**) Experiment 3, regarding the effect of increases in air temperature above the mean annual temperature (MAT). (**d**) Experiment 4, regarding the effect of soil type. Soil C values were for a 0–50 cm depth.

As in real-life practice in this region, the simulated annual crop failed to accumulate any biomass without OM addition and irrigation. Thus, we could make simulation comparisons only of the AFS 0 and Trees 0 with the Annual + system. The AFS 0 system had 270% more aboveground biomass C and 42% more soil C than the Annual + system (Figure 1a). The Trees 0 monoculture had 44% greater aboveground biomass and 14% more soil C than the AFS 0 system in these two scenarios without any additions. Thus, total C was 86% higher in the AFS 0 system than the Annual + monoculture, and increased 29% more in the transition from AFS 0 to the Tree 0 monoculture.

3.1.2. Experiment 2: Effects of Individual Tree Species

In this experiment, the five modeled tree "species" differed only in their lignin fraction parameters (WDLIG 1–5 in Table S3). Lignin fractions ranged from 0.05 to 0.25 in leaves, from 0.18 to 0.36 in fine roots, and from 0.25 to 0.45 in fine branches, large wood, and coarse roots (Table S3). Simulated C stocks in aboveground biomass increased with lignin content, approaching an asymptote by the intermediate lignin values, at which point biomass was 240% greater than at the lowest lignin contents (Tree "A" in Figure 1b). Soil C stocks doubled as lignin content increased. Thus, total C increased by 227% as lignin content increased among the experimental tree "species".

3.1.3. Experiment 3: Effects of Increasing Temperature

The experimental temperatures represented the current baseline mean monthly temperatures, plus increases of 1, 2, 4, and 6 °C above the current temperatures (Table S5). Simulated C stocks in aboveground biomass declined with an increase in temperature to 95% of the baseline biomass (Figure 1c). Soil C declined with an increase in temperature to 70% of the baseline. Thus, total C stocks decreased to 84% of the baseline run when there was no change in temperature.

3.1.4. Experiment 4: Effects of Soil Texture

To simulate different soil textures in this experiment, we ran the model at five different sand fractions, decreasing the parameter from 0.7 to 0.1, and increasing the clay fraction correspondingly from 0.1 to 0.6 (Table S6). The silt fraction was relatively similar across runs. Aboveground C stocks increased by only 2% as sand content declined, whereas soil C stocks doubled over this range of sand content (Figure 1d). Thus, total C stocks increased by 38% over this range of experimental soil types.

3.2. Field-Based Measurements of Soil Carbon

Soil C (0–10 cm) did not differ significantly between the sampling times which were 8 years apart ($p > 0.32$) (Table 1), and the interaction of sampling time with cropping system was not significant ($p > 0.79$). Soil C differed significantly among the cropping systems ($p = 0.0009$); soil C was higher in agroforestry systems and rubber tree monocultures than in palm monocultures. The modeled soil C for the agroforestry systems and rubber tree monocultures corresponded well with measured ranges of soil C.

Table 1. Soil carbon (Mg ha^{-1}) in Kerala agroforestry systems over an 8-year period.

Cropping System	Number of Plots	Measured Soil C [1]		Modeled Soil C
		Time 1	Time 2	
Monoculture: coconut, arecanut (palms) [a,2]	11	9.12 ± 1.21	10.25 ± 0.91	No data
Monoculture: rubber (tree) [b]	7	15.46 ± 2.58	15.19 ± 1.30	15.85
Agroforestry System [b]	21	14.54 ± 1.25	14.74 ± 0.92	13.83

[1] Means (± standard error) are for a 0–10 cm depth. [2] Different letters in this column denote significant differences at $\alpha \leq 0.05$.

In the field-based measurements, soil C increased rather than decreased in a greater number of plots in palm monocultures and agroforestry systems (Figure 2). Agroforestry systems had the largest relative increases in soil C, whereas tree monocultures had the lowest increases.

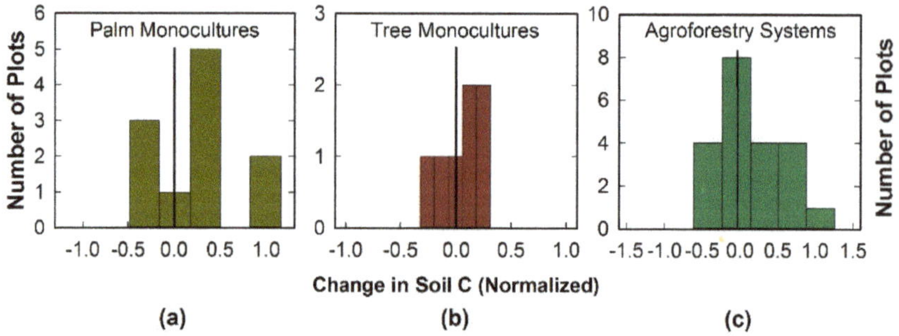

Figure 2. Changes in soil C in agroforestry systems in comparison with monocultures of palms and trees in Kerala. (**a**) Monocultures of palms (coconut and arecanut). (**b**) Monocultures of rubber trees. (**c**) Agroforestry systems. Soil (0–10 cm in depth) was sampled at two times over an 8 year period. Changes in soil C were calculated as: (Final minus Initial Soil C)/Initial Soil C.

4. Discussion

The model demonstrated a capacity to characterize measured surface soil C stocks and their changes in agroforestry systems and tree monocultures: modeled soil C values for these systems, ranging from 916 to 1585 g m^{-2}, fell within the variability measured in the field (Table 1). Moreover, the measured changes in soil C (Figure 2) were consistent with the variability in modeled predictions that arose from differences in management of OM additions and irrigation. Although we did not measure deeper soil (10–50 cm), the modeled C stocks for this depth were consistent with field studies by others in Kerala agroforestry systems [45]. In addition, the modeled aboveground C stocks were within ranges reported in the literature [40,42,46]. Although true validation of the model was not possible, the congruence of the modeled output with measured soil C stocks in this and other studies indicated that the model captured the C dynamics of these systems sufficiently well for us to evaluate the outcome of our modeling experiments. In contrast with field-based observational studies in which one factor may be confounded with other factors, these modeling experiments allowed for the evaluation of a single factor at a time within a holistic framework. This provided a means for direct comparison of the effects of individual factors on C storage in agricultural systems under various conditions in central Kerala.

4.1. Biomass C Stocks

4.1.1. Effects of Growth Form(s)

Of the four simulated factors, growth form, as modeled in Experiment 1, had the greatest effect on C stocks in aboveground biomass in Kerala agricultural systems (Figure 1). Changing from a monoculture of annual non-woody crops to an agroforestry system containing trees and annuals more than quadrupled C stocks in aboveground biomass when the systems received irrigation and OM additions.

Annual cropping systems generally necessitate OM additions and irrigation in some locations, but management of tree crops is more flexible. Even without OM additions and irrigation, aboveground C stocks in the agroforestry system were nearly triple that of the annual monoculture that did receive OM additions and irrigation. These results highlight the effectiveness of agroforestry systems as a mitigation strategy that can store nearly as much C as a monoculture of trees, while at the same time providing annuals and cash crops and requiring fewer inputs.

The change from annual monoculture to agroforestry represents an increase in the number of functional groups. An increase in biomass C stocks was similarly associated with increased plant functional diversity in other studies in agroforestry systems [14] and experiments [33,47,48]. Various mechanisms could explain the effect of this increase on biomass C stocks. As the number of plant functional groups increases, canopy structure can diversify, thus increasing light use, thereby increasing photosynthesis. Similarly, the stratified belowground architecture of such systems [49] can enhance water and nutrient uptake [16]. Tropical trees can tap into water and nutrient supplies at greater depths than most annual crops [50,51] and they exhibit considerable phenotypic plasticity in this respect [49]. As such, a higher proportion of trees in the system will promote biomass C accumulation.

4.1.2. Effects of Tree Species

Differences among individual tree species in the lignin content of their tissues (Experiment 2) had the next biggest impact on biomass C: stocks increased nearly 2.5 fold across the range of lignin contents (Figure 1b). It is notable that this increase occurred without any inputs of fertilizers or irrigation—it arose simply by changing to a different "species" of tree. These results align with field-based experimental results in which the identity of a species played a significant role in C accumulation because tree species differ in their traits, which influenced C cycling differentially [17,33,52,53].

Deforestation in the tropics increased from 6.9 to >7.9 million ha over the period 2000 to 2012 [54]. Thus, the dilemma of how to restore the landscape with trees, while providing land for agriculture continues to grow. Agroforestry presents an optimal solution for this. To maximize this land use, however, our modeling findings highlight the importance of evaluating the tree species traits that influence their capacity to sequester C. Researchers have evaluated species-specific traits for reforestation efforts with plantations, focusing on traits such as survival, wood density, height, diameter, crown cover, shade or sun tolerance, seed dispersal mechanism, photosynthetic characteristics and water use [55–58]. Our findings indicate that this same approach could advance the potential for sequestering biomass C in agroforestry systems. Evaluating more types of traits and incorporating results from this research into the design of agroforestry systems could further improve C storage in agroforests.

4.1.3. Increasing Temperatures

In the simulated warming of 6 °C above the current MAT, from 26.4 to 32.4 °C, biomass C stocks decreased by only 5% (Figure 1c). If the magnitude of this temperature increase did not surpass optimal thresholds for photosynthesis, increase respiration, or increase mortality rates, one would not expect a direct effect of increased temperature on biomass C. However, one would expect to see indirect effects of increased temperature via increased evapotranspiration, leading to increased soil moisture stress, especially in this rain-fed agroforestry system modeled, thereby leading to reduced plant growth and biomass C stocks. Indeed, modeled potential evapotranspiration (PET) increased with MAT (Figure 3a). However, as MAT increased, soil water available for crop/tree survival (parameter "avh2o(2)" in CENTURY) remained stable at 12.9 cm of water for a 0–50 cm depth and streamflow decreased by only 7% (Figure 3a). The relatively high clay content used in this experiment, 41%, allowed for relatively high soil water availability. The critical means by which the modeled trees avoided water stress was that the tree "LUQD" was deciduous, thereby allowing it to avoid water stress during the dry season.

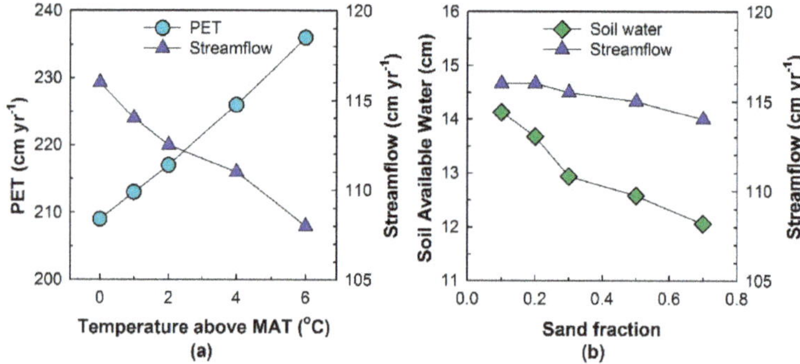

Figure 3. (**a**) Simulated changes in potential evapotranspiration (PET) and streamflow with increasing temperature. (**b**) Simulated changes in soil water available for crop/tree survival and streamflow with changes in soil sand fraction.

There are many uncertainties in modeled predictions of how climate will change [22], and our results suggest that there are also uncertainties in modeling how cropping systems will respond to increases in temperature, and the corresponding changes in plant water use. Although CENTURY contains nearly 100 tree parameters, only two parameters relate to water use: evergreen versus deciduous and the effect on transpiration rate of doubling the atmospheric CO_2 concentration. Another way to simulate the effect of water stress with this model is to change the soil texture, as in Experiment 4, given soil texture's effect on soil water-holding capacity. Future modeling could explore the effect of other tree traits on water use, including allocation to fine roots and evergreen leaves, to examine the indirect effects of increasing temperature on C stocks in agroforestry systems.

4.1.4. Soil Texture

The magnitude of the effect of soil texture on biomass C stocks was less than that of any of the other factors: biomass C declined only 2%–3% as the sand fraction increased from 0.1 to 0.7 (Figure 1c). As sand content increased, simulated available soil water declined by 17% (Figure 3b). One would have expected a greater reduction in biomass with this decline in soil water. The explanation is that in this modeled setting, soil nutrients were relatively more limiting to biomass accumulation than water limitation.

Because temperature parameters were held constant across levels of sand fractions in this Experiment 4, PET also remained constant, unlike in Experiment 3 on temperature. With no change in PET, simulated streamflow declined by only 2% (Figure 1c), even as soil water availability declined with increasing sand content. Future modeling could explore the relationships between water and nutrient limitations in agroforestry systems in conjunction with crop/tree traits and their sensitivity to drought and lower soil nutrient status. This research can inform practitioners about the impacts of the trade-offs involved in choosing species for agroforestry systems based on the environmental constraints within a site.

4.2. Soil C Stocks

4.2.1. Tree Species and Growth-Form Effects

Of the factors modeled, traits of individual tree species (i.e., species *identity*) had the biggest impact on soil C stocks (Figure 1b). The modeled species with the highest lignin content increased soil C more than two-fold (207%). This result aligns with the concept that lignin is relatively recalcitrant [18] and would thus accrue in soil, building up soil C stocks.

Manipulating the growth forms also had a large effect: soil C increased nearly two-fold (190%) with the addition of trees to an annual system (Figure 1a). These simulation results align well with multiple field-based studies in which soil C increased with plant functional diversity [44,54,59,60]. In home gardens in Kerala, Saha et al. [45] found that soil C stocks were directlyrelated to plant diversity.

4.2.2. Increasing Temperature

Whereas the effect of increasing temperature was minimal on biomass C storage, the effect on soil C stocks was much greater. Soil C declined to 90% of the original stocks under a 2 °C increase in temperature, and to 70% under the modeled 6 °C increase. Many other factors influence the decomposition of SOM, including water and O_2 availability, substrate quality, fire, physical and chemical protection, and enzymes and inhibitors. As such, it can be difficult to ascertain the sensitivity of SOM decomposition to temperature because these other factors may mask the effects of temperature on decomposition [61]. The CENTURY model addresses this inherent difficulty by compartmentalizing OM within three pools in mineral soil (i.e., active, slow, and passive), plus plant tissue pools, based on their mean residence time and chemistry. This provides a robust framework for modeling SOM decomposition and its controlling factors in a diverse array of soils. For future modeling, this represents a rich context for examining relationships between soil C and the crop/tree species planted, environmental conditions, and management.

4.2.3. Soil Texture

Similar to temperature, the simulated increase in sand content had a much stronger effect on soil C stocks than on biomass stocks. Soil C more than doubled over the range of sand/clay combinations, a response similar to that of the change in the modeled tree species. This strong response of modeled change in soil texture provides confidence in the model's sensitivity to environmentally important factors. The similar response to the effects of soil texture, a factor known to influence soil C and tree species, also indicates the relative strength of the effect of species traits on ecosystem processes.

This strong effect of modeled soil texture on soil C stocks is supported by field-based studies. As reviewed by Nair et al. [9], soil texture influences the formation of primary organo-mineral complexes. Clay soils enhance aggregation and the stability of soil aggregates and provides physical protection of SOM. In contrast, sands rely on physical binding by roots and hyphae, rather than organo-mineral complexes, such that aggregates in sandy soils tend to be weak. As such, clay soils retain more C than sandy soils.

4.3. Implications for Total (Biomass + Soil) C Storage in Tropical Agroforestry Systems

The potential for C storage in humid tropical systems is very high, ranking among the highest for agroforestry systems around the world [5]. In India, the potential C storage in agroforestry systems in the humid tropical lowlands is estimated to be 92–228 MgC ha^{-1} [62], on par with that of forest plantations and mature forest in the humid tropics [17]. With potential C sequestration rates of 1.1–2.2 Pg C/year in 585–1245 million ha in Africa, Asia, and the Americas, agroforestry systems can play an important role in mitigating climate warming [62]. A literature study found that soil C storage rates in agroforestry systems in sub-Saharan Africa were significantly higher than 4‰ [63]. Moreover, agroforestry systems can enhance pools of soil organic C that last for millennia [64].

Our field-based measurements of surface soil C indicated that soil C storage varied significantly among agroforestry systems, a finding supported by other soil C studies at deeper depths in sites in Kerala [45] and in sub-Saharan Africa [63]. Many factors, including the species composition and number and density of woody species, land use history, and soil type, can influence C storage in agroforestry systems [45,65]. Of the factors modeled in this study, manipulation of growth form had the largest effect on total C storage under management with OM additions and irrigation, with total C stocks 2.7 fold greater in the agroforestry system compared with the annual monoculture (Figure 1a). Without OM addition or irrigation, however, the traits of the individual tree species had the largest effect on simulated total C storage, with a 2.3 fold increase in total C across the range of lignin contents in trees (Figure 1b).

Management factors such as the type, quantity, and timing of organic amendments and inorganic fertilizers can also influence C storage. Across farms in this study, the type and quantity of organic amendments varied among farms and agricultural systems [14]. Organic amendments included dung, green manure, wood ash, neem (*Azadirachta indica* A. Juss.) extract, and finely ground rock phosphate. Farmers also used marketed fertilizers such as di-ammonium phosphate (18% N, 20% P), ammonium phosphate (16% N, 8.6% P), and "17-17-17" (percentages of N, P, and K on a mass basis). Agricultural land use has changed recently in Kerala [66], and that too could introduce variability in management as management options, policies, and socioeconomic factors have changed. Although best management practices (BMPs) for agroforestry systems involving tillage, green manure addition, and tree management have been described [65,67,68], future research is needed in helping farmers achieve BMPs for specific agroforestry systems.

Although these modeling experiments focused on the C storage capacity of agroforestry, these species-rich systems can enhance other ecosystem services, including provisioning of food, fibers, wood, and medicines; control of climate and disease; refugia for rare species; and cultural and spiritual benefits. Higher biodiversity of agroforests can result in a stable supply of ecosystem goods and services if their crop species respond differently to environmental and biotic perturbations and allow for a range of management options [13].

Our modeling results highlight the flexibility provided by agroforestry systems, i.e., one could increase C storage simply through judicious selection of tree species based on their traits, without economic investment in fertilizers and irrigation. This strategy of selecting species for their C-storage potential would require a better knowledge of tree traits that influence C dynamics, indicating future directions for research. From the standpoint of developing sustainable agricultural systems, the modeling results suggest the need to consider site-specific environmental conditions in conjunction with the water- and nutrient-use traits of the species to be planted. Research focused on evaluating the trade-offs between water use efficiency, nutrient use efficiency, and growth potential for common crops in agroforestry will provide valuable information for designing systems for maximizing C storage in a warming world and playing a role in mitigating climate change.

5. Conclusions

The integrated use of measurements from long-term plots in conjunction with process-based modeling allowed us to evaluate the effects of four factors on C storage in agroforestry systems in Kerala, India, a humid tropical environment. The type of growth form had the largest impact on C storage in aboveground biomass—inclusion of trees into an annual cropping system more than quadrupled C stocks. Species-specific traits such as lignin content had a greater effect on soil C stocks than did growth form, however. One of the most important implications from this modeling study is that inclusion of trees in the system promotes higher C storage in these agricultural systems without the economic costs of fertilizers or irrigation. Greater use and better management of trees within these agricultural systems thus provides a very low-cost means for mitigation of climate warming.

Supplementary Materials: The following are available online at http://www.mdpi.com/1999-4907/10/9/803/s1, Table S1: Climate data from the Vellanikkara, Thrissur Station. Latitude: 10°31'; Longitude: 76°13'. Height is 40 m above MSL. Monthly data from January 1983 to September 2005, Table S2: Schedule' file in CENTURY. Blocks #1–4 simulate the land-use scenario, i.e., the 'spin-up,' prior to the start of the experimental modeling. Block #5 simulates the first run in Experiment #1 on effects of cropping system for the treatment of 100% Annuals (Banana, plantain), Table S3: Tree parameters in the Tree.100 file. Table S4: Crop Parameters in the 'Crop.100' file, Table S5: Climate parameters used in Experiment 3 on effects of increase in temperature. Point # refers to the order of the temperature increase, from lowest to highest. Bolding denotes parameters that differed among the 'treatment' levels. Point #1 values are based on the actual climate data, i.e., no warming effect, Table S6: Site parameters used in Experiment 4 on the effect of soil type. Point # refers to the order of increase in sand content, from lowest to highest. Bolding denotes parameters that differed among the 'treatment' levels.

Author Contributions: Conceptualization, methodology, software, formal analysis, validation, data curation, visualization, writing—original draft preparation, funding acquisition, project administration: A.E.R. Resources, writing—review and editing: B.M.K.

Funding: This research was funded by the Fulbright Program through a Fellowship (to A.E.R.) in the Indo-American Environmental Leadership Program.

Acknowledgments: We thank Kerala Agricultural University (KAU) for use of facilities. G.S.L.H.V. Rao provided 20 years of meteorological data for parameterizing the model's climate variables. Cindy Keough provided assistance with model programming. We also thank the farmers who participated in this study and several research technicians at KAU for assistance in this research.

Conflicts of Interest: The authors declare no conflict of interest.

References

1. Le Quéré, C.; Moriarty, R.; Andrew, R.M.; Canadell, J.G.; Sitch, S.; Korsbakken, J.I.; Friedlingstein, P.; Peters, G.P.; Andres, R.J.; Boden, T.A. Global carbon budget 2015. *Earth Syst. Sci. Data* **2015**, *7*, 349–396. [CrossRef]

2. Houghton, R.A.; Nassikas, A.A. Negative emissions from stopping deforestation and forest degradation, globally. *Glob. Chang. Biol.* **2018**, *24*, 350–359. [CrossRef] [PubMed]

3. Minasny, B.; Malone, B.P.; McBratney, A.B.; Angers, D.A.; Arrouays, D.; Chambers, A.; Chaplot, V.; Chen, Z.-S.; Cheng, K.; Das, B.S. Soil carbon 4 per mille. *Geoderma* **2017**, *292*, 59–86. [CrossRef]

4. Nath, A.J.; Lal, R.; Sileshi, G.W.; Das, A.K. Managing India's small landholder farms for food security and achieving the "4 per Thousand" target. *Sci. Total Environ.* **2018**, *634*, 1024–1033. [CrossRef] [PubMed]

5. Kumar, B.M.; Nair, P.K.R. *Carbon Sequestration Potential of Agroforestry Systems: Opportunities and Challenges*; Springer Science & Business Media: Berlin, Germany, 2011; Volume 8.

6. Muschler, R.G. Agroforestry: Essential for Sustainable and Climate-Smart Land Use. In *Tropical Forestry Handbook*; Springer: Berlin, Germany, 2016; pp. 2013–2116.

7. MOSPI. Statistical Year Book India 2018. Available online: http://www.mospi.gov.in/statistical-year-book-india/2018/177 (accessed on 11 September 2019).

8. Lal, R. Soil carbon sequestration in India. *Clim. Chang.* **2004**, *65*, 277–296. [CrossRef]

9. Nair, P.K.R.; Nair, V.D.; Kumar, B.M.; Showalter, J.M. Carbon sequestration in agroforestry systems. In *Advances in Agronomy*; Elsevier: Amsterdam, The Netherlands, 2010; Volume 108, pp. 237–307.

10. Kumar, B.M.; Singh, A.K.; Dhyani, S.K. South Asian agroforestry: Traditions, transformations, and prospects. In *Agroforestry-The Future of Global Land Use*; Springer: Berlin, Germany, 2012; pp. 359–389.

11. McGuire, A.; Sitch, S.; Clein, J.S.; Dargaville, R.; Esser, G.; Foley, J.; Heimann, M.; Joos, F.; Kaplan, J.; Kicklighter, D. Carbon balance of the terrestrial biosphere in the twentieth century: Analyses of CO_2, climate and land use effects with four process-based ecosystem models. *Glob. Biogeochem. Cycles* **2001**, *15*, 183–206. [CrossRef]

12. Metherell, A. A Model for Phosphate Fertiliser Requirements for Pastures—Incorporating Dynamics and Economics. In *The Efficient Use of Fertilisers in a Changing Environment: Reconciling Productivity and Sustainability*; Occasional Report; Massey University: Palmerston North, New Zealand, 1994.

13. Hooper, D.U.; Chapin, F.S.; Ewel, J.J.; Hector, A.; Inchausti, P.; Lavorel, S.; Lawton, J.H.; Lodge, D.; Loreau, M.; Naeem, S. Effects of biodiversity on ecosystem functioning: A consensus of current knowledge. *Ecol. Monogr.* **2005**, *75*, 3–35. [CrossRef]

14. Russell, A.E. Relationships between crop-species diversity and soil characteristics in southwest Indian agroecosystems. *Agric. Ecosyst. Environ.* **2002**, *92*, 235–249. [CrossRef]

15. Fisher, R.F. Amelioration of degraded rain forest soils by plantations of native trees. *Soil Sci. Soc. Am. J.* **1995**, *59*, 544–549. [CrossRef]

16. Gliessman, S. *Agroecology: Ecological Processes in Sustainable Agriculture*; CRC Press: Boca Raton, FL, USA, 1998.

17. Russell, A.E.; Raich, J.W.; Arrieta, R.B.; Valverde-Barrantes, O.; González, E. Impacts of individual tree species on carbon dynamics in a moist tropical forest environment. *Ecol. Appl.* **2010**, *20*, 1087–1100. [CrossRef]

18. Melillo, J.M.; Aber, J.D.; Muratore, J.F. Nitrogen and lignin control of hardwood leaf litter decomposition dynamics. *Ecology* **1982**, *63*, 621–626. [CrossRef]

19. Hancock, J.E.; Loya, W.M.; Giardina, C.P.; Li, L.; Chiang, V.L.; Pregitzer, K.S. Plant growth, biomass partitioning and soil carbon formation in response to altered lignin biosynthesis in Populus tremuloides. *New Phytol.* **2007**, *173*, 732–742. [CrossRef] [PubMed]

20. Novaes, E.; Kirst, M.; Chiang, V.; Winter-Sederoff, H.; Sederoff, R. Lignin and biomass: A negative correlation for wood formation and lignin content in trees. *Plant Physiol.* **2010**, *154*, 555–561. [CrossRef] [PubMed]

21. Pachauri, R.K.; Allen, M.R.; Barros, V.R.; Broome, J.; Cramer, W.; Christ, R.; Church, J.A.; Clarke, L.; Dahe, Q.; Dasgupta, P. *Climate Change 2014: Synthesis Report. Contribution of Working Groups I, II and III to the Fifth Assessment Report of the Intergovernmental Panel on Climate Change*; IPCC: Geneva, Switzerland, 2014.

22. Rosenzweig, C.; Elliott, J.; Deryng, D.; Ruane, A.C.; Müller, C.; Arneth, A.; Boote, K.J.; Folberth, C.; Glotter, M.; Khabarov, N. Assessing agricultural risks of climate change in the 21st century in a global gridded crop model intercomparison. *Proc. Natl. Acad. Sci. USA* **2014**, *111*, 3268–3273. [CrossRef] [PubMed]

23. O'Neill, B.C.; Oppenheimer, M.; Warren, R.; Hallegatte, S.; Kopp, R.E.; Pörtner, H.O.; Scholes, R.; Birkmann, J.; Foden, W.; Licker, R. IPCC reasons for concern regarding climate change risks. *Nat. Clim. Chang.* **2017**, *7*, 28. [CrossRef]

24. Lambers, H.S.; Chapin, F.; Pons, T.L. *Plant Physiological Ecology*; Springer: New York, NY, USA, 1998.

25. Raich, J.W.; Parton, W.J.; Russell, A.E.; Sanford, R.L.; Vitousek, P.M. Analysis of factors regulating ecosystemdevelopment on Mauna Loa using the Century model. *Biogeochemistry* **2000**, *51*, 161–191. [CrossRef]

26. Sanchez, P.A. Soil productivity and sustainability in agroforestry systems. *Agrofor. A Decade Dev.* **1987**, 205–223.

27. Quesada, C.; Phillips, O.; Schwarz, M.; Czimczik, C.; Baker, T.; Patiño, S.; Fyllas, N.; Hodnett, M.; Herrera, R.; Almeida, S. Basin-wide variations in Amazon forest structure and function are mediated by both soils and climate. *Biogeosciences* **2012**, *9*, 2203–2246. [CrossRef]

28. Coleman, D.C.; Callaham, M.A.; Crossley, D., Jr. *Fundamentals of Soil Ecology*; Academic Press: Cambridge, MA, USA, 2017.

29. Raychaudhuri, S.; Kaw, R.; RAG-HAVAN, D. *Agriculture in Ancient India*; Indian Council of Agricultural Research: New Delhi, India, 1964.

30. Bhaskaran, B.; Mitchell, J.; Lavery, J.; Lal, M. Climatic response of the Indian subcontinent to doubled CO_2 concentrations. *Int. J. Climatol.* **1995**, *15*, 873–892. [CrossRef]

31. Pascal, J.-P. *Wet Evergreen Forests of the Western Ghats of India*; Institut Français de Pondichéry: Pondicherry, India, 1988.

32. Bourgeon, G. *Explanatory Booklet on the Reconnaissance Soil Map of Forest Area–Western Karnataka and Goa*; Institut Français de Pondichéry: Pondicherry, India, 1989.

33. Russell, A.E.; Cambardella, C.A.; Ewel, J.J.; Parkin, T.B. Species, rotation, and life-form diversity effects on soil carbon in experimental tropical ecosystems. *Ecol. Appl.* **2004**, *14*, 47–60. [CrossRef]

34. Blake, G.R.; Hartge, K. *Bulk density 1. Methods Soil Anal. Part 1—Phys. Mineral. Methods*; American Society of Agronomy: Madison, WI, USA, 1986; pp. 363–375.

35. SAS Institute. *SAS User's Guide: Statistics*; SAS Institute: Cary, NC, USA, 1985; Volume 2.

36. Motavalli, P.; Palm, C.; Parton, W.; Elliott, E.; Frey, S. Comparison of laboratory and modeling simulation methods for estimating soil carbon pools in tropical forest soils. *Soil Biol. Biochem.* **1994**, *26*, 935–944. [CrossRef]

37. Probert, M.; Keating, B.; Thompson, J.; Parton, W. Modelling water, nitrogen, and crop yield for a long-term fallow management experiment. *Aust. J. Exp. Agric.* **1995**, *35*, 941–950. [CrossRef]
38. Vitousek, P.M.; Turner, D.R.; Parton, W.J.; Sanford, R.L. Litter decomposition on the Mauna Loa environmental matrix, Hawai'i: Patterns, mechanisms, and models. *Ecology* **1994**, *75*, 418–429. [CrossRef]
39. Parton, W.; Tappan, G.; Ojima, D.; Tschakert, P. Ecological impact of historical and future land-use patterns in Senegal. *J. Arid Environ.* **2004**, *59*, 605–623. [CrossRef]
40. Kumar, B.M.; George, S.J.; Jamaludheen, V.; Suresh, T. Comparison of biomass production, tree allometry and nutrient use efficiency of multipurpose trees grown in woodlot and silvopastoral experiments in Kerala, India. *For. Ecol. Manag.* **1998**, *112*, 145–163. [CrossRef]
41. Kumar, B.M.; Thomas, J.; Fisher, R.F. Ailanthus triphysa at different density and fertiliser levels in Kerala, India: Tree growth, light transmittance and understorey ginger yield. *Agrofor. Syst.* **2001**, *52*, 133–144. [CrossRef]
42. Kumar, B.M.; Kumar, S.S.; Fisher, R.F. Galangal growth and productivity related to light transmission in single-strata, multistrata and 'no-over-canopy' systems. *J. New Seeds* **2005**, *7*, 111–126. [CrossRef]
43. Raich, J.W.; Russell, A.E.; Bedoya-Arrieta, R. Lignin and enhanced litter turnover in tree plantations of lowland Costa Rica. *For. Ecol. Manag.* **2007**, *239*, 128–135. [CrossRef]
44. Russell, A.E.; Raich, J.W.; Valverde-Barrantes, O.J.; Fisher, R.F. Tree species effects on soil properties in experimental plantations in tropical moist forest. *Soil Sci. Soc. Am. J.* **2007**, *71*, 1389–1397. [CrossRef]
45. Saha, S.K.; Nair, P.K.R.; Nair, V.D.; Kumar, B.M. Soil carbon stock in relation to plant diversity of homegardens in Kerala, India. *Agrofor. Syst.* **2009**, *76*, 53–65. [CrossRef]
46. Kumar, B.M.; George, S.; Suresh, T. Fodder grass productivity and soil fertility changes under four grass+ tree associations in Kerala, India. *Agrofor. Syst.* **2001**, *52*, 91–106. [CrossRef]
47. Tilman, D.; Knops, J.; Wedin, D.; Reich, P.; Ritchie, M.; Siemann, E. The influence of functional diversity and composition on ecosystem processes. *Science* **1997**, *277*, 1300–1302. [CrossRef]
48. Fornara, D.; Tilman, D. Plant functional composition influences rates of soil carbon and nitrogen accumulation. *J. Ecol.* **2008**, *96*, 314–322. [CrossRef]
49. Kumar, B.M.; Jose, S. Phenotypic plasticity of roots in mixed tree species agroforestry systems: Review with examples from peninsular India. *Agrofor. Syst.* **2018**, *92*, 59–69. [CrossRef]
50. Jobbágy, E.G.; Jackson, R.B. The distribution of soil nutrients with depth: Global patterns and the imprint of plants. *Biogeochemistry* **2001**, *53*, 51–77. [CrossRef]
51. Stahl, C.; Hérault, B.; Rossi, V.; Burban, B.; Bréchet, C.; Bonal, D. Depth of soil water uptake by tropical rainforest trees during dry periods: Does tree dimension matter? *Oecologia* **2013**, *173*, 1191–1201. [CrossRef]
52. Hobbie, S.E.; Ogdahl, M.; Chorover, J.; Chadwick, O.A.; Oleksyn, J.; Zytkowiak, R.; Reich, P.B. Tree species effects on soil organic matter dynamics: The role of soil cation composition. *Ecosystems* **2007**, *10*, 999–1018. [CrossRef]
53. Montagnini, F. Accumulation in above-ground biomass and soil storage of mineral nutrients in pure and mixed plantations in a humid tropical lowland. *For. Ecol. Manag.* **2000**, *134*, 257–270. [CrossRef]
54. Austin, K.G.; González-Roglich, M.; Schaffer-Smith, D.; Schwantes, A.M.; Swenson, J.J. Trends in size of tropical deforestation events signal increasing dominance of industrial-scale drivers. *Environ. Res. Lett.* **2017**, *12*, 054009. [CrossRef]
55. Butterfield, R.P. Promoting biodiversity: Advances in evaluating native species for reforestation. *For. Ecol. Manag.* **1995**, *75*, 111–121. [CrossRef]
56. Hooper, E.; Condit, R.; Legendre, P. Responses of 20 native tree species to reforestation strategies for abandoned farmland in Panama. *Ecol. Appl.* **2002**, *12*, 1626–1641. [CrossRef]
57. Dierick, D.; Hölscher, D. Species-specific tree water use characteristics in reforestation stands in the Philippines. *Agric. For. Meteorol.* **2009**, *149*, 1317–1326. [CrossRef]
58. Craven, D.; Dent, D.; Braden, D.; Ashton, M.; Berlyn, G.; Hall, J. Seasonal variability of photosynthetic characteristics influences growth of eight tropical tree species at two sites with contrasting precipitation in Panama. *For. Ecol. Manag.* **2011**, *261*, 1643–1653. [CrossRef]
59. Vitousek, P.; Hooper, D. Biological diversity and terrestrial ecosystem biogeochemistry. In *Biodiversity and Ecosystem Function*; Springer: Berlin, Germany, 1994; pp. 3–14.
60. Finzi, A.C.; Van Breemen, N.; Canham, C.D. Canopy tree–Soil interactions within temperate forests: Species effects on soil carbon and nitrogen. *Ecol. Appl.* **1998**, *8*, 440–446.

61. Davidson, E.A.; Janssens, I.A. Temperature sensitivity of soil carbon decomposition and feedbacks to climate change. *Nature* **2006**, *440*, 165. [CrossRef]

62. Dixon, R. Agroforestry systems: Sources of sinks of greenhouse gases? *Agrofor. Syst.* **1995**, *31*, 99–116. [CrossRef]

63. Corbeels, M.; Cardinael, R.; Naudin, K.; Guibert, H.; Torquebiau, E. The 4 per 1000 goal and soil carbon storage under agroforestry and conservation agriculture systems in sub-Saharan Africa. *Soil Till. Res.* **2019**, *188*, 16–26. [CrossRef]

64. Lorenz, K.; Lal, R. Soil organic carbon sequestration in agroforestry systems. A review. *Agron. Sustain. Dev.* **2014**, *34*, 443–454. [CrossRef]

65. Kunhamu, T.; Kumar, B.M.; Samuel, S. Does Tree Management Affect Biomass and Soil Carbon Stocks of Acacia mangium Willd. Stands in Kerala, India. In *Carbon Sequestration Potential of Agroforestry Systems*; Springer: Berlin, Germany, 2011; pp. 217–228.

66. Guillerme, S.; Kumar, B.; Menon, A.; Hinnewinkel, C.; Maire, É.; Santhoshkumar, A. Impacts of public policies and farmer preferences on agroforestry practices in Kerala, India. *Environ. Manag.* **2011**, *48*, 351–364. [CrossRef]

67. Jarecki, M.K.; Lal, R. Crop management for soil carbon sequestration. *Crit. Rev. Plant Sci.* **2003**, *22*, 471–502. [CrossRef]

68. Lorenz, K.; Lal, R. Carbon dynamics and pools in major forest biomes of the world. In *Carbon Sequestration in Forest Ecosystems*; Springer: Berlin, Germany, 2010; pp. 159–205.

Article

Effects of Changing Temperature on Gross N Transformation Rates in Acidic Subtropical Forest Soils

Xiaoqian Dan [1], Zhaoxiong Chen [1], Shenyan Dai [1,*], Xiaoxiang He [1], Zucong Cai [1,2], Jinbo Zhang [1,3] and Christoph Müller [4,5]

[1] School of Geography, Nanjing Normal University, Nanjing 210023, China; 181302119@stu.njnu.edu.cn (X.D.); 181302002@stu.njnu.edu.cn (Z.C.); 191301031@stu.njnu.edu.cn (X.H.); zccai@njnu.edu.cn (Z.C.); zhangjinbo@njnu.edu.cn (J.Z.)

[2] Key Laboratory of Virtual Geographic Environment, Nanjing Normal University, Ministry of Education, Nanjing 210023, China

[3] Jiangsu Center for Collaborative Innovation in Geographical Information Resource Development and Application, Nanjing 210023, China

[4] Institute of Plant Ecology, Justus-Liebig University Giessen, Heinrich-Buff-Ring 26, 35392 Giessen, Germany; Christoph.Mueller@bot2.bio.uni-giessen.de

[5] School of Biology and Environmental Science and Earth Institute, University College Dublin, Belfield, D04V1W8 Dublin, Ireland

* Correspondence: shenyandai.nnu@gmail.com; Tel.: +86-25-8589-1203; Fax: +86-25-8589-1745

Received: 16 September 2019; Accepted: 6 October 2019; Published: 10 October 2019

Abstract: Soil temperature change caused by global warming could affect microbial-mediated soil nitrogen (N) transformations. Gross N transformation rates can provide process-based information about abiotic–biotic relationships, but most previous studies have focused on net rates. This study aimed to investigate the responses of gross rates of soil N transformation to temperature change in a subtropical acidic coniferous forest soil. A ^{15}N tracing experiment with a temperature gradient was carried out. The results showed that gross mineralization rate of the labile organic N pool significantly increased with increasing temperature from 5 °C to 45 °C, yet the mineralization rate of the recalcitrant organic N pool showed a smaller response. An exponential response function described well the relationship between the gross rates of total N mineralization and temperature. Compared with N mineralization, the functional relationship between gross NH_4^+ immobilization and temperature was not so distinct, resulting in an overall significant increase in net N mineralization at higher temperatures. Heterotrophic nitrification rates increased from 5 °C to 25 °C but declined at higher temperatures. By contrast, the rate of autotrophic nitrification was very low, responding only slightly to the range of temperature change in the most temperature treatments, except for that at 35 °C to 45 °C, when autotrophic nitrification rates were found to be significantly increased. Higher rates of NO_3^- immobilization than gross nitrification rates resulted in negative net nitrification rates that decreased with increasing temperature. Our results suggested that, with higher temperature, the availability of soil N produced from N mineralization would significantly increase, potentially promoting plant growth and stimulating microbial activity, and that the increased NO_3^- retention capacity may reduce the risk of leaching and denitrification losses in this studied subtropical acidic forest.

Keywords: temperature change; gross N transformation rates; subtropical acidic forest soil; China; ^{15}N tracing experiment

1. Introduction

Available soil nitrogen (N) affects the growth of both plants and microorganisms. In natural forest ecosystems, available N is supplied via litter and organic matter degradation, plant N_2 fixation, and atmospheric N deposition processes; the interplay among these soil N transformations governs the availability of N which is also affected by soil physicochemical and microbial properties as well as local environmental factors (e.g., temperature, moisture) [1,2], where temperature is considered to be a key factor determining microbial activity levels in soils [3–5].

To better understand soil N transformation rates, both net and gross soil N transformation rates should be determined. Although net N mineralization and nitrification rates do provide an indication of N availability in ecosystems, they are not geared towards understanding the dynamics of specific soil N processes [6,7]. When testing the effects of temperature, net rate studies were mainly carried out, often reporting highly variable responses of soil N to temperature changes. This most likely arose because of different temperature responses among transformation processes that are grouped together when considered in the examination of net rates. For instance, net mineralization rates are the outcome of combined effects of mineralization rates—these often derived from different soil organic matter (SOM) fractions which themselves are characterized by different temperature responses—and also the individual consumption processes such as autotrophic nitrification, immobilization, and so forth. Thus, net N mineralization may increase, decrease, or even be non-responsive to changing temperatures [5,8,9]. Only gross rates of soil N transformations can provide the crucial information for insight into dynamics of the internal N cycle between the organic and mineral N pools [10,11].

Some previous studies found differing responses between the net and gross rates of soil N transformation to temperature changes. For example, recent work by Cheng et al. (2015) had suggested that gross rates of N mineralization and immobilization increased in equal proportions with temperature rising from 5 °C to 25 °C in their studied forest soil, leaving the net rate of N mineralization unaltered by temperature change [9]. Earlier, Zaman and Chang (2004) had reported that gross rates of nitrification were balanced by NO_3^- immobilization rates operating at 5 °C to 40 °C, which led to a negligible response of net rate of nitrification to temperature change [12]. Hence, to gain robust process-based insights into soil N availability, how the temperature change affects the individual gross transformation rate must be explicitly taken into account [13,14].

In this study, we investigated the temperature response of acidic forest soils in subtropical China using a ^{15}N tracing approach. Current ^{15}N tracing methods can simultaneously quantify the gross rates of multiple processes separately, thereby conveying the dynamics of soil N transformation processes more comprehensively and realistically [15,16]. Our aim was to explore temperature response functions for individual gross N rates and to provide an understanding of possible ecological implications in face of ongoing global change.

2. Materials and Methods

2.1. Soil Sampling

The sampling site was located at the Shuangzhen Forestry Center, Jiangxi Province, China (27°59′ N, 117°25′ E), which is characterized by a typical subtropical monsoon climate. The mean annual temperature is 17.6 °C, with a minimum and maximum monthly average temperatures respectively of 5.6 °C in January and 29.3 °C in July (30-year averages). The mean annual precipitation is 1778 mm (30-year average), of which approximately half occurs from April to June. Dominant vegetation at the site are *Pinus massoniana* Lamb. and *Cunninghamia lanceolata* R.Br. trees The soils here originated from granite and are classified as Hapludults according to the USDA soil taxonomy [17].

Soil sampling was carried out in November 2018. The soil's pH was 5.07, and its organic carbon (C), total N, and water-soluble organic C contents were 19.78 g kg^{-1} soil, 1.92 g kg^{-1} soil, and 236.5 mg kg^{-1} soil, respectively. At the site, five plots (1 m × 1 m) were randomly positioned, in which surface soil (0–20 cm depth layer) was collected after carefully removing the O horizon. All samples were pooled together, passed through a 2-mm sieve, and then divided into two sub-samples. One was stored at 4 °C for the incubation experiments within two weeks and the other was air-dried to determine soil physical and chemical properties.

2.2. ^{15}N tracing Experiment

A temperature gradient was set up that consisted of five temperature treatments: 5 °C, 15 °C, 25 °C, 35 °C, and 45 °C. There were two ^{15}N labeled treatments (^{15}N-labelled NH_4^+ and ^{15}N-labelled NO_3^-) each with three replicates. A total of 120 flasks (250 mL), each containing 20 g (oven-dry basis) fresh soil, were prepared. These flasks were divided into five subgroups (each with 24 flasks) and pre-incubated at 5 °C, 15 °C, 25 °C, 35 °C, and 45 °C, respectively, for 1 day. Then, either the ^{15}NH$_4$NO$_3$ (^{15}N at 10.20 atom%) or NH$_4$^{15}NO$_3$ (^{15}N at 10.25 atom%) (2 mL) was added evenly to the surface of the fresh soil in each flask, to give respective final concentrations of 30 mg NH_4^+-N kg^{-1} soil and 30 mg NO_3^--N kg^{-1} soil. All soil samples were adjusted to 60% water hold capacity (WHC), and all flasks were sealed with a perforated preservative film, after which they were placed into five different incubators for the 6-day incubation. Soil water content in the flasks was maintained by adding water every 2 days to compensate for lost water, especially under the 35 °C and 45 °C treatments. Soil inorganic N was extracted with 2 M KCl (potassium chloride) (soil: solution, 1:5) at 0.5 h, 48 h, 96 h, and 144 h after adding the 15N labeled solution, to determine the concentrations and ^{15}N abundances of NH_4^+ and NO_3^-.

2.3. Analysis of Soil Properties

Soil pH was determined using a DMP-2 mV/pH detector (soil:water, 1:1.25) (Quark Ltd., Nanjing, China) [18]. Soil organic C was measured using wet digestion with H_2SO_4-$K_2Cr_2O_7$ and soil total N by semi-micro Kjeldahl digestion with Se, $CuSO_4$ and K_2SO_4 as catalysts. Soil water-soluble organic C was measured with a TOC instrument (Multi N/C, Jena, Germany). Concentrations of NH_4^+ and NO_3^- were measured by colorimetry in a continuous flow analyzer (Skalar San^{++}, Breda, Netherlands) and their respective ^{15}N abundances determined by isotope ratio mass spectrometry (Europa Scientific Integra, Crewe, UK) via a modified diffusion method following Brooks et al. (1989) [19].

2.4. Calculations and Statistical Analyses

The ^{15}N tracing model Ntrace was used to quantify the simultaneously occurring multiple gross soil N transformation rates [16]. The data inputted to the model were the concentrations and ^{15}N excess values—measured ^{15}N abundance values minus ^{15}N natural abundance values of NH_4^+ and NO_3^-. All values entered were means ± standard deviations. Net rates of N mineralization were calculated as the change in inorganic N (NH_4^+ + NO_3^-) concentration from 0.5 h to 144 h divided by the incubation time, and likewise for net rates of nitrification but using the change in NO_3^- concentration.

Significant differences in the means among the five temperature treatments were tested by one-way ANOVA, carried out separately for each N transformation rate. Based on the actual experimental repetitions, the least significant differences at the 5% significance level (LSD0.05) were calculated for each N transformation rate, which presents the most conservative way to calculate LSDs [20]. The observed error in the observations is linked to the number of actual repetitions and is reflected in the probability density function (PDF) of each parameter [16]. Statistical analyses were performed in SigmaStat 4.0 Analysis.

3. Results

3.1. Changes in Concentrations and ^{15}N Enrichments of Mineral N

The modeled and observed concentrations and ^{15}N enrichments matched well, with R^2 values ranging from 0.91 to 0.98 for all treatments (Figure 1). The NH4+ concentration increased with longer incubation times and the higher treatment temperatures. It was significantly highest at 45 °C, indicating the large effect of temperature on NH_4^+ production (Figure 1a). The NO_3^- concentrations remained mostly unchanged, except at 45 °C, where they declined (Figure 1b).

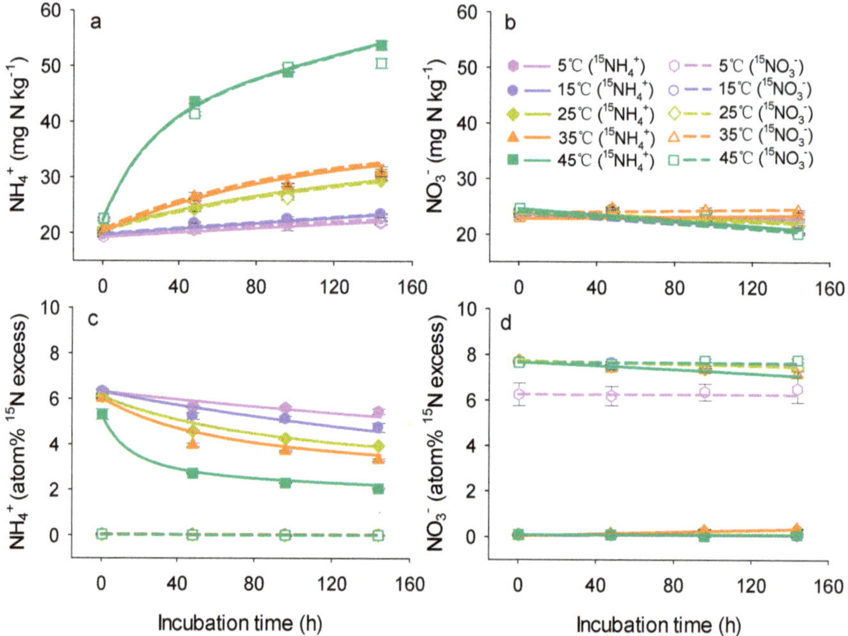

Figure 1. Measured (*symbols*) and modeled (*lines*) concentrations of NH_4^+ (**a**) and NO_3^- (**b**), and ^{15}N enrichments of NH_4^+ (**c**) and NO_3^- (**d**). Vertical bars indicate standard deviations of the mean value.

Under the $^{15}NH_4NO_3$ treatment, the ^{15}N enrichment in the NH_4^+ pool decreased with rising temperature, whereas that of the NO_3^- pool was nearly unchanged (Figure 1c). In the $NH_4{}^{15}NO_3$ treatment, however, the ^{15}N enrichment in both NH_4^+ and NO_3^- pools was only slightly changed (Figure 1d).

3.2. Influences of Temperature on Net N Transformation Rates

The net rate of N mineralization increased as the temperature rose from 5 °C to 45 °C. The highest rate, 4.3 mg N kg^{-1} day^{-1} soil, was observed at 45 °C, or ca. 25 times that of the 5 °C treatment. Net rates of nitrification in all treatments were negative, and they tended to decline with higher temperatures (Figure 2).

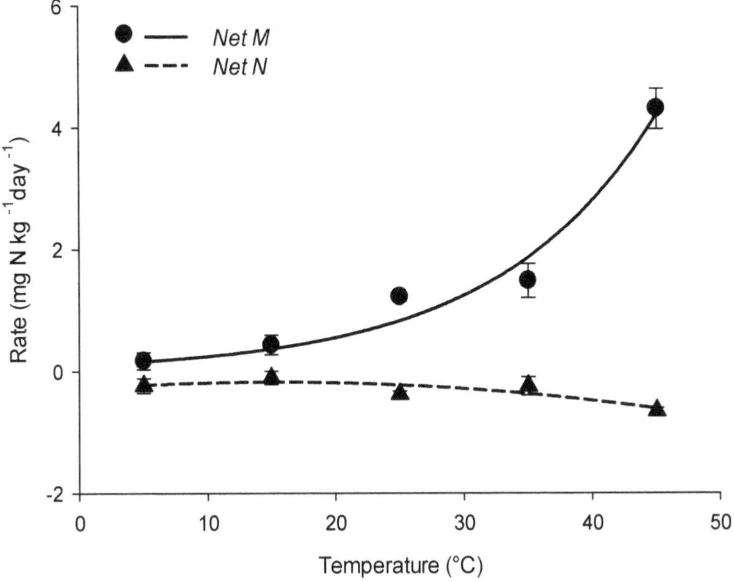

Figure 2. The non-linear trends in net rates of N mineralization (*Net M*) and nitrification (*Net N*) as function of imposed incubation temperatures. Vertical bars indicate standard deviations of the mean value.

3.3. Influences of Temperature on Gross N Transformation Rates

The gross rates of most soil N transformation processes were significantly influenced by temperature change, except for the dissimilatory NO_3^- reduction to NH_4^+, adsorption of NH_4^+ on cation exchange sites, and the release of adsorbed NH_4^+ (data was not shown, because they were near to zero). The responses of gross mineralization rates of soil recalcitrant organic N and labile organic N to temperature change were significantly different (Figure 3a). The mineralization rate of soil labile organic N pool was evidently promoted by a higher temperature, whereas that of the soil recalcitrant organic N pool was nearly stable from 5 °C to 35 °C but it significantly increased at 45 °C. Consequently, gross rate of total N mineralization exponentially increased with increasing temperature from 5 °C to 45 °C, reaching 5.27 mg kg^{-1} day^{-1} soil (i.e., at 45 °C). Gross rates of NH_4^+ immobilization—I_{NH4}, i.e., immobilization rate of NH_4^+ to recalcitrant organic N + immobilization rate of NH_4^+ to labile organic N—were much lower than those of total N mineralization, leading to a net production of NH_4^+. The response of gross NH_4^+ immobilization rates to temperature change was not as distinctive as the response of total gross N mineralization.

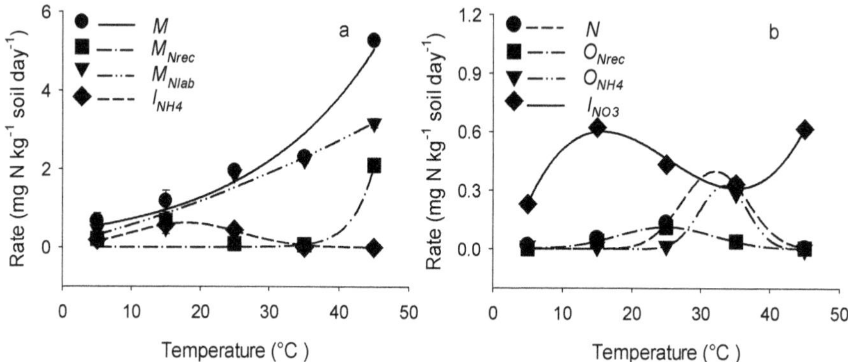

Figure 3. Non-linear trends in the gross rates of soil N transformation processes as a function of imposed incubation temperatures. Vertical bars indicate standard deviations of the mean value. (**a**) M (M_{Nrec} + M_{Nlab}), gross rates of total N mineralization; M_{Nrec}, mineralization rate of recalcitrant organic N pool to NH_4^+; M_{Nlab}, mineralization of labile organic N pool to NH_4^+; I_{NH4}, gross rates of NH_4^+ immobilization; (**b**) N (O_{Nrec} + O_{NH4}), gross rates of nitrification; O_{Nrec}, oxidation rate of recalcitrant organic N to NO_3^- (heterotrophic nitrification rate); O_{NH4}, oxidation rate of NH_4^+ to NO_3^- (autotrophic nitrification rate); I_{NO3}, rate of NO_3^- immobilization to recalcitrant organic N.

The heterotrophic nitrification rate was close to zero at 5 °C, but increased from 5 °C through the 25 °C treatment, where it was maximal (0.12 mg N kg^{-1} day^{-1} soil), but then decreased with increasing temperatures (Figure 3b). Autotrophic nitrification was negligible at 5 °C, 25 °C, and 45 °C, peaking at a rate of 0.29 mg N kg^{-1} day^{-1} soil at 35 °C. Therefore, the gross rate of nitrification continuously increased to 0.33 mg N kg^{-1} day^{-1} soil as the temperature rose from 5 °C to 35 °C, but it rapidly decreased to 0.01 mg N kg^{-1} day^{-1} soil at 45 °C. The NO_3^- immobilization rate was significantly higher than the gross rate of nitrification, leading to net NO_3^- consumption. The NO_3^- immobilization rates increased from 5 °C to 15 °C, and decreased thereafter from 15 °C to 35 °C, yet rapidly increased again to 0.616 mg N kg^{-1} day^{-1} soil at 45 °C. Response functions were fitted to describe the how the above dynamics for gross rates of soil N transformation processes were altered along the experimental incubation temperature gradient (Table 1).

Table 1. Equations of fitted curves of gross rates of soil N transformation processes as a function of incubation temperature shown in Figure 3.

Parameter [1]	Equations
M	$y = 0.4196\exp(0.0553x), R^2 = 0.9574, P < 0.05$
M_{Nlab}	$y = 0.0917 + 0.0432x + 0.0006x^2, R^2 = 0.9509, P < 0.05$
M_{Nrec}	$y = 7.2897 \times 10^{-7}\exp(0.3305x), R^2 = 0.8262, P < 0.05$
I_{NH4}	$y = 0.6310\exp[-0.5(\frac{x-17.9259}{7.9126})^2], R^2 = 0.9800, P < 0.05$
N	$y = 0.4001\exp[-0.5(\frac{x-32.0251}{4.7458})^2], R^2 = 0.9578, P < 0.05$
O_{Nrec}	$y = 0.1150\exp[-0.5(\frac{x-24.9095}{6.9720})^2], R^2 = 0.9983, P < 0.05$
O_{NH4}	$y = 0.0093 + 0.7553\exp[-0.5(\frac{x-31.7954}{2.2733})^2], R^2 = 0.9991, P < 0.05$
I_{NO3}	$y = -0.2863 + 0.1327x - 0.0073x^2 + 8.0250 \times 10^{-5}x^3, R^2 = 0.9834, P < 0.05$
Net M	$y = 0.1084\exp(0.0813x), R^2 = 0.9709, P < 0.05$
Net N	$y = -0.2894 + 0.0145x - 0.0005x^2, R^2 = 0.7406, P < 0.05$

[1] M (M_{Nrec} + M_{Nlab}), gross rates of total N mineralization; M_{Nrec}, mineralization rate of recalcitrant organic N pool to NH_4^+; M_{Nlab}, mineralization of labile organic N pool to NH_4^+; I_{NH4}, gross rates of NH_4^+ immobilization; N (O_{Nrec} + O_{NH4}), gross rates of nitrification; O_{Nrec}, oxidation rate of recalcitrant organic N to NO_3^- (heterotrophic nitrification rate); O_{NH4}, oxidation rate of NH_4^+ to NO_3^- (autotrophic nitrification rate); I_{NO3}, rate of NO_3^- immobilization to recalcitrant organic N.

4. Discussion

Because temperature is a key factor affecting soil microbial activities, it influences microbial-mediated soil N transformations [21–24]. Our results showed that gross rates of individual soil N transformation processes to temperature change responded differently in subtropical acid forest soil; for instance, the mineralization rate of labile organic N pool that significantly increased with temperature rising from 5 °C to 45 °C while the rate of recalcitrant SOM more or less remained stable. Soil enzyme activities are considered limited only by temperature when the supply rate of a substrate exceeds its reaction rate [25]. Therefore, the observation that soil enzyme activities increased with increasing temperature could explain the enhanced mineralization rate of labile organic N pool with temperature. Conversely, the mineralization rate of the recalcitrant organic N pool only increased significantly in the 45 °C treatment, a result possibly explained by the low temperature sensitivity of mineralization of the recalcitrant organic N pool [9]. Previous work done at our study site suggested a low amount of nutrients and a high proportion of recalcitrant compounds characterized the litter of coniferous trees [26]. The temperature sensitivity of microbial-mediated mineralization processes depends on the temperature sensitivity of soil organic matter decomposition [4,6]. Low litter quality likely resulted in low temperature sensitivity for decomposition, thereby requiring a higher temperature for decomposition to finally occur, as it did at 45 °C. The different responses of the gross mineralization rates of soil recalcitrant organic N and labile organic N to temperature change drove the gross N mineralization rates to increase exponentially with rising temperature, a trend that is consistent with many other studies [9,27–29].

Our study found that the gross rate of NH_4^+ immobilization gradually increased from 5 °C to 15 °C, a result in line with work by Binkley et al. (1994) and Cheng et al. (2015), yet it declined in the interval of 15 °C to 45 °C, which is likely regulated by the soil C content as observed in other studies [9,30–32]. Higher temperatures could cause labile C to be rapidly assimilated, which resulted in decreasing in C supply to heterotrophic NH_4^+ immobilizing microorganisms, thereby further reducing the gross rates of soil NH_4^+ immobilization [33–35]. Compared with the gross rate of soil total N mineralization, that of soil NH_4^+ immobilization had a less pronounced response to temperature change. With increasing temperature, an exponentially increasing gross rate of soil total N mineralization coupled to the slight response of soil NH_4^+ immobilization generated a significant increase net rate of soil N mineralization. Therefore, soil N availability is expected to increase markedly under future global warming conditions, promoting the growth of plants and microorganisms in this subtropical acidic forest soil. Autotrophic nitrification and heterotrophic nitrification are pathways of NO_3^- production in soils [36,37]. Our results revealed functionally different responses of these two processes to temperature change. While autotrophic nitrification is mainly driven by ammonia-oxidizing bacteria (AOB) and ammonia-oxidizing archaea (AOA), heterotrophic nitrification is mainly carried out by heterotrophic fungi or bacteria [37–43]. Previous investigations have suggested heterotrophic nitrification is an important and even dominant pathway for NO_3^- production in acidic forest soil [43–45]. Our results confirm this view, in showing that heterotrophic nitrification exceeded autotrophic nitrification at all temperatures except 35 °C. The former's rate first increased, peaking at 0.12 mg N kg^{-1} day^{-1} soil at 25 °C, then declined with higher temperatures, indicating that heterotrophic nitrifying bacteria or fungi in acidic forest soil preferred a more moderate temperature (e.g., 25 °C in this study), such that their activities may decrease at lower or higher temperatures. Liu et al. (2015) reported an optimum temperature for heterotrophic nitrification of ca. 15 °C in an acidic cropping soil, which contrasted with forest soil (coastal western hemlock) studied by Grenon et al. (2004), for which the heterotrophic nitrification rate was augmented by higher temperature [4,43]. Hence, an optimum temperature for heterotrophic nitrification may be ecosystem dependent and further depend on its organic substrates and microorganisms [46,47]. Our acidic coniferous forest soil featured a high content of complex organic matter, which can stimulate the growth of fungi [9]. Furthermore, Zhang et al. (2019) reported recently that refractory organic C content and fungal gene copy numbers of soil were each positively correlated with heterotrophic nitrification rates [47]. In our study, mineralization rate of the soil recalcitrant organic N pool significantly increased as the temperature rose from 35 °C to 45 °C, indicating that refractory organic matter content was unlimited, precluding its influence upon heterotrophic nitrification's response to temperature change. The responses of fungi, which are arguably the main divers of heterotrophic nitrification, can determine how heterotrophic nitrification responds to temperature change [39–41,43]. We found a maximal heterotrophic nitrification rate at 25 °C, which is in line with the optimum temperature for fungal growth, 25–30 °C [48,49].

Soil pH is another critical factor affecting the substrate (NH_3) and thus capable of greatly influencing the rate of autotrophic nitrification, which research suggests is positively correlated with pH [38,50–52]. In our study, the soil autotrophic nitrification rate was very low and it barely responded to the applied temperature gradient, except for the 35 °C treatment, in which it significantly increased. In acidic forest soils, autotrophic nitrification is carried out by AOA [52,53]. Thus, we suggest the response of autotrophic nitrification to temperature change may have been linked to the temperature sensitivity of AOA in the acid coniferous tree soil we studied. The optimum temperature for AOA is 35–40 °C, so greater autotrophic nitrification at 35 °C is expected, as found in our study [54–57]. Disparate responses of autotrophic vis-à-vis heterotrophic nitrification rates to temperature change led to gross rates of nitrification increasing from 5 °C to its peak at 35 °C (0.329mg N kg^{-1} day^{-1} soil).

The NO_3^- immobilization rate was responsible for net NO_3^- consumption and its response to temperature change substantially affected the available N. It is widely thought that NH_4^+ is more easily immobilized by microorganisms than is NO_3^-, even if the size of the NH_4^+ pool is very small [58–60]. However, our results showed that NO_3^- immobilization rates actually exceeded gross rates of nitrification, especially at 15 °C and 45 °C. This may be a unique feature of these acidic forest soils given similar observations for this kind of system [9,45,60–62]. We showed that NO_3^- immobilization rates accelerated, to a maximum of 0.621 mg N kg^{-1} day^{-1} soil at 15 °C, yet decreased at 15 °C to 35 °C but sharply rebounded from 0.233 to 0.616mg N kg^{-1} day^{-1} under 35 °C to 45 °C. This dynamic response of NO_3^- immobilization rates at different temperatures may be linked to specific responses of different microorganisms (e.g., fungi or bacteria) to temperature change [63].

Since the NO_3^- immobilization rate outpaced the gross rate of nitrification, the net NO_3^- immobilization rates decreased with increasing temperature. This result points to the NO_3^- retention capacity in the studied soil being enhanced by warming, which should reduce the risks of leaching and denitrification losses in this acidic subtropical forest under global warming conditions. However, Jansen-Willems et al. (2016) reported that increased soil temperatures could augment both total inorganic N and NO_3^- pools, mainly due to the gross rates of released stored NO_3^- and total nitrification (autotrophic and heterotrophic nitrification) being promoted by warming [64]. Yet more inorganic N did not markedly enhance N_2O emissions under higher temperatures, mainly because of a reduction in the rates of denitrification and the oxidation of organic N [64]. In our study, however, the responses of N_2O emissions to temperature were not determined. We recommend that future research investigate the effects of changing temperature on N_2O emissions and production pathways and the related microbial mechanisms in acidic subtropical forest soils.

5. Conclusions

Our results for a subtropical acidic coniferous forest soil showed that the responses of separate gross N transformation rates to temperature can vary. Because the response of gross rates of NH_4^+ immobilization to temperature change was weaker than that of total N mineralization, the net rate of N mineralization significantly increased with rising temperature. The higher NO_3^- immobilization rates than gross rates of nitrification led to higher net NO_3^- consumption that itself decreased with increasing temperatures. These results suggest that, under global warming conditions, soil N availability would significantly increase, which ought to promote the growth of plants and microorganisms, while retention capacity of soil NO_3^- would also increase, which should reduce the risks of leaching and denitrification losses in subtropical acidic forest soil.

Author Contributions: Conceptualization, Z.C. (Zucong Cai) and J.Z.; Data curation, X.D., Z.C. (Zhaoxiong Chen), S.D., and X.H.; Formal analysis, X.D.; Funding acquisition, S.D.; Investigation, X.D., Z.C. (Zhaoxiong Chen), and X.H.; Methodology, Z.C. (Zucong Cai), J.Z., and C.M.; Project administration, S.D.; Supervision, Z.C. (Zucong Cai) and J.Z.; Validation, X.H.; Writing—original draft, X.D.; Writing—review and editing, C.M.; All authors discussed the results and commented on the manuscript.

Funding: This research was funded by the National Natural Science Foundation of China, grant number 41830642; the CAS Interdisciplinary Innovation Team, grant number JCTD-2018-06; and the "Double World-classes" Development of Geography.

Acknowledgments: The study was carried out as part of the IAEA-funded Coordinated Research Project "Minimizing farming impacts on climate change by enhancing carbon and nitrogen capture and storage in Agro-Ecosystems (D1.50.16)" and it was carried out in close collaboration with the German Science Foundation research unit DASIM (FOR 2337). We would like to thank the native English speaking scientists of Elixigen Company (Huntington Beach, California) for editing our manuscript.

Conflicts of Interest: The authors declare no conflict of interest.

References

1. Compton, J.E.; Boone, R.D. Soil nitrogen transformations and the role of light fraction organic matter in forest soils. *Soil Biol. Biochem.* **2002**, *34*, 933–943. [CrossRef]
2. Templer, P.; Findlay, S.; Lovett, G. Soil microbial biomass and nitrogen transformations among five tree species of the Catskill Mountains, New York, USA. *Soil Biol. Biochem.* **2003**, *35*, 607–613. [CrossRef]
3. Joergensen, R.G.; Brookes, P.C.; Jenkinson, D.S. Survival of the soil microbial biomass at elevated temperatures. *Soil Biol. Biochem.* **1990**, *22*, 1129–1136.
4. Grenon, F.; Bradley, R.L.; Titus, B.D. Temperature sensitivity of mineral N transformation rates, and heterotrophic nitrification: Possible factors controlling the post-disturbance mineral N flush in forest floors. *Soil Biol. Biochem.* **2004**, *36*, 1465–1474. [CrossRef]
5. Lang, M.; Cai, Z.C.; Mary, B.; Hao, X.; Chang, S.X. Land-use temperature affect gross nitrogen transformation rates in Chinese and Canadian soils. *Plant Soil* **2010**, *334*, 377–389. [CrossRef]
6. Magill, A.H.; Aber, J.D. Variation in soil net mineralization rates with dissolved organic carbon additions. *Soil Biol. Biochem.* **2000**, *32*, 597–601. [CrossRef]
7. Hart, S.C.; Nason, G.E.; Myrold, D.D.; Perry, D.A. Dynamics of gross nitrogen transformations in an old-growth forest: The carbon connection. *Ecology* **1994**, *75*, 880–891. [CrossRef]
8. Guntiñas, M.E.; Leirós, M.C.; Trasar-Cepeda, C.; Gil-Sotres, F. Effects of moisture and temperature on net soil nitrogen mineralization: A laboratory study. *Eur. J. Soil Biol.* **2012**, *48*, 73–80. [CrossRef]
9. Cheng, Y.; Wang, J.; Zhang, J.B.; Wang, S.Q.; Cai, Z.C. The different temperature sensitivity of gross N transformations between the coniferous and broad-leaved forests in subtropical China. *Soil Sci. Plant Nutr.* **2015**, *61*, 506–515. [CrossRef]
10. Huygens, D.; Rütting, T.; Boeckx, P.; Van Cleemput, O.; Godoy, R.; Müller, C. Soil nitrogen conservation mechanisms in a pristine south Chilean Nothofagus forest ecosystem. *Soil Biol. Biochem.* **2007**, *39*, 2448–2458. [CrossRef]
11. Rütting, T.; Huygens, D.; Müller, C.; Van Cleemput, O.; Godoy, R.; Boeckx, P. Functional role of DNRA and nitrite reduction in a pristine south Chilean *Nothofagus* forest. *Biogeochemistry* **2008**, *90*, 243–258. [CrossRef]
12. Zaman, M.; Chang, S.X. Substrate type, temperature, and moisture content affect gross and net N mineralization and nitrification rates in agroforestry systems. *Biol. Fertil. Soils* **2004**, *39*, 269–279. [CrossRef]
13. Stark, J.M.; Hart, S.C. High rates of nitrification and nitrate turnover in undisturbed coniferous forests. *Nature* **1997**, *385*, 61–64. [CrossRef]
14. Han, W.Y.; Xu, J.M.; Yi, X.; Lin, Y.D. Net and gross nitrification in tea soils of varying productivity and their adjacent forest and vegetable soils. *Soil Sci. Plant Nutr.* **2012**, *58*, 173–182. [CrossRef]
15. Mary, B.; Recous, S.; Robin, D. A model for calculating nitrogen fluxes in soil using 15N tracing. *Soil Biol. Biochem.* **1998**, *30*, 1963–1979. [CrossRef]
16. Müller, C.; Rutting, T.; Kattge, J.; Laughlin, R.J.; Stevens, R.J. Estimation of parameters in complex ^{15}N tracing models by Monte Carlo sampling. *Soil Biol. Biochem.* **2007**, *39*, 715–726. [CrossRef]
17. Zhao, Q.G.; Xie, W.M.; He, X.Y.; Wang, M.Z. *Red Soils of Jiangxi Province*; Jiangxi Science and Technology Publishing House: Nanchang, China, 1988.

18. Lu, R.K. *Soil Agro-Chemical Analyses*; Agricultural Technical Press of China: Beijing, China, 2000.
19. Brooks, P.D.; Stark, J.M.; McInteer, B.B.; Preston, T. Diffusion method to prepare soil extracts for automated nitrogen-15 analysis. *Soil Sci. Soc. Am. J.* **1989**, *53*, 1707–1711. [CrossRef]
20. Müller, C.; Laughlin, R.J.; Christie, P.; Watson, C.J. Effects of repeated fertilizer and slurry applications over 38 years on N dynamics in a temperate grassland soil. *Soil Biol. Biochem.* **2011**, *43*, 1362–1371. [CrossRef]
21. Parton, W.J.; Schimel, D.S.; Cole, C.V.; Ojima, D.S. Analysis of factors controlling soil organic matter levels in Great Plains Grasslands. *Soil Sci. Soc. Am. J.* **1987**, *51*, 1173–1179. [CrossRef]
22. Insam, H. Are the soil microbial biomass and basal respiration governed by the climatic regime? *Soil Biol. Biochem.* **1990**, *22*, 525–532. [CrossRef]
23. Dalias, P.; Anderson, J.M.; Bottner, P.; Coûteaux, M.M. Temperature responses of net nitrogen mineralization and nitrification in conifer forest soils incubated under standard laboratory conditions. *Soil Biol. Biochem.* **2002**, *34*, 691–701. [CrossRef]
24. Wang, C.; Wan, S.; Xing, X.; Zhang, L.; Han, X. Temperature and soil moisture interactively affected soil net N mineralization in temperate grassland in Northern China. *Soil Biol. Biochem.* **2006**, *38*, 1101–1110. [CrossRef]
25. Giardina, C.P.; Ryan, M.G. Evidence that decomposition rates of organic carbon in mineral soil do not vary with temperature. *Nature* **2000**, *404*, 858–861. [CrossRef] [PubMed]
26. Reich, P.B.; Oleksyn, J.; Modrzynski, J.; Mrozinski, P.; Hobbie, S.E.; Eissenstat, D.M.; Tjoelker, M.G.; Chorover, J.; Chadwick, O.A.; Hale, C.M. Linking litter calcium, earthworms and soil properties: A common garden test with 14 tree species. *Ecol. Lett.* **2005**, *8*, 811–818. [CrossRef]
27. Cookson, W.R.; Cornforth, I.S.; Rowarth, J.S. Winter soil temperature (2–15 °C) effects on nitrogen transformations in clover green manure amended or unamended soils: A laboratory and field study. *Soil Biol. Biochem.* **2002**, *34*, 1401–1415. [CrossRef]
28. Cheng, Y.; Wang, J.; Wang, S.; Cai, Z.; Wang, L. Effects of temperature change and tree species composition on N_2O and NO emissions in acidic forest soils of subtropical China. *J. Environ. Sci.* **2014**, *26*, 617–625. [CrossRef]
29. Miller, K.S.; Geisseler, D. Temperature sensitivity of nitrogen mineralization in agricultural soils. *Biol. Fertil. Soils* **2018**, *54*, 853–860. [CrossRef]
30. Binkley, D.; Stottlemyer, R.; Saurez, F.; Cortina, J. Soil nitrogen availability in some arctic ecosystems in northwest Alaska: Responses to temperature and moisture. *Ecoscience* **1994**, *1*, 64–70. [CrossRef]
31. Bengtson, P.; Falkengren-Grerup, U.; Bengtsson, G. Relieving substrate limitation-soil moisture and temperature determine gross N transformation rates. *Oikos* **2005**, *111*, 81–90. [CrossRef]
32. Booth, M.S.; Stark, J.M.; Rastetter, E. Controls on nitrogen cycling in terrestrial ecosystems: A synthetic analysis of literature data. *Ecol. Monogr.* **2005**, *75*, 139–157. [CrossRef]
33. Hoyle, F.C.; Murphy, D.V.; Fillery, I.R.P. Temperature and stubble management influence microbial CO2-C evolution and gross N transformation rates. *Soil Biol. Biochem.* **2006**, *38*, 71–80. [CrossRef]
34. Cookson, W.R.; Osman, M.; Marschner, P.; Abaye, D.A.; Clark, I.; Murphy, D.V.; Stockdale, E.A.; Watson, C.A. Controls on soil nitrogen cycling and microbial community composition across land use and incubation temperature. *Soil Biol. Biochem.* **2007**, *39*, 744–756. [CrossRef]
35. Rennenberg, H.; Dannenmann, M.; Gessler, A.; Kreuzwieser, J.; Simon, J.; Papen, H. Nitrogen balance in forest soils: Nutritional limitation of plants under climate change stresses. *Plant Biol.* **2009**, *11*, 4–23. [CrossRef] [PubMed]
36. Pedersen, H.; Dunkin, K.A.; Firestone, M. The relative importance of autotrophic and heterotrophic nitrification in a conifer forest soil as measured by N tracer and pool dilution techniques. *Biogeochemistry* **1999**, *44*, 135–150. [CrossRef]
37. De Boer, W.; Kowalchuk, G.A. Nitrification in acid soils: Micro-organisms and mechanisms. *Soil Biol. Biochem.* **2001**, *33*, 853–866. [CrossRef]
38. Liu, R.; Suter, H.; He, J.; Hayden, H.; Chen, D. Influence of temperature and moisture on the relative contributions of heterotrophic and autotrophic nitrification to gross nitrification in an acid cropping soil. *J. Soils Sediments* **2015**, *15*, 2304–2309. [CrossRef]
39. Hirsch, P.; Overrein, L.; Alexander, M. Formation of nitrite and nitrate by actinomycetes and fungi. *J. Bacteriol.* **1961**, *82*, 442–448. [PubMed]
40. Doxtader, K.G.; Alexander, M. Nitrification by heterotrophic soil microorganisms. *Soil Sci. Soc. Am. J.* **1966**, *30*, 351–355. [CrossRef]

41. Kreitinger, J.P.; Klein, T.M.; Novick, N.J.; Alexander, M. Nitrification and characteristics of nitrifying microorganisms in an Acid Forest Soil. *Soil Sci. Soc. Am. J.* **1985**, *49*, 1407–1410. [CrossRef]

42. Zhu, T.B.; Meng, T.Z.; Zhang, J.B.; Yin, Y.; Cai, Z.; Yang, W.; Zhong, W. Nitrogen mineralization, immobilization turnover, heterotrophic nitrification, and microbial groups in acid forest soils of subtropical China. *Biol. Fertil. Soils* **2013**, *49*, 323–331. [CrossRef]

43. Zhu, T.; Meng, T.; Zhang, J.; Zhong, W.; Müller, C.; Cai, Z. Fungi-dominant heterotrophic nitrification in a subtropical forest soil of China. *J. Soils Sediments* **2015**, *15*, 705–709. [CrossRef]

44. Killham, K. Nitrification in coniferous forest soils. *Plant Soil* **1990**, *128*, 31–44. [CrossRef]

45. Zhang, J.B.; Müller, C.; Zhu, T.B.; Cheng, Y.; Cai, Z. Heterotrophic nitrification is the predominant NO_3^- production mechanism in coniferous but not broad-leaf acid forest soil in subtropical China. *Biol. Fertil. Soils* **2011**, *47*, 533–542. [CrossRef]

46. Zhang, J.; Sun, W.; Zhong, W.; Cai, Z. The substrate is an important factor in controlling the significance of heterotrophic nitrification in acidic forest soils. *Soil Biol. Biochem.* **2014**, *76*, 143–148. [CrossRef]

47. Zhang, Y.; Liu, S.; Cheng, Y.; Cai, Z.; Müller, C.; Zhang, J. Composition of soil recalcitrant C regulates nitrification rates in acidic soils. *Geoderma* **2019**, *337*, 965–972. [CrossRef]

48. Pietikäinen, J.; Pettersson, M.; Bååth, E. Comparison of temperature effects on soil respiration and bacterial and fungal growth rates. *FEMS Microbiol. Ecol.* **2005**, *52*, 49–58. [CrossRef] [PubMed]

49. Wood, T.G. Termites and the soil environment. *Biol. Fertil. Soils* **1988**, *6*, 228–236. [CrossRef]

50. Zhao, W.; Cai, Z.C.; Xu, Z.H. Does ammonium-based N addition influence nitrification and acidification in humid subtropical soils of China? *Plant Soil* **2007**, *297*, 213–221. [CrossRef]

51. Zhang, J.; Zhu, T.; Cai, Z.; Müller, C. Nitrogen cycling in forest soils across climate gradients in Eastern China. *Plant Soil* **2011**, *342*, 419–432. [CrossRef]

52. Jia, Z.J.; Conrad, R. Bacteria rather than Archaea dominate microbial ammonia oxidation in an agricultural soil. *Environ. Microbiol.* **2009**, *11*, 1658–1671. [CrossRef] [PubMed]

53. Zhang, L.M.; Offre, P.R.; He, J.Z.; Verhamme, D.T.; Nicol, G.W.; Prosser, J.I. Autotrophic ammonia oxidation by soil thaumarchaea. *Proc. Natl. Acad. Sci. USA* **2010**, *107*, 17240–17245. [CrossRef] [PubMed]

54. Ouyang, Y.; Norton, J.M.; Stark, J.M. Ammonium availability and temperature control contributions of ammonia oxidizing bacteria and archaea to nitrification in an agricultural soil. *Soil Biol. Biochem.* **2017**, *113*, 161–172. [CrossRef]

55. Taylor, A.E.; Giguere, A.T.; Zoebelein, C.M.; Myrold, D.D.; Bottomley, P.J. Modeling of soil nitrification responses to temperature reveals thermodynamic differences between ammonia-oxidizing activity of archaea and bacteria. *ISME J.* **2017**, *11*, 896–908. [CrossRef] [PubMed]

56. Duan, P.; Wu, Z.; Zhang, Q.; Fan, C.; Xiong, Z. Thermodynamic responses of ammonia-oxidizing archaea and bacteria explain N_2O production from greenhouse vegetable soils. *Soil Biol. Biochem.* **2018**, *120*, 37–47. [CrossRef]

57. Jones, J.M.; Richards, B.N. Effect of reforestation on turnover of ^{15}N-labelled nitrate and ammonia in relation to changes in soil microflora. *Soil Biol. Biochem.* **1977**, *9*, 383–392. [CrossRef]

58. Jackson, L.E.; Schimel, D.S.; Firestone, M.K. Short-term partitioning of ammonium and nitrate between plants and microbes in an annual grassland. *Soil Biol. Biochem.* **1989**, *21*, 409–415. [CrossRef]

59. Recous, S.; Machet, J.M.; Mary, B. The partitioning of fertiliser-N between soil and crop: Comparison of ammonium and nitrate applications. *Plant Soil* **1992**, *144*, 101–111. [CrossRef]

60. Zak, D.R.; Groffman, P.M.; Pregitzer, K.S.; Christensen, S.; Tiedje, J.M. The vernal dam: Plant-microbe competition for nitrogen in northern hardwood forests. *Ecology* **1990**, *71*, 651–656. [CrossRef]

61. Davidson, E.A.; Stark, J.M.; Firestone, M.K. Microbial production and consumption of nitrate in an annual grassland. *Ecology* **1990**, *71*, 1968–1975. [CrossRef]

62. Davidson, E.A.; Hart, S.C.; Firestone, M.K. Internal cycling of nitrate in soils of a mature coniferous forest. *Ecology* **1992**, *73*, 1148–1156. [CrossRef]

63. Ziegler, S.E.; Billings, S.A.; Lane, C.S.; Li, J.; Fogel, M.L. Warming alters routing of labile and slower-turnover carbon through distinct microbial groups in boreal forest organic soils. *Soil Biol. Biochem.* **2013**, *60*, 23–32. [CrossRef]

64. Jansen-Willems, A.B.; Lanigan, G.J.; Clough, T.J.; Andresen, L.C.; Müller, C. Long-term elevation of temperature affects organic N turnover and associated N_2O emissions in a permanent grassland soil. *Soil* **2016**, *2*, 601–614. [CrossRef]

 *forests*

Review

Above- and Below-Ground Carbon Sequestration in Shelterbelt Trees in Canada: A Review

Rafaella C. Mayrinck *, Colin P. Laroque, Beyhan Y. Amichev and Ken Van Rees

School of Environment and Sustainability, University of Saskatchewan, Maintenance Road, 110, Saskatoon, SK S7N 5C5, Canada; colin.laroque@usask.ca (C.P.L.); beyhan.amichev@vt.edu (B.Y.A.); kcv903@mail.usask.ca (K.V.R.)
* Correspondence: rcm786@mail.usask.ca; Tel.: +1-306-850-2717 or +55-15-9-9798-5054

Received: 26 August 2019; Accepted: 17 October 2019; Published: 19 October 2019

Abstract: Shelterbelts have been planted around the world for many reasons. Recently, due to increasing awareness of climate change risks, shelterbelt agroforestry systems have received special attention because of the environmental services they provide, including their greenhouse gas (GHG) mitigation potential. This paper aims to discuss shelterbelt history in Canada, and the environmental benefits they provide, focusing on carbon sequestration potential, above- and below-ground. Shelterbelt establishment in Canada dates back to more than a century ago, when their main use was protecting the soil, farm infrastructure and livestock from the elements. As minimal-and no-till systems have become more prevalent among agricultural producers, soil has been less exposed and less vulnerable to wind erosion, so the practice of planting and maintaining shelterbelts has declined in recent decades. In addition, as farm equipment has grown in size to meet the demands of larger landowners, shelterbelts are being removed to increase efficiency and machine maneuverability in the field. This trend of shelterbelt removal prevents shelterbelt's climate change mitigation potential to be fully achieved. For example, in the last century, shelterbelts have sequestered 4.85 Tg C in Saskatchewan. To increase our understanding of carbon sequestration by shelterbelts, in 2013, the Government of Canada launched the Agricultural Greenhouse Gases Program (AGGP). In five years, 27 million dollars were spent supporting technologies and practices to mitigate GHG release on agricultural land, including understanding shelterbelt carbon sequestration and to encourage planting on farms. All these topics are further explained in this paper as an attempt to inform and promote shelterbelts as a climate change mitigation tool on agricultural lands.

Keywords: agroforestry systems; carbon sequestration; climate change mitigation; windbreaks; shelterbelts

---

1. Overview: Shelterbelt Qualities and Their Role around the World

Shelterbelts, also known as windbreaks, are agroforestry systems that can be defined as barriers of trees, or trees combined with shrubs, that are planted to reduce wind speed [1–4]. Hedges are a similar feature, defined as a narrow row of a low and dense shrub species used to separate fields [4]. Sometimes shelterbelt and hedge concepts can be interchangeable, because shelterbelts are also used to separate fields and hedges end up reducing wind speed. However, in this paper, we will be focusing on the aspects of shelterbelts only.

Shelterbelts have been established all over the world to protect soil, crops, homes, farm infrastructure, livestock, and pastures. In Britain, shelterbelts were largely planted in the mid-18th century for crop protection and to keep farm pollution away from busy roads [5]. In the United States (U.S.), a shelterbelt-incentive program was carried out by the Prairie States Forestry Project (PSFP), which resulted in nearly 30,000 km of shelterbelts planted from 1935 to 1942 across six Great Plains states [6]. In China, shelterbelts have been used to isolate the coastal zone from sea and land disturbances [7], to protect agricultural systems from dry winds and sandstorms [8], and to stabilize sand dunes [9]. In 1950, an extensive shelterbelt planting took place, aimed at defeating agricultural lands from erosion [7]. Later, the "Three-North Shelterbelt Project" started, and has increased treed land area from 5%, in 1978, to 10%, in 2008 [9]. In New Zealand, landowners have planted shelterbelts since 1850, when the settlers arrived, totaling more than 300,000 km in length [10]. In Australia, shelterbelts were planted on treeless areas such as the western plains of Victoria [11]. In Argentina, there are more than 1500 km of windbreaks planted to protect crops, cattle and homes from wind [12].

Shelterbelts can be composed of perennial and or annual trees and shrubs [1–3]. The species chosen should be adapted to local climate, topography, and soil [13]. To make it sustainable through time, it is recommended to alternate rows of fast and slow-growing species [14], creating a forest-like dynamic. Fast-growing species start protecting the area earlier allowing the slow-growing species to reach maturity when the fast-growing species are in decline, thus always maintaining an effective shelter. This system enriches biodiversity, while producing wood that can be harvested periodically for fencing, furniture and housing, as well as increasing carbon residence time in the system. Combining species within the overall design makes the shelterbelt system sustainable through time, as well as making the system more resistant to pests and disease, diversifying shelterbelt structure and assisting to mitigate any of its vulnerabilities [3,13].

Ideal shelterbelt structure and design depend on its function [2]. For example, for wind protection, it should have multiple rows (usually 2–3) of trees to achieve high shelterbelt density, located at 2–5 times the shelterbelt height (H) from the edge of the field, to increase the amount of land protected and amplify economic returns [1,2,13]. For snow management, normally, the ideal design is planting one single row with a tall deciduous tree species using wide spacing (5 to 7 m between trees to achieve a medium shelterbelt density), perpendicular to the prevailing winds [2]. In many cases, there are other directions that winds can put crops in danger, rather than the prevailing winds, so it is beneficial to have two right-angle oriented shelterbelt rows [13]. For severe winters, as in the Canadian Prairies, five to seven rows may provide the ideal protection from weather events [13]. For livestock systems, shelterbelt rows should be dense, planted at narrow spacing (2–3 m between trees), so that the animals are protected from associated wind chill [2]. Normally, one windbreak is not enough to protect a whole field, so more rows need to be added, within a distance of 10–20 times the shelterbelt height, depending on the level of protection desired, size of equipment used, and degree of crop tolerance to wind [2,15]. For example, Helmers and Brandle [16] recommended to add a shelterbelt every 13 H for corn and soybean production in a 70-year planning horizon.

An important factor on establishing shelterbelts is row spacing indicated by the distance between planted trees within a shelterbelt row. If narrow spacing is adopted, the trees will shade the soil beneath much sooner, which can reduce the costs of weed control; however, the disadvantage is that trees will be competing earlier for resources, and if not managed properly, could lead to reducing their health and growth [13]. Wider spacing also has disadvantages, since the trees will take longer to shade the soil to avoid weed competition, so the landowner will have to combat weeds; the trees will develop larger crowns, demanding more water; greater tree and soil exposure to the sun and wind between the rows will also increase evapotranspiration. Between shrubs, conifers and deciduous trees the minimal recommended spacing is 0.3–1, 2.0–2.5 and 3 m, respectively [17].

Shelterbelt maintenance is similar to forest maintenance. At their establishment, weeding and pest control should be conducted, to help seedlings get established. If the shelterbelt is planted in a pasture or an area with wild animals, building a fence should be considered to protect the seedlings from animal injuries [14,17]. Later, branch pruning and thinning of densely planted rows may be required in some cases, to boost height and diameter growth, respectively [14].

The efficiency of a shelterbelt as a wind barrier is determined by its external and internal structure. Its external structure is related to its height, length, orientation, continuity, width, and cross-sectional shape; its internal structure is related to the amount and structure of open and solid spaces in the tree crowns, plant shape, and surface area [1–3]. Shelterbelt height and length determine the extent of windbreak protection [1,3]. Shelterbelt length should be 10 [1–3] to 20 [15] times its height, to reduce wind flow around the ends of the shelterbelt [1–3].

As shelterbelt use and importance become more popular around the world, more research on the topic has been published. Figure 1A illustrates the percentage of existing literature published per year, available on the Web of Science, on the topic of shelterbelts and the respective percentages of published papers per year on the topics of agriculture and forestry. It was expected that agriculture and forestry publication rates would surpass shelterbelt publication rates for three main reasons. First, agriculture and forestry are crucial to feed the increasing demand of the growing global population. Second, shelterbelts are normally not established with the intention of producing material goods as they are for agriculture and forestry. Third, the environmental benefits provided by shelterbelts are only mostly experienced in the long run and are difficult to be translated into monetary values, unlike the products from agriculture and forestry. However, shelterbelt publications were increasing at a pace similar to agriculture and forestry rates, rising sharply after the 90s (Figure 1A). This may be attributed to the increasing access to computers, making it easy to process studies and publish more, or perhaps to the increasing awareness on climate change and environmental issues. In either case, awareness on the importance of shelterbelts and their environmental services was increasing, regardless of their smaller role on providing material goods. Similar searches were made using the keyword windbreak, instead of shelterbelt which yielded the same trend as shown in Figure 1A. The total contribution to shelterbelt research by country varied (Figure 1B). The top three leading countries on shelterbelt research are (in decreasing order) China, U.S., and Canada.

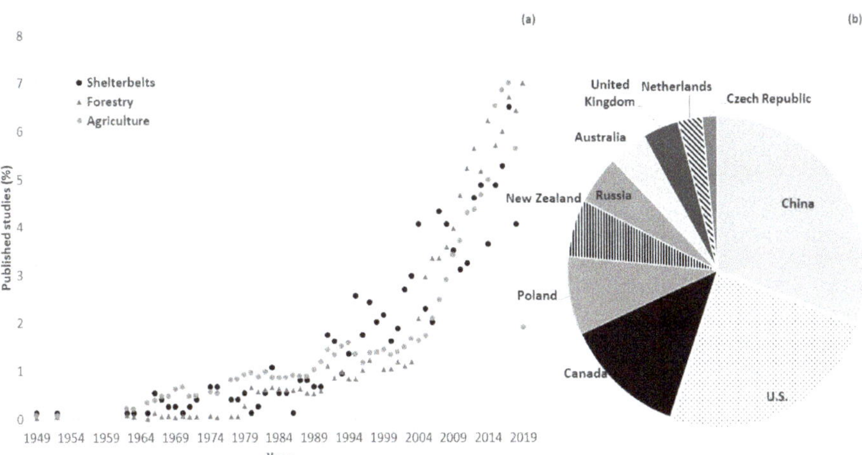

Figure 1. Number of journal publications on shelterbelt agroforestry systems found on Web of Science (**a**), and the relative contribution by country (**b**).

Shelterbelt researchers in Canada have contributed much to the literature due to the great number of shelterbelts planted over the years. According to the Government of Canada, farmers in Western Canada have planted more than 600 million trees during the past century [18]. Just in Saskatchewan alone, the total length of shelterbelts equals to 60,633 km [19], sequestering around 4.85 Tg C in the past nine decades, more than half of which were sequestered since 1990 (3.77 Tg C) [20].

The motivations driving farmers around the world to plant shelterbelts have been in general agreement: to protect farm yards, crops, infrastructure, and livestock from the harsh environment, and for aesthetics [1,2,5]. However, this general motivation has changed over the years to address the needs of each generation of farmers; thus, new reasons for planting and maintaining shelterbelts were added. For example, in New Zealand, shelterbelts planting was also focused on timber production [21]. In Canada, as described above, shelterbelts have been planted mainly to protect soil from wind erosion;

however, because of new agricultural techniques, such as the no-till system, wind erosion is less of an issue than in past decades, and some farmers are currently removing their shelterbelts [1,2]. In a recent study, it was determined that 29.8% of farmers in a 1400 km^2 study area in Saskatchewan, Canada, removed their shelterbelts between 2008 and 2016 [21]. The main reasons that farmers were removing their existing shelterbelts were: (1) shelterbelts required additional labor for maintenance; (2) shelterbelts made it difficult to operate large farming equipment in crop fields; and (3) shelterbelts reduced the land area available for crop production [21–24].

However, the benefits provided by the shelterbelt may be able to overcome even these new disadvantages. Some researchers have suggested that farmers are not aware of the advantages of having shelterbelts, so, they are more prone to remove them [23–25]. It is very likely that if education on the environmental benefits that shelterbelts provide, mainly their carbon sequestration, were projected into the long term, it would help farmers to better value them. Rewarding farmers for keeping existing, and or, planting new shelterbelts, by for example granting them a tax reduction or tax credit in a carbon trading market place [21], would make landowners more prone to stop removing them. If a carbon market were implemented, any carbon sequestered by shelterbelt trees would have a monetary value, so trees would become a part of the farm's overall budget, which could motivate future shelterbelt planting and better maintenance of existing ones.

2. Shelterbelt History in the Canadian Prairies

When European settlers first started to populate Western Canada in the late 1890s, they were encouraged to settle in various regions, including the driest area, known as the Palliser Triangle [24,26]. This area encompasses southern Saskatchewan, into Alberta, and Manitoba including the Canada–U.S. border, covering more than 200,000 square kilometers [24,26]. There is even a dryer area inside the Palliser Triangle, named the Dry Belt, located in southern Alberta and Saskatchewan [24,26]. This land is extremely arid, and has had many cycles of extreme dry years through time [26].

Between the 1880s and 1980s, several droughts occurred in the Canadian Prairies: in 1910, 1914, 1917–20, 1924, 1929–30 [6,26]. During these dry periods, shelterbelts were useful in capturing snow, and helped increase and maintain soil moisture. Planted trees improved the soil water regime by shading and reducing soil surface temperatures [6], all of which reduce soil evapotranspiration. At the time, there was also a false belief that planting trees would increase local rainfall [6,26], which contributed to increases in local shelterbelt planting. The 1930s drought occurred during one of the worst crises in North America, known as the Great Depression. It was a combination of drought, insect infestation, and dropping of global commodities prices [26]. The Great Depression was so brutal that, in 1936, about 14,000 people left their farms, totaling around 12,140 km^2 of abandoned land within the Palliser Triangle alone [6,26].

To aid farmers in Western Canada facing Depression-like conditions, the Government of Canada established the Prairie Farm Rehabilitation Administration (PFRA) in 1935, in Indian Head, Saskatchewan, where 200 hundred agrologists, engineers, field husbandmen, inspectors and administrators focused on rehabilitating farms in the Prairies after the Great Depression [26]. The PFRA received $750,000 for the first year, and $1 million annually for the next four years, to help combat the consequences of drought [26]. By 1939, the PFRA assisted with the construction of thousands of dugouts and earthen dams designed for stocking water [26]. Additionally, the PFRA assisted farms in many ways, including the management of community pastures in Saskatchewan and Manitoba, and by conducting a soil survey covering 90% of the Palliser Triangle, at a one-mile resolution [26].

One of the most notable legacies of the PFRA activity included providing free seedlings and assistance for shelterbelt establishment to farmers, known as the Prairie Shelterbelt Program (PSP) [27,28]. A nursery in Indian Head was created to function as a live demonstration farm to show planting options, and serve as a reference for farmers displaying the diversity of shelterbelt agroforestry systems that could be implemented on the Prairies [6]. Initially, seedling demand was low, but this increased through time, from 1000 to 9.2 million trees per year, peaking in the particular years of 1961, 1970, 1981 and 1991, probably due to landowners receiving shelterbelt-related information from PFRA (i.e., hand-outs, talks at agricultural fairs and shows) related to droughts that occurred in the previous years [28]. As the demand for seedlings changed, the number of species offered also changed, increasing from five, at the beginning of the program, to 37 shelterbelt species in the 2000s [28]. It is calculated that in total, over 600 million shelterbelt trees/shrubs were provided by the PSP program [28,29]. Towards the end of the program, the demand for shelterbelt trees dropped, and in 2006 and 2009, 3.7 million and 2.5 million seedlings were ordered, respectively [28]. In the end, the PSP program was shut down in 2013.

The work promoted by PFRA's PSP program on implementing shelterbelts for the past nine decades showed Canada's commitment to sustainable development in the Prairies. This commitment was reaffirmed in 2009 by signing the Copenhagen Accord, when Canada committed to reduce GHG emissions by 17% by 2020, based on 2005 levels. To help with reaching the goal, the Federal government launched the Agricultural Greenhouse Gases Program (AGGP) in 2013. The goal of AGGP is to mitigate GHG emissions in the agricultural sector by creating technologies and practices that promote carbonless agriculture [30]. One of the aims of AGGP is to study shelterbelt carbon sequestration and their potential for climate change mitigation. This branch financed the majority of studies on shelterbelts in Canada since its inception. In 2016, the Paris Accord was signed by 170 countries, including Canada, which required committed countries to join efforts to keep global temperature increases under 2 °C, based on pre-industrial levels, with further efforts aimed at limiting warming to 1.5 °C [31]. As stipulated by the AGGP program, and in the context of global efforts towards a carbonless economy, the carbon sequestration potential of shelterbelts remains a viable research priority for the Canadian Prairies.

3. Environmental Services Provided by Shelterbelts

Shelterbelts provide both public and private benefits, commonly referred to as environmental services [32]. Private benefits include protecting soil, homes, farm infrastructure, and livestock from the elements [23,28,32–34], reducing animal odor from livestock systems, lowering the risk of crop environmental damage due to pesticide spray-drift [24,28,35], reducing noise [13], and heating costs for households and livestock operations [7,28]. It is estimated that shelterbelts can save up to 18% in energy costs for heating homes [36].

Public benefits provided by shelterbelts include reducing soil runoff into rivers, streams and creeks, sequestering carbon dioxide from the atmosphere, enhancing and protecting animal and plant diversity, including pollinators, as well as improving water quality [18,32,37]. Some shelterbelt benefits can be categorized as both private and public. For example, improving water quality and protecting biodiversity are more commonly classified as public environmental services; however, these can also be considered as private benefits since it would also be beneficial on the land where the shelterbelt is located.

These benefits are hard to quantify in terms of monetary value [18,24,32], due to the complexity of the factors affecting, and being affected by, shelterbelt systems, but there are some studies that address this question. Kulshreshtha et al. [32] assessed public service worth provided by shelterbelt seedlings given by PFRA's PSP program from 1981 to 2001. They found that those benefits were worth $140 million, the majority provided by carbon sequestration ($73 million), and soil erosion reduction ($15 million). Similarly, Amichev et al. [20] studied carbon sequestration of six common shelterbelt species in Saskatchewan, planted from 1925 to 2009, and estimated that the carbon additions (through CO_2 sequestration) in shelterbelt systems since 1990 (equal to 3.77 Tg C) would be worth $208 million dollars at current carbon prices.

Because shelterbelts can improve their surrounding conditions through environmental services, they can also impact crop production by retaining soil moisture, slowing wind speed, shading areas beside the trees, and reducing soil loss [1,15,38]. Shelterbelts can increase monetary gains for landowners by increasing crop yield and/or saving on chemical applications [20,24,32]. Normally, these benefits can span up to a distance of 10 H on the leeward and 0–3 H on the windward side of the trees [38]. For example, it was calculated that shelterbelts planted from 1981 to 2001 in the Prairie Provinces in Canada, prevented soil erosion, equal to a benefit of $15 million [32]. Shelterbelts were also calculated to reduce crop production costs related to the use of pesticides, and reduce overall crop loss due to pest damage [39,40]. Gámez-Viruez et al. [40], studying feces from birds inhabiting shelterbelts in Australia, found that birds fed on crop pests, helping as a biological control. Shelterbelts can also help to keep a more stable soil temperature range in its surroundings. During the night, soil temperatures near shelterbelts are from 1–2 °C higher than in open fields, which can assist crops to germinate and grow faster in colder environments [1].

Baldwin [38] conducted a review on the effect of shelterbelts on crop production and concluded that gains can be up to 50%. Kort [41] completed an extensive review on how the crops respond to shelterbelt trees and found that in most cases, shelterbelts increased crop yield, and that this can be further maximized by choosing adequate shelterbelt species and designs, based on specific crops. The author concluded that wheat (*Triticum aestivum* L.), barley (*Hordeum vulgare* L.), rye (*Secale cereal* L.), millet (*Pennisetum americanum* L.), alfalfa (*Medicago sativa* L.), and hay (mixed grass and legumes) yields are more responsive to shelterbelt presence, and that oats (*Avena sativa* L.), and maize (*Zea may* L.) are affected as well, though they are less responsive. Hawke and Tombleson [10] observed a 15% increase in pasture production on both sides of shelterbelts at distances equal to 70% of the tree height.

The general rule is that shelterbelts have positive impacts on adjacent crop yield [1,15,38,42], which are more pronounced during short-duration or more intense droughts [25,41,43,44]. However, there are cases where shelterbelts impact yield negatively. Land trade-off is one factor leading to decreasing yield since shelterbelts take out land area from crop production [41]. To make shelterbelts economically viable, crop yield increases by a shelterbelt's presence should be more than compensated for the yield lost in the area that is used to plant the shelterbelt. In the long run, the benefits from the shelterbelt will therefore be felt economically [44]. According to Brandle et al. [2] shelterbelts are economically viable if less than 6% of the land is occupied by trees.

Another factor that may decrease crop yield is allelopathy on the adjacent crop, as it can inhibit seed germination and overall crop growth. Another factor is shading, created by the trees, that can reduce crop photosynthesis and growth from reduced sunlight [6,15,18,37]. For example, Kowalchuk and Jong [45,46] assessed the effect of shelterbelts on wheat yield and soil erosion for three years and found that when environmental conditions were dry, trees and crops competed for moisture, and yield was reduced at distances up to 10 m from the shelterbelt edge. Singh and Kohli [47] studied the effect of an eight-year-old *Eucalyptus tereticornis* Sm. shelterbelt on yields of chickpea (*Cicer arietinum* L.), lentil (*Lens culinaris* (LENCU) wheat, cauliflower (*Brassica oleracea* L.) and barseem (*Trifolium alexandrinum* L.) and concluded that yield was always reduced by shelterbelts and that chickpeas were the most affected crop from the list.

Table 1 shows the effect of shelterbelts on crops reported in the literature. It varies from case to case and can be positive or negative for the same crop, depending on many environmental factors, varying from year-to-year environmental inputs. In most cases, shelterbelt impact was positive. In the cases of decreasing yields, the results were mainly attributed to below-ground tree/crop competition for soil moisture and nutrients. One alternative to reducing below-ground competition is pruning the lateral tree roots [2,48,49]. The frequency of root pruning was dependent on the shelterbelt and crop species and spacing, but is normally done every one to five years [2,42]. It was calculated that in North America, root pruning can decrease competition from 10 to 44% in the adjacent shelterbelt-influenced area [2,38]. Onyewotu et al. [49] found that millet yield increased after pruning roots at 0.25 H distance from *Eucalyptus camaldulensis* Dehnh shelterbelt roots beside the field. Lyles et al. [42] also found that yield from 1 to 2 H distance were 1.6 higher than on the unpruned zone. However, root pruning is an expensive operation. To avoid root pruning, it is important to choose species with deep root systems, so they do not spread laterally and compete with crop root systems [2,42,43]. Greb and Black [43] found that American elm (*Ulmus Americana* L.), black walnut (*Juglans nigra* L.), ponderosa pine (*Pinus ponderosa* Douglas ex P. Lawson & C. Lawson, and Siberian elm (*Ulmus pumila* L.) have shallow roots, so they spread laterally in search of nutrients, and compete more with crops. Thevs [50] found that among the shelterbelt species tested (tamarack (*Tamarix*), Siberian elm, Russian-olive (*Elaeagnus angustifolia* L.), honeysuckle (*Lonicera*), and caragana (*Caragana arborescens* Lam., tamarack had the highest soil water uptake and caragana had the lowest, and therefore was a better suited species to be used in a shelterbelt composition.

Table 1. Effect of shelterbelts on adjacent crop yield from studies around the world.

Shelterbelt Species	Crop Species	Influence	Yield (%)	Reference	Location
	Winter wheat	+	23	Kort [a] [41]	
	Spring wheat	+	8	Kort [41]	
	Barley	+	25	Kort [41]	
	Oat	+	6	Kort [41]	
	Rye	+	19	Kort [41]	
	Millet	+	44	Kort [41]	
	Alfalfa	+	99	Kort [41]	
Ponderosa pine	Winter wheat	+	4.19	Greb and Black [43]	Colorado, U.S.
Caragana, Chokeckerry (*Prunus virginiana* L.)	Winter wheat	+	12.25	Greb and Black [43]	Colorado, U.S.
Black walnut, Black locust (*Robinia pseudoacacia* L.)	Winter wheat	−	12.89	Greb and Black [43]	Colorado, U.S.
Ponderosa pine	Sorghum	−	2.4	Greb and Black [43]	Colorado, U.S.
Siberian pea, Chokeckerry	Sorghum	−	13.5	Greb and Black [43]	Colorado, U.S.
Black walnut, Black locust	Sorghum	−	3.75	Greb and Black [43]	Colorado, U.S.
(*Populus* × *euramericana*)	Soybeans (*Glycine max* L.)	+	23	Qi et al. [25]	Great Plains, U.S.
Green ash, Austrian pine (*Pinus nigra*), Eastern red cedar (*Juniperus virginiana* L.)	Soybeans	+	26	Ogbuehi and Brandle [51]	Nebraske, U.S.
Eastern red cedar	Beans	+	21	Rosenber [52]	Nebraske, U.S.
Indian rosewood (*Dalbergia sissoo*)	Cotton (*Gossypium hirsutum* L.)	+	10	Puri et al. [53]	Dhiranvas, India
Corymbia intermedia (R.T.Baker) K.D.Hill & L.A.S.Johnson, *Corymbia tessellaris* (F.Muell.) K.D.Hill & L.A.S.Johnson	Potato (*Solanum tuberosum* L.)	+	6.7	Sun and Dickinson [54]	Atherton Tablelands, Australia
Arizona cypress (*Cupressus arizonica*)		−	25	Campi et al. [55]	Rutigliano, Italy
Aleppo pine (*Pinus halepensis* Miller)	Wheat	+	44	Nuberg et al. [56]	Southern Australia
Aleppo pine	Faba beans (*Vicia faba* L.)	+	49	Nuberg et al. [56]	Southern Australia
Aleppo pine	Oat	+	25	Nuberg et al. [56]	Southern Australia
Populus canadensis Mönch, *P. Beijingensis* W.Y. Hsu, *P. xiaozuanrica*, *P. Simonii* Carrière, *P. pseudo-simonii*	Mayze	+	6.13	Zheng et al. [57]	Western China, Heilongjiang, Jilin, Liaoning and Inner Mongolia

[a] Shelterbelt species and location are not provided because this numbers were summarized by Kort [41] after an extensive literature review from variety of studies (including a variety of locations and shelterbelt species).

Even though shelterbelt tree roots can compete with crops, they impact the overall ecosystem in a positive manner, due to their significant role on soil health and structure. It is calculated that global land degradation annually costs about $300 billion U.S. dollars [58]. A study conducted in England and Wales illustrated that soil compaction alone was responsible for 39% of all costs of soil recovery [59]. Soil compaction is one of the most serious issues faced by agricultural producers today [34]. Soil compaction is caused by a variety of factors, such as overuse of heavy machinery and short crop rotations, which reduces crop yield and soil health in the long term. In contrast, shelterbelt systems can ameliorate crop growing conditions by improving soil structure and adding organic matter into the soil by means of growing extensive and deep tree root systems, thus enhancing soil porosity, which increases soil water infiltration and recharge, and improving the overall soil health. Carrol et al. [34] studied the effect of shelterbelts in pastures and observed that water infiltration under and near shelterbelts was 60 times greater compared to open areas on the pasture, and that significant changes in the rate of soil water infiltration happened soon after planting, as early as two years after shelterbelt establishment.

Besides water infiltration, crop water-use efficiency is also affected by shelterbelts. The treed barrier reduces wind speed, thus slowing heat transfers from the crops to the air, and slowing down evapotranspiration. Davis and Norman [60] reviewed the effects of shelterbelts on crop water-use and found a significant reduction in turbulent air which decreased evapotranspiration, improving water-use efficiency. Similarly, Ogbuehi [51] found that sheltered (shaded) soybeans had greater photosynthesis rate, stomatal conductance and deeper light penetration than non-sheltered plants. Thevs et al. [50], studying corn, potato, wheat, and barley production, found that crop water consumption was 10–12% lower in areas in the proximity of shelterbelts, compared to open field conditions.

Environmental services provided by shelterbelts, as previously discussed, modify the micro environment, and can be seen as a useful tool to mitigate the effects of climate change that will impact agricultural production worldwide in the near future [1,44]. For example, Easterling et al. [44] used a model to simulate the effects of climate change related stress on maize planted in dry environments in Nebraska, comparing yield on sheltered and unsheltered crops. They concluded that sheltered crops yields were greater, and that shelterbelts are an important tool to ameliorate global warming consequences.

3.1. Shelterbelt Carbon Sequestration Potential

Shelterbelts are useful not just to mitigate local weather extremes caused by climate change, but are also useful tools to mitigate global warming, thanks to their carbon sequestration potential. The Inter-Governmental Panel on Climate Change indicated that there were about 630 million hectares of unproductive land on the planet in 2000, and they suggested that if it was used for agroforestry, it would sequester 1.43 and 2.15 Tg CO_2 every year by 2010 and 2040, respectively [61].

Carbon sequestration in agroforestry varies among the different types of agroforestry practiced, ecological regions where it takes place, and soil type, ranging from 0.29 to 15.21 Mg ha^{-1} year^{-1} for above-ground, and 30–300 Mg C ha^{-1} year^{-1} up to 1 m depth in the soil [62]. Currently, the global area under agroforestry systems is 1023 million ha [63] and as previously stated, there is approximately 630 million ha of unproductive lands in the world that could be used to promote carbon sequestration through agroforestry practices [61]. Carbon sequestration potential tends to be greatest in natural forests, then in agroforestry systems, followed by tree plantations, and finally in cropped lands [49,63]. Table 2 shows the land use and its carbon sequestration potential reported in the literature around the world. Land use systems that include trees have a great potential in comparison with agriculture alone. Schoeneberger [64] and Wang and Feng [65] values are already in hectares, and they considered that shelterbelts occupied 5 and 2.5% of the cropland area, respectively.

Table 2. Carbon sequestration potential of diverse land use classes using different tree species around the world.

Land Use	Species	Total Mg C (km^{-1} year^{-1})	Above-ground Mg C (ha^{-1} year^{-1})	Below-ground Mg C (ha^{-1} year^{-1})	Total Mg C (ha^{-1} year^{-1})	Location	Reference
Shelterbelt	Hybrid Poplar	6.03–6.54	0.79*Total	0.21*Total	3.3–5.2	Saskatchewan, Canada	Amichev et al. [a] [20]
Shelterbelt	Scots Pine (*Pinus sylvestris* L.)	1.90–2.17	0.90*Total	0.10*Total	1.4–3.3	Saskatchewan, Canada	Amichev et al. [20]
Shelterbelt	Manitoba Maple	2.39–2.60	0.80*Total	0.20*Total	2.8–5.3	Saskatchewan, Canada	Amichev et al. [20]
Shelterbelt	White Spruce (*Picea glauca* Moench)	2.43–2.75	0.81*Total	0.19*Total	2.2–4.1	Saskatchewan, Canada	Amichev et al. [20]
Shelterbelt	Green Ash	1.78–1.98	0.77*Total	0.23*Total	2.0–3.9	Saskatchewan, Canada	Amichev et al. [20]
Shelterbelt	Caragana	1.73–2.03	0.74*Total	0.26*Total	1.3–2.7	Saskatchewan, Canada	Amichev et al. [20]
Shelterbelt			0.58-1.17			Nebraska	Schoeneberger [b] [64]
Shelterbelt	*Poplar canadensis* Mönch, *Paulownia elongala*		0.38			China	Wang and Feng [c] [65]
Woodlots for firewood, fodder, land reclamation			1.0–5.0	1.0–6.0	2.0–11	Asia/Africa	Nair et al. [63]
Shade tree system	*Cordia alliodora* (Ruiz & Pav.) Oken, *Theobroma cacao* L.		3			Costa Rica	Montagnini et al. [66]
Shade tree system	*Erythrina poeppigiana* (Walp.) O.F. Cook), *Theobroma cacao* L.		4.4			Costa Rica	Montagnini et al. [66]
Forest	*Dipteryx panamensis* (Pitt.) Rec. & Mell		20.26			Costa Rica	Montagnini et al. [66]
Monoculture	Cooffe (*Coffea arabica* L.)		2.14		2.14	Brazil	Palm et al. [67]
Imporved fallow	*Crotalaria grahamiana* Wight & Arn		8.5	2.7	11.2	Kenya	Albrecht et al. [68]
Imporved fallow	*Eucalyptus saligna* Sm.		21.7	9.55	31.25	Kenya	Albrecht et al. [68]
Silvipasture			6.1		6.1	North America	Udawatta and Jose [69]
Alley crop			3.4		3.4	North America	Udawatta and Jose [69]
agriculture	Fallow, Soybean, Maize		0.3		0.3	Goias, Brazil	Bayer et al. [70]
Intercroping	Gliricidia sepium ((Jacq.) Steud)			12.3	12.3	Zomba, Malawi	Makumba et al. [71]

[a] Reported per-area data (Mg ha^{-1} year^{-1}) in Amichev et al. [20] represent the area directly underneath the live shelterbelt tree crowns. For comparison purposes in this study, we assumed that the area of the shelterbelts in Amichev et al. [20] represented 5% of the total farm area, similar to Schoeneberger [64]. [b] It was estimated that the area of shelterbelts represented 5% of the total farm area. [c] It was estimated that the area of shelterbelts represented 2.5% of the total farm area. #*Total: value times the total biomass (Total Mg C (ha^{-1} year^{-1})).

3.1.1. Shelterbelt Carbon Sequestration Potential and Stocks Above-Ground

Carbon sequestration has been extensively discussed as one of the main strategies to keep atmospheric carbon dioxide at acceptable levels, and minimize environmental risks from climate change effects. Given the increased awareness about shelterbelt carbon sequestration, many studies have been conducted in this regard, across a variety of climatic and edaphic regions, and across different shelterbelt designs, and species mixes [9,10,20,26,29,30,33,34,39,45,72]. The means to quantify carbon sequestration in shelterbelt systems have improved over the years, starting with the use of simple linear relationships and yield tables [73], to more sophisticated, and more accurate methods, using complex modelling frameworks, such as Holos, CBM-CFS3 (Carbon Budget Model for the Canadian Forest Sector) [37,74–76], and 3PG (Physiological Principles Predicting Growth) models [74].

The earliest research on shelterbelt carbon stocks in Canada was carried out by Kort and Turnock [73]. They used destructive sampling techniques to measure shelterbelt tree biomass, and fitted linear models to predict above-ground biomass for different shelterbelt species, ranging from 17 to 90 years, across all soil zones in the Saskatchewan Prairies. They reported an average biomass of 79 kg tree^{-1} (32 Mg km^{-1}) for green ash, 263 kg tree^{-1} (105 Mg km^{-1}) for hybrid poplar, and 144 kg tree^{-1} (41 Mg km^{-1}) for white spruce. Some early work was also done to understand the interaction between shelterbelts and the adjacent crops in terms of C sequestration. Peichl et al. [75] compared carbon sequestration within three systems in Ontario, Canada: 13-year-old hybrid poplar shelterbelt plus barley; a 13-year-old Norway spruce (*Picea abies*) shelterbelt plus barley; and a barley-only crop system. Total carbon sequestration was 15.1 and 6.4 Mg C ha^{-1} higher than the barley-only system for hybrid poplar and Norway spruce, respectively. Carbon stock in the soil was also significantly different between the systems: 78, 66, and 65 Mg C ha^{-1} for hybrid poplar, spruce, and barley-only systems, respectively.

Amichev et al. [74] and Amichev et al. [77] used the 3PG and CBM-CFS3 models to quantify tree growth and carbon stocks of shelterbelts. The CBM-CFS3 model was originally developed for the Canadian forest industry sector and has been used at various scales of analysis, from stand to landscape levels, to simulate forest stand growth and carbon dynamics. Similarly, 3PG is a hybrid model, designed to model forest growth, which was also adapted for use in shelterbelt systems [73,76]. Amichev et al. [74] parametrized 3PG to quantify carbon stocks of white spruce shelterbelts in a large scale study extending across five soil zones in Saskatchewan and spanning several decades of planting, from 1925 to 2009. They estimated the total above-ground biomass at 117.6 Mg C km^{-1}, ranging from 106 to 195 Mg C km^{-1}, depending on the soil zone. The total ecosystem carbon flux increased from 0.33 to 4.4 Mg C km^{-1} year^{-1}, from year 1 to 25, reaching a peak of 5.5 C km^{-1} year^{-1} Mg 39 years after planting. Average-stand biomass at age 60 was 241.3, 238.6, and 227.3 Mg km^{-1} at 2.0, 3.5, and 5.0 m tree spacing design, respectively. Overall, total carbon stocks for all white spruce shelterbelts in the province, planted over the span of eight decades, was 50,440 Mg C, sequestered in over 991 km of planted shelterbelts.

Using the same methodology, Amichev et al. [20] estimated the growth of five additional common shelterbelt species planted across Saskatchewan—hybrid poplar, Manitoba maple, Scots pine, green ash, and Caragana planted between 1925 to 2009, in three spacing designs (2.0, 3.5, and 5.0 m), and at four different mortality rates (0, 15, 30, and 50%). They estimated total shelterbelt carbon stocks for the province at 10.8 Tg C. Overall, the average carbon sequestration rate on a length basis (per km) was estimated 1.73 to 6.54 Mg C km^{-1} year^{-1}. The carbon sequestration rates for the individual species were 6.03–6.54 Mg C km^{-1} year^{-1} for hybrid poplar, 1.73–2.03 Mg C km^{-1} year^{-1} for caragana, 1.90–2.17 Mg C km^{-1} year^{-1} for Scots pine, 2.43–2.75 Mg C km^{-1} year^{-1} for white spruce, 1.78–1.98 Mg C km^{-1} year^{-1} for green ash, and 2.39–2.60 Mg C km^{-1} year^{-1} for Manitoba maple shelterbelts. The per-unit-area C rates (Mg C ha^{-1} year^{-1}) represent the C sequestration rate across 1-ha cumulative land area located directly underneath the shelterbelt tree crowns. These C sequestration rates included the C locked in live and dead above- and below-ground biomass (i.e., stems, branches, leaves, roots), as well as litter layer (i.e., decomposing tree branches and leaves) on the soil surface, and soil organic matter added into the soil.

The Holos model is an empirical, process-based, farm-scale model that estimates GHGs emissions from farms based on site-specific input information [37,76]. The model relies on details such as enteric fermentation, manure management, cropping systems, energy use, and presence of planted trees [76]. Researchers have used the Holos model for exploration of diverse farming scenarios through many simulations, aiming for minimal GHGs releases from a farm. For example, Holos was used by Amadi et al., [38] to calculate the potential of hybrid poplar, white spruce and caragana shelterbelts to offset GHGs emissions from cereal (*Triticum aestivum* and *Avena sativa*) production for a 60-year simulation period, at five shelterbelt planting densities. At the highest planting density (i.e., 5% of total farm area occupied by trees), hybrid poplar, white spruce, and caragana shelterbelts reduced farm GHGs emissions by 23%, 18% and 8%, respectively. The majority of this GHGs offset (95%) was attributed to C sequestered in wood biomass and the soil. The rest was attributed to lower N_2O emissions and CH_4 oxidation, commonly observed within the shelterbelt zone. For a 60-yr simulation, the estimated carbon stocks were 8712, 5581, and 1705 Mg C (at the most-dense spacing) for hybrid poplar, white spruce, and caragana, respectively.

Likewise, statistical models also have been used to estimate carbon sequestration in shelterbelts [10,50]. Possu et al. [78] assessed 15 allometric models on Ponderosa pine windbreaks and used the best model to estimate carbon sequestration for 16 shelterbelt tree species in Nebraska, projected over 50 years in nine areas of the U.S. They found that carbon sequestration potential ranged from 1.07 ± 0.21 to 3.84 ± 0.04 Mg C ha^{-1} year^{-1} for conifer species and from 0.99 ± 0.16 to 13.6 ± 7.72 Mg C ha^{-1} year^{-1} for broadleaved deciduous species. Zhou et al. [9] assessed carbon stocks in shelterbelts in Montana, U.S., with two types of statistical models: precise preferred models which required more variables from expensive data inventories; and cost preferred models, which were simpler to fit, but were less precise. The authors concluded that both sets of equations were effective to estimate shelterbelt biomass and that the precision preferred models were between 0.8 and 1.2% more precise than the cost preferred models. They found that above-ground biomass for a single row of Russian-olive shelterbelt was 110 metric tonnes km^{-1} (110 Mg km^{-1}), that converted to approximately 55 Mg C km^{-1}, 60 years after planting (equal to sequestration rate of 0.91 Mg C km^{-1} year^{-1}).

3.1.2. Carbon Stocks Below-Ground

Soil is the biggest organic carbon pool on Earth [58]. It is calculated that the world's agricultural and degraded soil are able to sequester 50 to 66% of all carbon released, equivalent to 42–78 gigatons of carbon [79]. The two major below-ground carbon pools are the soil organic carbon (SOC) and below-ground biomass (i.e., fine and coarse roots). However, even though the below-ground carbon sequestration potential is known, the methods to quantify it are still in their infancy and there is no established standard protocol to follow when measuring it worldwide.

For determining SOC, the two main problems are the lack of standards for soil aggregate class definitions and what soil depth to sample [62]. Aggregates are often classified according to their ability to resist slaking in water, and fortunately, a trend in new studies is starting to follow a standard (<53 μm, 53–250 μm, and <250μm) [62]. Soil depth is the most serious issue on assessing and comparing underground carbon sequestration/stocks [62]. Most studies sample up to 20 or 30 cm depth [62]. For carbon stock studies in agroforestry, assessing soil to a greater depth is extremely important, since the sub-soil is a crucial part of carbon stabilization [62]. A general technique that is simple and practical for any situation is needed in order to allow for comparisons among studies, facilitating better whole system shelterbelt carbon estimation. This is extremely important, given that above-ground biomass alone represents just one pool of the carbon sequestered in shelterbelt agroforestry systems and all ecosystem components need to be considered. Soil carbon sequestration is estimated to be around two-thirds of the whole carbon sequestered in the ecosystem [80,81]. For example, Chu et al. [72], studying shelterbelt trees planted by the Three-North Shelterbelt Program, found that 67% of the carbon was stored in the soil, with roots representing 13%, and above-ground biomass representing only 10%.

The SOC pool is in constant interaction with other C pools in shelterbelt systems, receiving inputs from above- and below-ground system components. For example, above-ground inputs include litter fall (i.e., fallen leaves and branches), animal excrements, and decomposed biomass. Below-ground carbon inputs include root litter and rhizosphere depositions [35]. Higher SOC inputs help to maintain soil moisture and fertility, and is strongly affected by precipitation, temperature, soil texture, average stem and crown diameter, tree height, amount of surface litter, and shelterbelt species and age [28,35].

Below-ground biomass measurement techniques for shelterbelt systems also have not been thoroughly explored, since they tend to be resource demanding and time consuming, and there is no well-established methodology to sample the below-ground system. Because of this, comparison between studies is problematic [37,62,66,74,80]. Aiming to facilitate below-ground carbon estimation, the IPCC recommended below-ground biomass estimations using established relationships with above-ground biomass [55]. Similarly, Kort and Turnock [73] recommended root biomass estimations to be done by considering constant ratios of 40%, 30% and 50% of above-ground biomass for deciduous, coniferous, and shrub shelterbelts, respectively, which were also prescribed by Freedman et al. [79] and Grier et al. [82]. However, this method is problematic, since root systems vary with species, climatic zone, and environmental conditions within the region [62]. To be more comprehensive, more methods need to be tested for many species across many site conditions. For example, a dry environment would stimulate a deeper root system, while a less dry environment would produce shallower roots for the same tree species.

Numerous studies have observed a trend of soil carbon loss occurring when a new shelterbelt is first planted, most likely due to site preparation and land use change, which is offset years later, as the trees grow more extensive roots systems [28,35,74,80]. This is a natural process that usually takes place when land use is changed. For example, when a natural ecosystem is replaced by agriculture, around 60% and 75% of SOC is lost in temperate and tropical climate, respectively [80]. Amichev et al. [74] found that soil carbon stocks decreased during the first 10 years following shelterbelt implementation, losing about 3.5% within the first five years. Their model simulations illustrated that carbon emissions due to land cover change were completely offset by the ages of 17, 18, and 21 for shelterbelts planted at 2.0, 3.5, and 5.0 m spacing, respectively.

Even though it takes years to compensate carbon loss due to shelterbelt planting, in dry environments, such as the Prairies, where biomass production is not very high, shelterbelts can be an important source of organic matter for the soil. Shelterbelts increase SOC, moisture, and fertility, and consequently, increase carbon sequestration into the soil pool. Research has demonstrated that SOC under shelterbelt trees canopies is greater than SOC under crops, and that this difference tends to be less pronounced in deeper layers of the soil [28,83], but vary according to the species considered [36]. For example, Amadi et al. [28] studying the carbon sequestration potential of different shelterbelt species in Saskatchewan in the 0–7.5 cm and 7.5–15 cm soil layers, reported higher SOC stocks within the top soil layer. Similar were the results reported by Dhillon and Van Rees [35], who studied the SOC sequestration potential of six common shelterbelt species planted in Saskatchewan (green ash, hybrid poplar, Manitoba maple, white spruce, Scots pine and caragana), ranging from five to 63 years of age. Their results showed that soil organic matter concentration was 30% greater under shelterbelts than adjacent cropped fields, and that the SOC stock was 19% greater under shelterbelts than under crops. This difference is due to lower bulk density of soils under shelterbelts than under cropped fields. This lower bulk density is attributed to the presence of organic matter, extensive tree root systems, and due to the absence of heavy machinery traffic for the years since the shelterbelt was implemented. This finding was corroborated by Sauer et al. [80] who also reported lower bulk density under shelterbelts than under the adjacent cropped field. The bulk density in this case was 13% lower in the 0–10 cm layer, and 7% lower in the 10–30 cm layer. They also found that soil under shelterbelts had 18.6 Mg C ha^{-1} more SOC than the soil under crop production within the top 50 cm of soil, and that litter under shelterbelts contained an additional 3–8 Mg C ha^{-1}. The SOC stocks vary greatly by species because of differences in litter composition leading, to differences in litter decomposition rates. For example, these authors found that white spruce shelterbelts had 20.8 g C kg^{-1} more SOC in the 0–5 cm soil layer than the adjacent cropped field, while green ash had only 0.8 g C kg^{-1} more SOC than the adjacent cropped field.

An important aspect of the shelterbelt SOC pool is the long residence time, which emphasizes the shelterbelts' role as an effective climate mitigation tool from a global perspective. Needless to say, the longer the added soil carbon remains in the soil pool, the better. Organic carbon compounds in the soil can be classified as either labile, with residence time in the soil of a few months, or as recalcitrant, with residence time in the soil of a few decades. Dhillon and Van Rees [35] analyzed the effect of shelterbelts and cropped fields in Saskatchewan on the distribution of soil organic carbon density fractions. They found an increase in the SOC labile light fraction (71%) and the stable heavy fraction (22%) for soils under shelterbelts compared to cropped fields. The majority of SOC added in the 0–10 cm layer belonged to the labile light fraction, and the majority of the SOC added in the 10–30 cm layer belonged to the heavy fraction. The SOC light fraction was generally associated with conifer shelterbelts, whereas the SOC heavy fraction was associated with deciduous trees. For example, Manitoba maple litter was abundant with more resistant forms of soil organic matter (i.e., needing more time to decompose) [61].

Shelterbelt management practices also influence SOC stocks, and more specifically, the type of soil organic matter compounds, and consequently their residence time within the soil pool. The chemical composition of these soil compounds affects microorganisms-enzymes interactions and determine their stability and residence time in the soil [83]. The soil under shelterbelts have more processed C forms, such as aliphatic C, aromatic C, and ketones, that are harder for microbes to break down, while the soil under cropped fields have more sugars and alcohols [62,63].

Soil greenhouse gases flux studies are also important to better understand below-ground carbon dynamics. Amadi et al., [37] compared soil CO_2, CH_4 and N_2O fluxes in shelterbelt systems with adjacent cropped field in Prince Albert, Saskatchewan, using non-steady state vented chambers. Even though they found greater CO_2 fluxes under shelterbelts than in crop fields (probably due to higher microbial activity, root respiration and litter decomposition), soil organic carbon under the shelterbelts was 27% greater than under the adjacent cropped field. Shelterbelts contributed to the offset of other GHGs released from farming activities, not just CO_2, by enlarging the CH_4 soil sink, absorbing 58% and 81% more CH_4 than soil in cropped fields, in 2013 and 2014, respectively. Cumulative seasonal N_2O emissions from shelterbelt areas were two- to five-times lower than emissions by cropped fields nearby. A similar study by Amadi et al. [28] investigated whether a two-row 31-year-old hybrid poplar-caragana shelterbelt influenced the soil organic carbon and GHG flux dynamic on its surrounding area, compared to an adjacent cropped field. They found that soil organic carbon concentration in the soil was greater in the proximity to shelterbelts, and that CH_4 uptake decreased with increasing distance away from the shelterbelt. The N_2O release was smaller under the shelterbelt and increased towards the cropped field, and the CO_2 flux between soil and atmosphere was more intense in the proximity to shelterbelts, which was in agreement with Amadi et al. [37]. Higher fluxes were attributed to higher organic matter concentration, microbial activity, and tree root respiration in the proximity of the shelterbelts [28,37].

4. Conclusions

This review paper summarized the currently available research-based knowledge surrounding shelterbelt agroforestry systems, and aimed to increase the awareness of researchers, farmers, industry, and policymakers of the climate change mitigation potential of planted shelterbelts throughout Canada and the world. The current knowledge-base clearly indicates that shelterbelts have a great potential for carbon sequestration, both in above- and below-ground pools of the system. As the trees in the shelterbelts continue to grow, they are able to reduce and offset a significant portion of the carbon released from agricultural practices, while providing several other social and environmental benefits, both for the public and private sectors.

In order to preserve existing shelterbelts, and promote the planting of new ones, new effective policies are needed in Canada that would provide farmers with the necessary economic incentives, and cost recovery for shelterbelt establishment and maintenance. This is especially important in the post-Paris Accord era, as the Government of Canada is steadily transitioning towards a carbonless economy, at the forefront of which are the farming communities in the Canadian Prairies. To better understand the budgetary impact of a carbonless economy on the Canadian farmer, whole-farm cost analysis studies, and shelterbelt decision-support tools for farmers that account for shelterbelt establishment and decades-long maintenance costs (i.e., time, machinery, and labor), will be needed. New policies that will help farmers meet shelterbelt-related costs will likely have a significant impact on the carbon mitigation potential of these systems in the long term. All these actions, if executed and coordinated effectively, can provide a major step for Canada, and for the world, towards a truly decarbonized economy.

Author Contributions: Conceptualization, R.C.M., C.P.L., B.Y.A., K.V.R.; writing—original draft preparation, R.C.M.; writing—review and editing, B.Y.A., K.V.R., R.C.M., C.P.L.; supervision, C.P.L.

Funding: The authors gratefully acknowledge the financial support given by the Agricultural Greenhouse Gases Program, Government of Canada (Grant #AGGP2-017).

Conflicts of Interest: The authors declare no conflicts of interest.

References

1. Brandle, J.R.; Hodges, L.; Zhou, X.H. Windbreaks in North American agricultural systems. *Adv. Agrofor.* **2004**, *1*, 65–78.
2. Brandle, J.R.; Laurie, H.; Wight, B. Windbreak Practices. In *North American Agroforestry: An Integrated Science and Practice*; American Society of Agronomy, Inc.: Madison, WI, USA, 2000; pp. 79–118.
3. Zhu, J.J. Wind Shelterbelts. In *Encyclopedia of Ecology*; Elsevier: Copenhagen, Denmark, 2008; pp. 3803–3812.
4. Price, L.W. Hedges and Shelterbelts on the Canterbury Plains, New Zealand: Transformation of an Antipodean landscape. *Ann. Assoc. Am. Geogr.* **1993**, *83*, 119–140. [CrossRef]
5. Heath, B.; Maughan, J.; Morrison, A.; Eastwood, I.; Drew, I.; Lofkin, M. The influence of wooded shelterbelts on the deposition of copper, lead and zinc at Shakerley Mere, Cheshire, England. *Sci. Total. Environ.* **1999**, *235*, 415–417. [CrossRef]
6. Dunlop, A. Spatial and Temporal Aspects of Saskatchewan Field Shelterbelts. Ph.D. Thesis, University of Saskatchewan, Saskatoon, SK, Canada, 2000.
7. Wu, L.; He, D.; Ji, Z.; You, W.; Tan, Y.; Zhen, X.; Yang, J. Protection efficiency assessment and quality of coastal shelterbelt for Dongshan Island at the coastal section scale. *J. For. Res.* **2017**, *28*, 577–584. [CrossRef]
8. Xiao, Q.; Huang, M. Fine root distributions of shelterbelt trees and their water sources in an oasis of arid northwestern China. *J. Arid. Environ.* **2016**, *130*, 30–39. [CrossRef]
9. Zhou, X.; Brandle, J.R.; Schoeneberger, M.M.; Awada, T. Developing above-ground woody biomass equations for open-grown, multiple-stemmed tree species: Shelterbelt-grown Russian-olive. *Ecol. Model.* **2007**, *202*, 311–323. [CrossRef]
10. Hawke, M.F.; Tombleson, J.D. Production and interaction of pastures and shelterbelts in the central North Island. *Proc. N. Z. Grassl. Assoc.* **1993**, *197*, 193–197.
11. Burke, S. *Windbreaks*; Inkata Press: Port Melbourne, VIC, Australia, 1998; 68p.
12. Peri, P.L.; Bloomberg, M. Windbreaks in southern Patagonia, Argentina: A review of research on growth models, windspeed reduction, and effects on crops. *Agrofor. Syst.* **2002**, *56*, 129–144. [CrossRef]
13. Wight, B. 15. Farmstead windbreaks. *Agric. Ecosyst. Environ.* **1988**, *22*, 261–280. [CrossRef]
14. FAO—Special Forest Plantations. Available online: http://www.fao.org/3/t0122e/t0122e0a.htm (accessed on 6 October 2019).
15. Cleugh, H.A. Effects of windbreaks on airflow, microclimates and crop yields. *Agrofor. Syst.* **1988**, *41*, 55–84. [CrossRef]

16. Helmers, G.A.; Brandle, J. Optimum windbreak spacing in Great plains agriculture. *Great Plains Res. A J. Nat. Soc. Sci.* **2005**, *15*, 1–22.

17. Agriculture and Agri-Food Canada. Shelterbelt Planning and Establishment. 2015. Available online: http://www.agr.gc.ca/eng/science-and-innovation/agricultural-practices/agroforestry/shelterbelt-planning-and-establishment/?id=1344636433852 (accessed on 6 October 2019).

18. Piwowar, J.M.; Amichev, B.Y.; Van Rees, K.C.J. The Saskatchewan shelterbelt inventory. *Can. J. Soil Sci.* **2017**, *438*, 433–438. [CrossRef]

19. Amichev, B.Y.; Bentham, M.J.; Cerkowniak, D.; Kort, J.; Kulshreshtha, S.; Laroque, C.P.; Piwowar, J.M.; Van Rees, K.C.J. Mapping and quantification of planted tree and shrub shelterbelts in Saskatchewan, Canada. *Agrofor. Syst.* **2015**, *89*, 49–65. [CrossRef]

20. Amichev, B.Y.; Bentham, M.; Kulshreshtha, S.; Laroque, C.P.; Piwowar, J.M.; Van Rees, K. Carbon sequestration and growth of six common tree and shrub shelterbelts in Saskatchewan, Canada. *Can. J. Soil Sci.* **2016**, *97*, 368–381. [CrossRef]

21. Ha, T.V.; Amichev, B.Y.; Belcher, K.W.; Bentham, M.J.; Kulshreshtha, S.N.; Laroque, C.P.; Van Rees, K.C. Inventory and removal analyzed by object-based classification of satellite data in Saskatchewan, Canada. *Can. J. Remote Sens.* **2019**, *45*, 246–263. [CrossRef]

22. Rempel, C.J. Costs, benefits, and barriers to the adoption and retention of shelterbelts in prairie agriculture as identified by Saskatchewan producers. Master's Thesis, University of Saskatchewan, Environment and Sustainability, Saskatoon, SK, Canada, 2013.

23. Rempel, J.; Kulshreshtha, S.; Amichev, B.Y.; Van Rees, K. Costs and benefits of shelterbelts: A review of producers' perception and min map analysis for Saskatchewan, Canada. *Can. J. Soil Sci.* **2017**, *97*, 341–352. [CrossRef]

24. Rempel, J.; Kulshreshtha, S.; Van Rees, K.; Amichev, B. Factors that Influence Shelterbelt Retention and Removal in Prairie Agriculture as Identified by Saskatchewan Producers. In Proceedings of the Soil and Science Workshop, Saskatoon, SK, Canada, 11 March 2014.

25. Qi, X.; Mize, C.W.; Batchelor, W.D.; Takle, E.S.; Litvina, I.V. SBELTS: A model of soybean production under tree shelter. *Agrofor. Syst.* **2001**, *52*, 53–61. [CrossRef]

26. Marchildon, G. The Prairie Farm Rehabilitation Administration: Climate Crisis and Federal – Provincial Relations during the Great Depression. *Can. Hist. Rev.* **2014**, *90*, 275–301. [CrossRef]

27. Maillet, J.; Laroque, C.; Bonsal, B. A dendroclimatological assessment of shelterbelt trees in a moisture limited environment. *Agric. For. Meteorol.* **2017**, *237*, 30–38. [CrossRef]

28. Amadi, C.C.; Van Rees, K.C.; Farrell, R.E. Soil–atmosphere exchange of carbon dioxide, methane and nitrous oxide in shelterbelts compared with adjacent cropped fields. *Agric. Ecosyst. Environ.* **2016**, *223*, 123–134. [CrossRef]

29. Wiseman, G.; Kort, J.; Walker, D. Quantification of shelterbelt characteristics using high-resolution imagery. *Agric. Ecosyst. Environ.* **2009**, *131*, 111–117. [CrossRef]

30. Agriculture and Agri-Food Canada Agricultural Greenhouse Gases Project. Available online: http://www.agr.gc.ca/eng/programs-and-services/agricultural-greenhouse-gases-program/?id=1461247059955 (accessed on 19 July 2019).

31. Rogelj, J.; Den Elzen, M.; Höhne, N.; Fransen, T.; Fekete, H.; Winkler, H.; Schaeffer, R.; Sha, F.; Riahi, K.; Meinshausen, M. Paris Agreement climate proposals need a boost to keep warming well below 2 °C. *Nature* **2016**, *534*, 631–639. [CrossRef] [PubMed]

32. Kulshreshtha, S.; Kort, J. External economic benefits and social goods from prairie shelterbelts. *Agrofor. Syst.* **2009**, *75*, 39–47. [CrossRef]

33. Czerepowicz, L.; Case, B.; Doscher, C. Using satellite image data to estimate aboveground shelterbelt carbon stocks across an agricultural landscape. *Agric. Ecosyst. Environ.* **2012**, *156*, 142–150. [CrossRef]

34. Carroll, Z.L.; Bird, S.B.; Emmett, B.A.; Reynolds, B.; Sinclair, L.N. Can tree shelterbelts on agricultural land reduce flood risk? *Soil Use Manag.* **2004**, *20*, 357–359. [CrossRef]

35. Dhillon, G.S.; Rees, K.C.J. Van. Soil organic carbon sequestration by shelterbelt agroforestry systems in Saskatchewan. *Can. J. Soil Sci.* **2017**, *16*, 1–16. [CrossRef]

36. Liu, Y.; Harris, D. Effects of shelterbelt trees on reducing heating-energy consumption of office buildings in Scotland. *Appl. Energy* **2008**, *85*, 115–127. [CrossRef]

37. Amadi, C.C.; Van Rees, K.C.J.; Farrell, R.E. Greenhouse gas mitigation potential of shelterbelts: Estimating farm-scale emission reductions using the Holos. *Can. J. Soil Sci.* **2017**, *367*, 353–367. [CrossRef]

38. Baldwin, C.S. 10. The influence of field windbreaks on vegetable and specialty crops. *Agric. Ecosyst. Environ.* **1988**, *22*, 191–203. [CrossRef]

39. Mize, C.W.; Brandle, J.R.; Schoeneberger, M.M.; Bentrup, G. Ecological Development and function of Shelterbelts in Temperate North America. In *Advances in Agroforestry*; Springer Science and Business Media LLC: Berlin, Germany, 2008; Volume 4, pp. 27–54.

40. Bonifacio, R.S.; Gurr, G.M.; Kinross, C.; Raman, A.; Nicol, H.I.; Gámez-Virués, S.; Gámez-Virués, S. Arthropod prey of shelterbelt-associated birds: Linking faecal samples with biological control of agricultural pests. *Aust. J. Èntomol.* **2007**, *46*, 325–331.

41. Kort, J. Benefits of Windbreaks to Field and Forage Crops. *Wind. Technol.* **1988**, *23*, 165–190.

42. Lyles, L.; Tatarko, J.; Dickerson, J.D. Windbreak effects on soil water and wheat yield. *Trans. Am. Soc. Agric. Eng.* **1984**, *27*, 69–0072. [CrossRef]

43. Greb, B.W.; Black, A.L. Effects of windbreak plantings on adjacent crops. *J. Soil Water Conserv.* **1961**, *16*, 223–227.

44. Easterling, W.E.; Hays, C.J.; Easterling, M.M.; Brandle, J.R. Modelling the effect of shelterbelts on maize productivity under climate change: An application of the EPIC model. *Agric. Ecosyst. Environ.* **1967**, *61*, 163–176. [CrossRef]

45. Li, M.-M.; Liu, A.-T.; Zou, C.-J.; Xu, W.-D.; Shimizu, H.; Wang, K.-Y. An overview of the "Three-North" Shelterbelt project in China. *For. Stud. China* **2012**, *14*, 70–79. [CrossRef]

46. Kowalchuk, T.E.; De Jong, E. Shelterbelts and their effect on crop yield. *Can. J. Soil Sci.* **1995**, *75*, 543–550. [CrossRef]

47. Singh, D.; Kohli, R.K.; Kohli, R. Impact of Eucalyptus tereticornis Sm. shelterbelts on crops. *Agrofor. Syst.* **1992**, *20*, 253–266. [CrossRef]

48. Zohar, Y. Root distribution of a eucalypt shelterbelt. *For. Ecol. Manag.* **1985**, *12*, 305–307. [CrossRef]

49. Onyewotu, L.O.Z.; Ogigirigi, M.A.; Stigter, C.J. A study of competitive effects between a Eucalyptus camaldulensis shelterbelt and an adjacent millet (Pennisetum typhoides) crop. *Agric. Ecosyst. Environ.* **1984**, *51*, 281–286. [CrossRef]

50. Thevs, N.; Strenge, E.; Aliev, K.; Eraaliev, M.; Lang, P.; Baibagysov, A.; Xu, J. Tree Shelterbelts as an Element to Improve Water Resource Management in Central Asia. *Water* **2017**, *9*, 842. [CrossRef]

51. Ogbuehi, S.N.; Brandle, J.R. Influence of Windbreak-Shelter on Light Interception, Stomatal Conductance, and CO 2-Exchange Rate of Soybeans, Glycine max (Linnaeus) Merrill. *Neb. Acad. Sci.* **1981**, *49*, 1–6.

52. Rosenberg, N.J. Microclimate, air mixing and physiological regulation of transpiration as influenced by wind shelter in an irrigated bean field. *Agric. Meteorol.* **1966**, *3*, 197–224. [CrossRef]

53. Puri, S.; Singh, S.; Khara, A. Effect of windbreak on the yield of cotton crop in semiarid regions of Haryana. *Agrofor. Syst.* **1992**, *18*, 183–195. [CrossRef]

54. Sun, D.; Dickinson, G.R. A case study of shelterbelt effect on potato (Solanum tuberosum) yield on the Atherton Tablelands in tropical north Australia. *Agrofor. Syst.* **1994**, *25*, 141–151. [CrossRef]

55. Campi, P.; Palumbo, A.; Mastrorilli, M. Effects of tree windbreak on microclimate and wheat productivity in a Mediterranean environment. *Eur. J. Agron.* **2009**, *30*, 220–227. [CrossRef]

56. Nuberg, I.K.; Mylius, S.J.; Edwards, J.M.; Davey, C. Windbreak research in a South Australian cropping system. *Aust. J. Exp. Agric.* **2002**, *42*, 781. [CrossRef]

57. Zheng, X.; Zhu, J.; Xing, Z. Assessment of the effects of shelterbelts on crop yields at the regional scale in Northeast China. *Agric. Syst.* **2016**, *143*, 49–60. [CrossRef]

58. Nkonya, E.; Mirzabaev, A.; Braun, J. Von *Economics of Land Assessment for Improvement—A Global Degradation and Sustainable Development*; Springer: Berlin, Germany, 2015; pp. 1–695.

59. Graves, A.; Morris, J.; Deeks, L.; Rickson, R.; Kibblewhite, M.; Harris, J.; Farewell, T.; Truckle, I. The total costs of soil degradation in England and Wales. *Ecol. Econ.* **2015**, *119*, 399–413. [CrossRef]

60. Davis, J.; Norman, J. 22. Effects of shelter on plant water use. *Agric. Ecosyst. Environ.* **1988**, *22*, 393–402. [CrossRef]

61. IPCC. *Land Use, Land Use Change, and Forestry*; A Special Report of the IPCC; Cambridge University Press: Cambridge, UK, 2000; 375p.

62. Nair, P.R.; Nair, V.D.; Kumar, B.M.; Showalter, J.M. Carbon Sequestration in Agroforestry Systems. *Adv. Agron.* **2010**, *108*, 237–307.

63. Nair, P.K.R.; Kumar, B.M.; Nair, V.D. Agroforestry as a strategy for carbon sequestration. *J. Plant Nutr. Soil Sci.* **2009**, *172*, 10–23. [CrossRef]

64. Schoeneberger, M.M. Agroforestry: Working trees for sequestering carbon on agricultural lands. *Agrofor. Syst.* **2008**, *75*, 27–37. [CrossRef]

65. Wang, X.; Feng, Z. Atmospheric carbon sequestration through agroforestry in China. *Energy* **1995**, *20*, 117–121. [CrossRef]

66. Montagnini, F.; Nair, P.K.R. Carbon sequestration: An underexploited environmental benefit of agroforestry systems. *Adv. Agrofor.* **2004**, *1*, 281–295.

67. Palm, C.A.; Woomer, P.L.; Alegre, J.; Arevalo, L.; Castilla, C.; Cordeiro, D.G.; Rodrigues, V. Carbon Sequestration and Trace Gas Emissions in Slash-and-Burn and Alternative Land Uses in the Humid Tropics. In Proceedings of the ASB Climate Change Working Group, October 2019.

68. Albrecht, A.; Kandji, S.T. Carbon sequestration in tropical agroforestry systems. *Agric. Ecosyst. Environ.* **2003**, *99*, 15–27. [CrossRef]

69. Udawatta, R.P.; Jose, S. Carbon Sequestration Potential of Agroforestry Practices in Temperate North America. *Adv. Agrofor.* **2011**, *8*, 17–42.

70. Bayer, C.; Martin-Neto, L.; Mielniczuk, J.; Pavinato, A.; Dieckow, J. Carbon sequestration in two Brazilian Cerrado soils under no-till. *Soil Tillage Res.* **2006**, *86*, 237–245. [CrossRef]

71. Makumba, W.; Akinnifesi, F.K.; Janssen, B.; Oenema, O. Long-term impact of a gliricidia-maize intercropping system on carbon sequestration in southern Malawi. *Agric. Ecosyst. Environ.* **2007**, *118*, 237–243. [CrossRef]

72. Chu, X.; Zhan, J.; Li, Z.; Zhang, F.; Qi, W. Assessment on forest carbon sequestration in the Three-North Shelterbelt Program region, China. *J. Clean. Prod.* **2019**, *215*, 382–389. [CrossRef]

73. Kort, J.; Turnock, R. Carbon reservoir and biomass in Canadian prairie shelterbelts. *Agrofor. Syst.* **1999**, *44*, 175–186. [CrossRef]

74. Amichev, B.Y.; Bentham, M.J.; Kurz, W.A.; Laroque, C.P.; Kulshreshtha, S.; Piwowar, J.M.; Van Rees, K.C. Carbon sequestration by white spruce shelterbelts in Saskatchewan, Canada: 3PG and CBM-CFS3 model simulations. *Ecol. Model.* **2016**, *325*, 35–46. [CrossRef]

75. Peichl, M.; Thevathasan, N.V.; Gordon, A.M.; Huss, J.; Abohassan, R.A. Carbon Sequestration Potentials in Temperate Tree-Based Intercropping Systems, Southern Ontario, Canada. *Agrofor. Syst.* **2006**, *66*, 243–257. [CrossRef]

76. Agriculture and Agri-Food Canada Holos: A Tool to Estimate and Reduce GHGs from Farms. Available online: http://publications.gc.ca/site/eng/9.691658/publication.html (accessed on 12 July 2019).

77. Amichev, B.Y.; Johnston, M.; Van Rees, K.C. Hybrid poplar growth in bioenergy production systems: Biomass prediction with a simple process-based model (3PG). *Biomass Bioenergy* **2010**, *34*, 687–702. [CrossRef]

78. Possu, W.B.; Brandle, J.R.; Domke, G.M.; Schoeneberger, M.; Blankenship, E.; Ballesteros, W. Estimating carbon storage in windbreak trees on U.S. agricultural lands. *Agrofor. Syst.* **2016**, *90*, 889–904. [CrossRef]

79. Keith, T.; Freedman, B. Planting trees for carbon credits: A discussion of context, issues, feasibility, and environmental benefits. *Environ. Rev.* **1996**, *4*, 100–111.

80. Sauer, T.J.; Cambardella, C.A.; Brandle, J.R. Soil carbon and tree litter dynamics in a red cedar–scotch pine shelterbelt. *Agrofor. Syst.* **2007**, *71*, 163–174. [CrossRef]

81. Lal, R. Soil Carbon Sequestration Impacts on Global Climate Change and Food Security. *Science* **2004**, *304*, 1623–1627. [CrossRef] [PubMed]

82. Grier, C.C.; Vogt, K.A.; Keyes, M.R.; Edmonds, R.L. Biomass distribution and above- and below-ground production in young and mature Abies amabilis zone ecosystems of the Washington Cascades. *Can. J. For. Res.* **1981**, *11*, 155–167. [CrossRef]

83. Dhillon, G.S.; Gillespie, A.; Peak, D.; Van Rees, K.C. Spectroscopic investigation of soil organic matter composition for shelterbelt agroforestry systems. *Geoderma* **2017**, *298*, 1–13. [CrossRef]

 *forests*

Article

Intercropping the Sharp-Leaf Galangal with the Rubber Tree Exhibits Weak Belowground Competition

Junen Wu [1,2], Huanhuan Zeng [1,2,3], Chunfeng Chen [1,2], Wenjie Liu [1,2,*] and Xiaojin Jiang [1,2]

[1] CAS Key Laboratory of Tropical Forest Ecology, Xishuangbanna Tropical Botanical Garden, Chinese Academy of Sciences, Menglun, Yunnan 666303, China; wujunen@xtbg.ac.cn (J.W.); zenghuanhuan@xtbg.ac.cn (H.Z.); chenchunfeng@xtbg.ac.cn (C.C.); jiangxiaojin@xtbg.ac.cn (X.J.)
[2] Center of Plant Ecology, Core Botanical Gardens, Chinese Academy of Sciences, Menglun, Yunnan 666303, China
[3] University of Chinese Academy of Sciences, Beijing 100049, China
* Correspondence: lwj@xtbg.org.cn; Tel.: +86-691-8715071; Fax: +86-691-8715070

Received: 3 September 2019; Accepted: 18 October 2019; Published: 20 October 2019

Abstract: Intercropping the sharp-leaf galangal with the rubber tree could help to improve the sustainability of the rubber tree planting industry. However, our understanding of belowground competition in such agroforestry systems is still limited. Therefore, we used stable isotope methods (i.e., water δ^2H and δ^{18}O and leaf δ^{13}C) to investigate plant water-absorbing patterns and water use efficiency (WUE) in a monocultural rubber plantation and in an agroforestry system of rubber trees and sharp-leaf galangal. We also measured leaf carbon (C), nitrogen (N), and phosphorus (P) to evaluate the belowground competition effects on plant nutrient absorption status. Through a Bayesian mixing model, we found that the monocultural rubber trees and the intercropped sharp-leaf galangal absorbed much more surface soil water at a depth of 0–5 cm, while the rubber trees in the agroforestry system absorbed more water from the shallow and middle soil layers at a depth of 5–30 cm. This phenomenon verified the occurrence of plant hydrologic niche segregation, whereas the WUE of rubber trees in this agroforestry system suggested that the competition for water was weak. In addition, the negative correlation between the leaf P concentration of the rubber trees and that of the sharp-leaf galangal demonstrated their competition for soil P resources, but this competition had no obvious effects on the leaf nutrient status of the rubber trees. Therefore, this study verified that the belowground competition between rubber trees and sharp-leaf galangal is weak, and this weak competition may benefit their long-term intercropping.

Keywords: *Alpinia oxyphylla*; interspecific competition; leaf nutrient diagnosis; plant water use; rubber-based agroforestry system; stable isotope

---

1. Introduction

Natural rubber is an indispensable and essential raw material for a variety of industrial applications and products [1], and it has brought huge economic benefits to the cultivated regions of the rubber tree (*Hevea brasiliensis* (Willd. ex A. Juss.) Müll. Arg.), especially in mainland Southeast Asia [2]. However, negative impacts from the large-scale monocultural cultivation of rubber trees on the ecological environment has engendered an adverse reputation for the rubber tree, thus demonstrating a severe hindrance to the sustainable development of the rubber planting industry [3,4]. Apart from environmental problems, the present serious issue is the continuous low price of natural rubber [4]. Due to the direct impact of this issue on the livelihoods of rubber smallholders, rubber plantations in Xishuangbanna have a tendency to convert to other monocultural plantations, especially aging rubber plantations (i.e., with a planting history of more than 30 years) that exhibit lower rubber yields [5,6]. It is noteworthy that nearly half of rubber plantations in Xishuangbanna have grown old after decades of a rubber boom [7], and the smallholders of these aged rubber plantations have shown their willingness to plant another cash crop instead of replanting rubber trees [6]. Therefore, the future of the rubber planting industry seems pessimistic.

Seriously speaking, before a suitable substitute is found, natural rubber is still an indispensable national strategic resource, and the products of natural rubber are still part of our daily lives [1,8]. Moreover, aged rubber plantations also display lots of benefits. For example, they can store much more carbon [9,10], increase biodiversity [11], and be of great benefit to forest restoration [12]. Therefore, planning for aged rubber plantations needs more scrupulous and farsighted consideration. From a conservative perspective, it is especially necessary to extend the planting period of aged rubber plantations to acquire time for in-depth studies and final decisions. Meanwhile, corresponding countermeasures for reversing the current adverse trends of the rubber planting industry are urgently needed.

As a promising approach to achieve the green and sustainable cultivation of rubber plantations, the use of rubber-based agroforestry systems is conducive to improving the ecological environment, making full use of environmental resources, promoting the productivity of a system, and creating a favorable environment for transforming an aged rubber plantation into a secondary forest or another plantation [12,13]. In order to increase and stabilize the income of rubber smallholders, the economic value of intercropping plants is always considered in the design of rubber-based agroforestry systems, such as their medicinal value, edible value, industrial value, and ornamental value [12]. Therefore, successful intercrops should always exhibit high economic value, such as sharp-leaf galangal (*Alpinia oxyphylla* Miq.), an important medicinal herb that is frequently adopted in the treatment of dyspepsia, diarrhea, abdominal pain, and poor memory in East and Southeast Asia [14]. In addition, it is also an excellent ornamental plant. Because of its convenient management and maintenance, low economic cost and risk, handsome profit, and long-time attraction in the market, sharp-leaf galangal has been used as a suitable cash crop for intercropping with rubber trees [15]. Therefore, the rubber tree and sharp-leaf galangal agroforestry system (RS-AFS), which may help rubber smallholders pass through hard times of economic fluctuations in theory, has become one of the most promising rubber-based intercropping systems in the rubber-cultivating regions of China [13,16].

However, in addition to economic considerations, the suitability of this intercropping species is mainly evaluated through the adaptive capacity of low light availability and the aboveground growth characteristics of the intercropping species [12,13]. There have been few studies on the belowground competition between the rubber tree and this intercropping medicinal herb despite economic studies on such rubber-based agroforestry systems having been reported on for decades. Some studies have pointed out that competition between rubber trees and intercrops can be avoided by selecting intercrops with a differing root strategy, but this argument still needs to be verified [12]. As many lessons from intercropping history have suggested, ignoring belowground competition often leads to the failure of many agroforestry systems [16]. Therefore, before a large-scale extension of the rubber-based agroforestry system, more field studies on belowground competition are needed to verify the suitability of the species combination.

In general, soil water and various soil nutrients are the main resources plants compete for underground. Although plants have unique ways of competing for soil water and nutrients due to the different properties of these two resources, there are still many similarities in water and nutrient competition because both kinds of competition are related to root characteristics, especially the feeder root, which is commonly regarded as the main organ that absorbs water and nutrients in soil. However, due to the invisibility of root interaction, the complexity of plant resource use strategies, and the temporal and spatial heterogeneity of environmental resources [17,18], traditional methods for the investigation of plant belowground competition have had many limitations. For example, a root-digging method or other direct investigation of underground biomass is destructive for the subsequent sampling. Comparatively, the stable isotope tracer method shows many advantages, such as nondestructive sampling and precise and quantitative estimations of sources [19]. Because the processes of water absorption by plants do not discriminate between the stable isotopes of hydrogen and oxygen (i.e., ^{2}H and ^{18}O) in soil water [20], an analysis of the stable isotopic compositions (i.e., δ^2H and δ^{18}O) of plant xylem water and soil water could help recognize plant water use patterns belowground, thus assisting in the investigation of the distribution of plant feeder roots and plant competition for belowground resources [21,22]. In addition, nutrients are transported to plant roots mainly through three mechanisms: root interception, diffusion, and mass flow. Relatively speaking, root interception contributes less of a nutrient supply than diffusion and mass flow because it is not a major mechanism that supplies nutrients directly to roots, and diffusion and mass flow must use soil water as a medium. Meanwhile, their occurrences depend on plant water uptake and transpiration, which result in a difference in soil water potential and the nutrient concentrations of soil solutions within soil [23]. Therefore, an investigation of plant water use based on the stable isotope method is also conducive to understanding plant nutrient competition, because soil water regulates the distribution of root hairs and the mobility and availability of soil nutrients [24].

In order to investigate the belowground competition between rubber trees and the sharp-leaf galangal in an RS-AFS, we adopted stable hydrogen and oxygen isotopes to investigate plant niche differentiation in an RS-AFS. We also adopted stable carbon isotopes (i.e., plant leaf δ^{13}C) to study water use efficiency (WUE) and the response to water stress of the rubber trees and the intercropped sharp-leaf galangal. In addition, plants can change the carbon (C), nitrogen (N), and phosphorus (P) concentrations and C/N/P ratios in their tissues through nutrient translocation and retranslocation immediately when the uptake of soil N and P is insufficient to support plant growth [23]. In order to understand the competition effects on plant leaf C/N/P stoichiometry, we also measured the leaf C, N, and P concentrations of rubber trees and intercropped sharp-leaf galangal in an RS-AFS to evaluate the competition effects on plant leaf nutrient status and related growth during the rainy and dry seasons of 2016. Then we put forward two main questions: (1) Does the intercropping of sharp-leaf galangal help to change the water use of rubber trees and improve the soil water condition of this agroforestry system? (2) Does the belowground competition between rubber trees and sharp-leaf galangal influence the nutrient absorption of the rubber trees? Due to the shallow root depth of sharp-leaf galangal within the soil [15], we hypothesized that competition between the sharp-leaf galangal and the rubber tree would not be intense enough to change the water use and nutrient absorption of rubber trees in an RS-AFS.

2. Materials and Methods

2.1. Study Site

All study sites were located in Xishuangbanna Tropical Botanical Garden, Chinese Academy of Sciences (XTBG; 21°55′39″ N, 101°15′55″ E), in Xishuangbanna Prefecture, Yunnan Province, Southwestern China. The rainy season and the dry season are apparent in this region because the local climate is mainly controlled by the tropical southern monsoon of the Indian Ocean. The annual mean temperature is 21.8 °C, and the annual average precipitation is about 1550 mm. Despite the precipitation of this region being seemingly abundant, over 80% of the precipitation is concentrated in the rainy season, especially from May to August, and the dry season in this region exhibits less precipitation and a higher air temperature (the maximum air temperature in the daytime always reaches above 30 °C, especially from March to April; data were provided from the Xishuangbanna Station for Tropical Rainforest Ecosystem Studies).

The studies were performed in a rubber monoculture ((RM), as the control) and in an agroforestry system of rubber trees (*H. brasiliensis*) and sharp-leaf galangal (*A. oxyphylla*) (i.e., RS-AFS) (Figure S1). The rubber trees in all study sites were planted in 1973, they were planted in sets of double rows (rows were spaced 2 m apart, and the trees within the rows were spaced approximately 3 m apart), and each set of double rows was separated by an 18-m-wide gap. The sharp-leaf galangals in the RS-AFS were cultivated in this wide gap in 2010, and the planting pattern of this intercropped species was about 80 cm × 120 cm. After six years of intercropping, the sharp-leaf galangals grew densely and gradually covered the rubber tree planting rows through lateral extensions of their creeping stems. They had well-developed and thick rhizomes and stolons, but their rooting depths were about 25.46 ± 2.25 cm ($n = 8$). As a perennial herbaceous plant, sharp-leaf galangal produces clumps of leafy stems from a large creeping rhizome, and the aboveground heights of the sharp-leaf galangals were approximately 1.6 m. In addition, the heights of the rubber trees were more than 30 m during the studying period. The altitude of these two study sites was approximately 500 m above sea level, and the distances between study sites were less than 50 m. The terrain of all investigated sites was flat.

2.2. Sampling and Measuring Methods

Each study site was divided into several 6 × 9 m blocks (Figure S2). In total, there were 20 blocks in the RM and 14 blocks in the RS-AFS. On each sampling date, three blocks were selected randomly for sampling.

The leaves of the rubber trees and sharp-leaf galangals were sampled from the extremely dry season to the pronounced rainy season (i.e., 12 March 2016; 21 May 2016; 22 June 22 2016; and 20 July 2016). At noon of each sampling date, we collected the leaves from one individual rubber tree and one clump of the intercropped herbs at these three randomly selected blocks (6 × 9 m) in each study site (Figure S2). Therefore, on each sampling date, three individual rubber trees were selected for sampling in each site, and three clumps of sharp-leaf galangal were selected for sampling in the RS-AFS. We cut the shoots of rubber trees from the sunny slope of their canopy in each site and collected the leaves from these shoots. For the sharp-leaf galangals, we sampled the leaves directly. All leaf samples were dried (65 °C, 48 h) to a constant weight and then crushed with a pulverizer and sifted to fine powder with a 100-mesh sieve.

The leaf $\delta^{13}C$ value was measured through an IsoPrime100 (Isoprime, Stockport, UK). The isotope ratios for ^{13}C were expressed in parts per thousand relative to Vienna Pee Dee Belemnite (i.e., V-PDB). The total C and N concentrations of all samples were measured using the Vario MAX cube (Elementar; Hanau, Germany) at the Central Laboratory, XTBG (Xishuangbanna Tropical Botanical Garden). The total P concentration of plant tissue was determined through inductively coupled plasma atomic-emission spectrometry (Thermo Fisher; Waltham, USA) after digestion in HNO_3–$HClO_4$.

For the study of plant interspecific competition for water, we collected soil and plant xylem samples separately during the extremely dry season (on 12 March 2016) and the pronounced rainy season (on 20 July 2016) for a comparison of plant water-absorbing patterns in water-limited and water-abundant environments. This design mainly considers the occurrence of plant competition for water when water availability is low, and the best way to understand plant competition for water is to compare the functional traits of water use when water is limited and abundant [17]. To be specific, March is the driest period in this region, and it is also the leaf-flushing period of the rubber trees (Figure S3a,b), which suggests that they rubber trees have enough reserved water for their upcoming leaf expansion [25]. Therefore, the competition for water during this period may be very intense in an RS-AFS. In contrast, due to adequate rainfall in July (Figure S3b), soil water would be abundant in July, which may result in weak competition for water in an RS-AFS. Therefore, exploring plant water-absorbing patterns in an RS-AFS under these two extremely contrary situations is beneficial for understanding the plant competition for water in an agroforestry system.

On each sampling day, we randomly selected three blocks (6 × 9 m) in the RM and RS-AFS for sampling. We used an increment borer to obtain xylem samples with lengths of 10–15 cm from 3-5 rubber tree trunks in each study site and collected four samples of mature rhizomes from sharp-leaf galangals in the RS-AFS at the same time. It is worth explaining that the sharp-leaf galangal has no xylem. Therefore, water within its belowground rhizome seemed more suitable for stable isotope analysis because the root water uptake of most plant species either does not lead to isotopic discrimination or the discrimination is too low to be observed [22].

All sampled rubber trees and intercropped herbs were selected randomly in the blocks, and the sampled materials of one plant species in one block were mixed to form one sample. Therefore, there were three samples for each plant species in each site. The phloem on the xylem samples of both the rubber trees and the intercropped herbs was removed, and the treated samples were stored in a 15-ml screw-cap glass vial, sealed with Parafilm, and frozen immediately (−20 °C). At the same time, we dug a straight hole through a soil auger (4.5 cm in diameter) in the middle location of each selected block in the RM and RS-AFS (three holes for each study site) and collected soil samples from the surface (0–5-cm depths), shallow (5–15-cm depths), middle (15–30-cm depths), and deep (30–80-cm depths) soil layers. This soil sampling design mainly considered the relatively shallow vertical root depths of sharp-leaf galangals at depths of less than 30 cm and considered that there is no significant variation in soil water isotope compositions below 80 cm of depth [26]. Therefore, there was no need to sample soil from deeper soil layers below 80 cm of depth because a 0-80-cm deep soil layer was enough for a comparison of the water-absorbing patterns of the rubber tree and the sharp-leaf galangal. Therefore, we regarded the 30–80-cm soil layer as a deep soil layer. A small part of each sample from each soil layer was collected into a 15-ml screw-cap glass vial and sealed with Parafilm immediately. The rest of the sampled soil was sealed in LDPE (Low-Density Polyethylene) zip-lock bags to measure the gravimetric soil water content (SWC) via oven-drying (105 °C, 48 h).

The water of all plant xylem and soil samples was extracted by means of cryogenic vacuum distillation with liquid nitrogen. The negative pressure was set to be 2 Pa, and the heating temperature was set at 80 °C for stems and 105 °C for soils. The average extraction time was 75 min for stems and 60 min for soils. The δ^2H and δ^{18}O values of each water sample were also measured through an IsoPrime100 (Isoprime, Stockport, UK), and the isotope ratios were expressed in parts per thousand relative to Vienna Standard Mean Ocean Water (V-SMOW).

2.3. Calculations and Statistical Analyses

We used general linear models (GLMs) to analyze the differences in the SWCs and soil water δ^2H and δ^{18}O values for different seasons, sites, and depths (i.e., soil layers of different depths). Similarly, the differences in leaf δ^{13}C and leaf nutrient (C, N, and P) concentrations and ratios (C/N, C/P, and N/P) for different seasons, sites, and species were also analyzed by GLMs. For other purposes, we selected various kinds of combinations of influencing factors as fixed factors (Table 1; Table 2). If the analyzed results attained a significant level ($p \leq 0.05$), differences between groups were compared using a post hoc Tukey's test.

MixSIAR, which is a Bayesian mixing model, was used to distinguish the water sources of the rubber tree and the sharp-leaf galangal quantitatively in R [27]. In this study, we used two isotopic values (i.e., δ^2H and δ^{18}O) to estimate the water-absorbing patterns of the rubber trees and the sharp-leaf galangal on the premise of considering individual effects. Because the process in which source water enters plant roots and differentiates very little in water ^{2}H and ^{18}O can be ignored totally [20,23], the isotopic discrimination of this mixing model was set at zero. The raw isotopic values in the xylem water were treated as the mixture data, and the mean ± SD of the soil water isotopic values in different soil layers was set as the source data. The MCMC (Markov chain Monte Carlo) run length was set as "very long".

In addition, the soil water evaporation line of each site in each season was generated by fitting a linear regression to the isotope data in a dual isotope plot (i.e., δ^2H and δ^{18}O), and a Pearson correlation analysis was performed to display the relationships between the leaf δ^{13}C values; the leaf C, N, and P concentrations; and the leaf C/N, C/P, and N/P ratios of the rubber trees and the intercropped species. Meanwhile, on the basis of foliar parameters, an analysis of Pearson correlation-based similarities between the plant species was conducted, and the statistical analyses were performed using R 3.5.1 [28].

3. Results

3.1. Plant Xylem Water and Soil Water

The xylem water δ^2H and δ^{18}O values of the rubber trees exhibited no significant differences between the sites either in the extremely dry season or in the pronounced rainy season (Figure 1). The xylem water δ^2H and δ^{18}O values of the rubber trees in the RS-AFS were significantly lower than the intercropped sharp-leaf galangal during the extremely dry season ($p < 0.01$, Figure 1), but the difference was not significant during the rainy season.

The δ^2H and δ^{18}O values of the soil water in both the RM and RS-AFS differed between the seasons and the sampling depths significantly ($p < 0.01$, Table 1). Similarly, the SWCs of each study site also exhibited significant variations between seasons and depths ($p < 0.01$, Table 1). In general, soil water δ^2H and δ^{18}O values were significantly higher in the extremely dry season than in the pronounced rainy season ($p < 0.01$), and the soil water isotope was significantly diminished from the surface soil layer to the deep soil layer during the extremely dry season, but the difference was not obvious in the pronounced rainy season (Figure 1c,d). In addition, the difference in the soil water isotope values between the sites was not significant (except for the δ^{18}O values of the soil water in the pronounced rainy season; Table 1). In both the RM and RS-AFS, there existed a good linear relation between the δ^2H and δ^{18}O values of soil water.

In the extremely dry season, the difference in the SWC between the sites was not significant (Table 1; Figure 2a). However, the SWC of the RS-AFS was significantly higher than that of the RM during the pronounced rainy season ($p < 0.01$; Figure 2b). In general, the SWC of both the RM and RS-AFS was significantly increased in the pronounced rainy season relative to the extremely dry season.

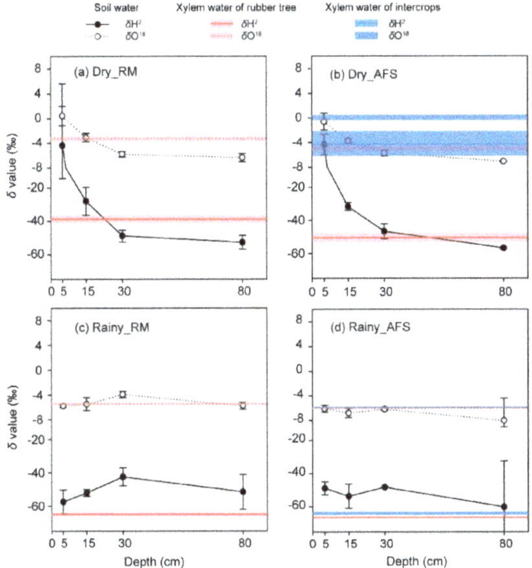

Figure 1. The δ^2H and δ^{18}O values of the consumers (i.e., the xylem water of rubber trees and intercrops) and their water sources (i.e., soil water within four different depths) at each site during the extremely dry season (**a,b**) and the pronounced rainy season (**c,d**). The bold lines inside of the shaded bars indicate mean values, and the width of the shaded bar indicates the SD. All data are expressed as the mean ± SD. "Dry_RM" and "Dry_AFS" indicate the rubber monoculture (RM) and the rubber tree and sharp-leaf galangal agroforestry system (RS-AFS) in the extremely dry season, respectively. Similarly, "Rainy_RM" and "Rainy_AFS" respectively indicate those two sites in the pronounced rainy season.

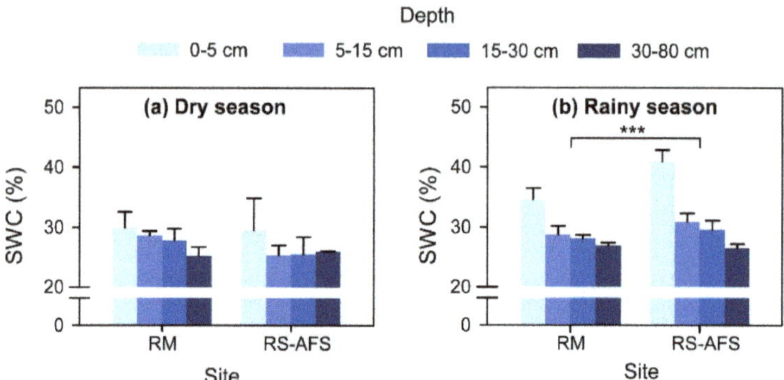

Figure 2. Soil water content (SWC) within the soil profiles at each site during (**a**) the dry season and (**b**) the rainy season. "RM" and "RS-AFS" indicate the rubber monoculture and the rubber tree and sharp-leaf galangal agroforestry system, respectively. Data are expressed as the mean ± SD. * $p < 0.05$, ** $p < 0.01$, *** $p < 0.001$.

Table 1. Results of a general linear model testing the effects on soil water content (SWC) and soil water δ^2H and δ^{18}O. F-values and significance are reported. The results included the following: (A) the effects of season, site, and depth; the effects of season and depth (B) in the RM and (C) in the RS-AFS; and the effects of site and depth (D) in the dry season and (E) in the rainy season. "*d.f.*" indicates the degree of freedom, "*F*" indicates *F* value and "*P*" indicates *P* value. * $p < 0.05$, ** $p < 0.01$, *** $p < 0.001$.

	Tested Effects	d.f.	δ^2H F	δ^2H P	δ^{18}O F	δ^{18}O P	SWC F	SWC P
(A)	Season	1	47.66	***	35.02	***	35.66	***
	Site	1	0.39		9.5	**	0.78	
	Depth	3	18.69	***	21.29	***	31.05	***
	Season * Site	1	0.01		2.02		9.48	**
	Season * Depth	3	21.22	***	18.52	***	6.73	**
	Site * Depth	3	0.74		0.25		1.92	
	Season * Site * Depth	3	0.52		0.95		2.11	
(B)	Season	1	37.21	***	21.43	***	7.19	*
	Depth	3	9.85	***	17.94	***	16.55	***
	Season * Depth	3	22.71	***	29.47	***	2.7	
(C)	Season	1	17.79	***	17.62	***	28.88	***
	Depth	3	9.66	***	8.56	**	16.46	***
	Season * Depth	3	5.52	**	3.64	*	5.12	*
(D)	Site	1	0.38		2.68		1.54	
	Depth	3	106	***	72.3	***	2.85	
	Site * Depth	3	0.47		0.45		0.74	
(E)	Site	1	0.16		6.83	*	18.13	***
	Depth	3	0.95		1.7		76.91	***
	Site * Depth	3	0.66		0.66		6.64	**

3.2. Plant Water Sources

As the MixSIAR revealed, rubber trees in the RM absorbed a certain proportion of the soil water from the shallow and middle soil layers (5–30-cm depths) in the extremely dry season (37.4%, on average; Figure 3a), but the water-absorbing proportion from these soil layers decreased (16.7%, on average), and the main area of absorption shifted to the surface soil layer (0–5-cm depths) in the pronounced rainy season (51.2%, on average; Figure 3b).

However, more than 50% of the absorbed water of rubber trees in the RS-AFS came from the shallow and middle soil layers (5–30-cm depths), both in the extremely dry season and the pronounced rainy season. Therefore, the water-absorbing patterns of rubber trees in the RS-AFS looked more stable relative to rubber trees in the RM (Figure 3).

Relative to the rubber trees in the RS-AFS, the absorbed water of the intercropped sharp-leaf galangal came mostly from the surface, shallow, and middle soil layers (close to 80%; Figure 3), and the water-absorbing proportions of each soil layer were also stable between the seasons.

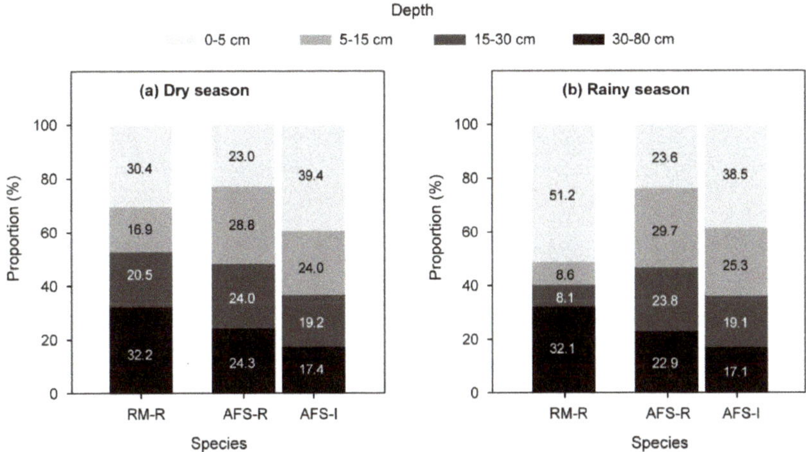

Figure 3. Mean water-absorbing proportion of each plant species in (**a**) the extremely dry season and (**b**) the pronounced rainy season. "RM-R" indicates rubber trees in the rubber monoculture. "AFS-R" and "AFS-I" indicate rubber trees and intercrops in the agroforestry system, respectively.

3.3. Leaf δ^{13}C, Nutrient Concentrations, and Ratios

In the RM and RS-AFS, the mean leaf δ^{13}C values of the rubber trees were −31.39‰ and −31.53‰, respectively. The mean leaf δ^{13}C value was −30.83‰ for the sharp-leaf galangals in the RS-AFS. The rubber trees in the RM and the RS-AFS exhibited no significant differences in their leaf δ^{13}C values (Table 2), but a significant decrease in the leaf δ^{13}C values of the rubber trees was found from the dry season to the rainy season. In addition, rubber trees had significantly lower leaf δ^{13}C values than did the intercropped sharp-leaf galangal in the RS-AFS ($p < 0.05$; Figure 4a). Furthermore, there was no significant difference in the leaf nutrient concentrations and ratios of the rubber trees between the two study sites (Table 2). In addition, in the RS-AFS, the leaf C, N, and P concentrations of the rubber trees were significantly higher than those of the sharp-leaf galangals, but the ratios of the leaf C/N, C/P, and N/P of the rubber trees were significantly lower than in the sharp-leaf galangals.

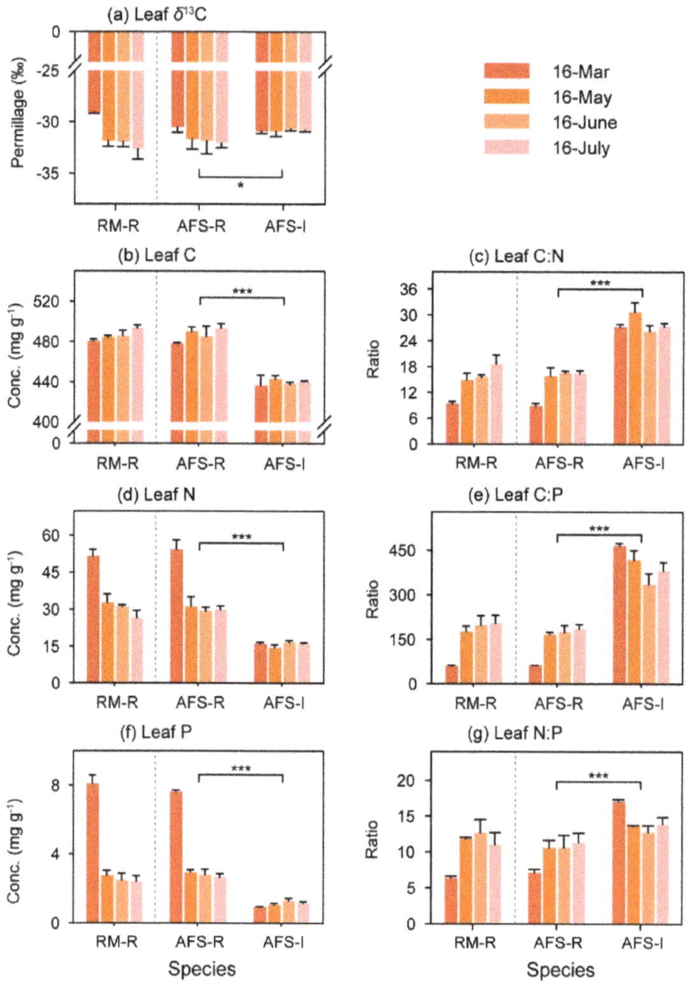

Figure 4. (**a**) Leaf $\delta^{13}C$; (**b**–**d**) leaf C, N, and P concentrations' and (**e**–**g**) leaf C/N, C/P, and N/P ratios of each plant species at each site. Data are expressed as the mean ± SD. * $p < 0.05$, ** $p < 0.01$, *** $p < 0.001$. See Figure 3 for abbreviations.

Table 2. Results of a general linear model testing the effects on leaf δ^{13}C, leaf nutrient (C, N, and P) concentration, and leaf ratios (C/N, C/P, and N/P). *F*-values and significance are reported. The results included the following: (A) time effects on the rubber trees in the RM; (B) time effects on the rubber trees in the RS-AFS; (C) time effects on the intercrops in the RS-AFS; (D) the effects of time and site on the rubber trees; and (E) the effects of time and species on the plants in the RS-AFS. "*d.f.*" indicates the degree of freedom, "*F*" indicates *F* value and "*P*" indicates *P* value. * $p < 0.05$, ** $p < 0.01$, *** $p < 0.001$.

	Tested Effects	d.f.	δ^{13}C		C		N		P		C/N		C/P		N/P	
			F	P	F	P	F	P	F	P	F	P	F	P	F	P
(A)	Time	3	19.86	***	7.66	**	50.44	***	162.91	***	24.62	***	25.82	***	15.21	**
(B)	Time	3	2.01		3.82		50.58	***	400.68	***	34.6	***	47.41	***	7.12	*
(C)	Time	3	0.39		0.79		4.01		8.24	**	5.57	*	10.81	**	22.85	***
(D)	Time	3	13.96	***	8.8	**	99.61	***	436.11	***	54.91	***	63.05	***	20.82	
	Site	1	0.21		0.15		0.38		0.28		0.17		2.47		1.3	
	Time * Site	3	2.06		0.77		1.41		2.06		2.23		0.57		1.65	
(E)	Time	3	1.58		3.79	*	47.78	***	298.91	***	18.42	***	3.37	*	0.88	
	Species	1	7.05	*	395.01	***	534.04	***	1893.82	***	665.24	***	731.62	***	119.17	***
	Time * Species	3	2.1		0.9		46.58	***	366.16	***	14.35	***	32.76	***	20.8	***

Furthermore, the leaf $\delta^{13}C$ of the rubber tree was significantly correlated with the leaf nutrient concentration and leaf nutrient ratio in the RM (Figure 5a), but the correlation coefficients declined, and the significant correlations with leaf C, leaf C/P, and leaf N/P disappeared in the RS-AFS (Figure 5b). Like the rubber trees in the RS-AFS, the leaf $\delta^{13}C$ of the intercrops was also significantly correlated with leaf N and P (Figure 5c). Except for leaf C, all leaf parameters of the rubber trees in the RM were significantly similar to those of rubber trees in the RS-AFS (Figure 6). However, in the RS-AFS, no significant similarity between the rubber trees and the intercropped sharp-leaf galangals was found in their leaf $\delta^{13}C$, leaf C and N, and leaf C/N (Figure 6a–c,e), but their leaf P concentrations and leaf C/P and N/P exhibited significant but opposite variations (Figure 6d,f,g).

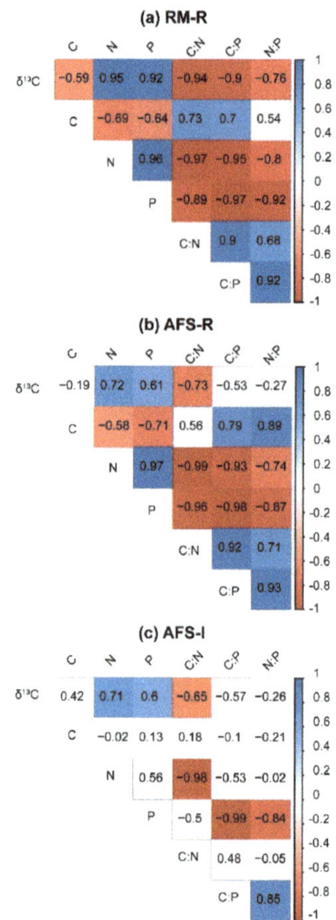

Figure 5. The Pearson correlations between all leaf parameters (i.e., leaf $\delta^{13}C$ values; leaf C, N, and P concentrations; and leaf C/N, C/P, and N/P ratios) of (**a**) the rubber trees in the RM and the leaves of (**b**) the rubber trees and (**c**) their intercrops in the RS-AFS. The color indicates the sign of the correlation (blue indicates a positive correlation, and red indicates a negative correlation). All the correlation coefficients in the red and blue boxes reached a significant level ($p < 0.05$). See Figure 3 for abbreviations.

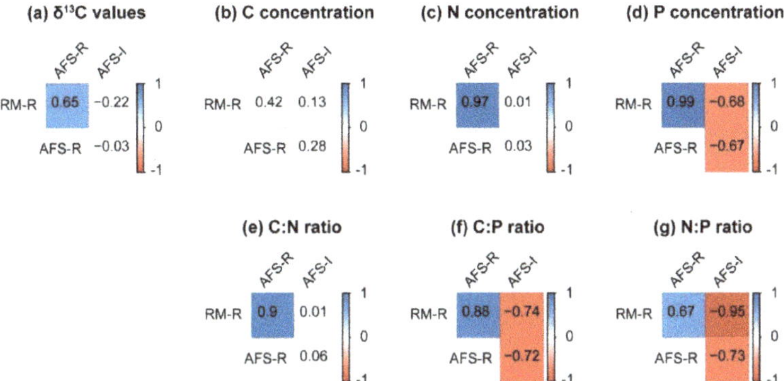

Figure 6. The Pearson correlation-based similarities between RM-R, AFS-R, and AFS-I, which were analyzed through the (**a**) leaf δ^{13}C values; (**b-d**) leaf C, N, and P concentrations; and (**e-g**) leaf C/N, C/P, and N/P ratios of the rubber trees and their intercrops. The color indicates the sign of the correlation (blue indicates a positive correlation, and red indicates a negative correlation). All the correlation coefficients in the red and blue boxes were significant ($p < 0.05$).

4. Discussion

4.1. Plant Water-Absorbing Patterns

Because of the significantly lower δ^2H and δ^{18}O values of the soil water in the deep soil layer in comparison to those in the shallow soil layers during the extremely dry season, the significantly higher xylem water δ^2H and δ^{18}O values of the sharp-leaf galangal relative to the rubber trees in the RS-AFS suggested that the sharp-leaf galangal absorbed much more water from the surface and shallow soil layers than the rubber tree did in the extremely dry season. Indeed, as demonstrated by the results of the MixSIAR model, the sharp-leaf galangals absorbed more water at a depth of 0-5 cm (Figure 3a), and their water-absorbing proportion from the surface soil layer to the deep soil layer gradually decreased in the extremely dry season. Obviously, this was mainly due to the shallow distribution of their roots (less than 30 cm of depth). Meanwhile, from the extremely dry season to the pronounced rainy season, the vertical water-absorbing patterns of this intercrop merely changed a little (Figure 3b). Relatively speaking, rubber trees in the RS-AFS seemed to absorb more water from the middle and deep soil layers than the sharp-leaf galangal did, both in the extremely dry season and in the pronounced rainy season (Figure 3).

Essentially, the monocultural rubber trees mainly depend on surface and shallow resources, because approximately 55%–86% of their feeder roots are concentrated in the soil layers above 20 cm of depth, and they are sensitive to a soil water deficit [29–32]. Therefore, we found that the rubber trees in the RM in this study mainly depended on the soil water from the soil layer above 30 cm of depth, and their absorption of the surface soil water (0-5-cm depths) was greatly enhanced from the extremely dry season to the pronounced rainy season (Figure 3). However, the main water-absorbing area of rubber trees shifted to the middle soil layer (i.e., 5–30-cm soil layers) when the sharp-leaf galangal was intercropped with them, whether it was in the extremely dry season or in the pronounced rainy season.

Despite the vertical rhizospheres of the sharp-leaf galangals being less than 30 cm, deep soil water (from 30–80-cm deep soil layers) was also taken as the water source of this intercropped species. This was mainly because rubber trees exhibit the function of hydraulic redistribution [33]. As a common phenomenon, hydraulic redistribution can be explained by the mechanism of vascular plant root systems transporting deep soil water to a shallow soil layer or transporting water from a water-abundant soil layer to a water-exhausted soil layer [34]. Therefore, as one possible water source, the deep soil water contributed a certain proportion to the absorbed water source of the sharp-leaf galangals in the RS-AFS (Figure 3). Such a phenomenon could not only benefit the water use of the intercropped sharp-leaf galangals in the extremely dry season, but also balance the SWCs of different soil layers, thus helping both the rubber trees and their intercrops form relatively stable water-absorbing patterns relative to those of rubber trees in the RM (Figure 3).

It is obvious that the flexible water uptake of the rubber trees (Figure 3) helped to form relatively complementary water-absorbing patterns in the RS-AFS. This was mainly because rubber roots demonstrate strong plasticity [30], and such plastic roots could detect and avoid neighboring roots [35]. Therefore, hydrological niche segregation, which is a phenomenon where coexisting plants use a water source through different strategies of water use [21], thus helped the rubber trees and the sharp-leaf galangals form relatively complementary water-absorbing patterns in the RS-AFS gradually. Therefore, the interspecific water competition between the rubber trees and the sharp-leaf galangals was particularly small in the RS-AFS, even in the extremely dry season. Moreover, the relatively deeper water-absorbing behavior of the rubber trees in the RS-AFS during the rainy season indicated a deeper rooting depth of rubber trees in the RS-AFS. Such deep rooting behavior (from rubber trees in an RS-AFS) benefits the hydrological processes within the soil in an RS-AFS. For example, a dye tracer experiment found that the deeper rooting phenomenon of rubber trees in rubber-based agroforestry systems could reduce soil compaction, optimize soil structure, improve soil water infiltration, enhance soil water-holding capacity, and facilitate soil water movement [36]. In addition, nutrient mass flow may also be enhanced because of enhanced infiltration, and rubber trees in an RS-AFS therefore may absorb more nutrients because deep roots imply the extension of root contact areas with soil nutrients [23]. Furthermore, the deep rooting depth of rubber trees in an RS-AFS could help improve their nutrient uptake efficiency from deep soil layers, thus effectively reducing nutrient losses through deep leaching [37] (Liu et al. found that a large proportion of P compounds were accumulated in the deep soil layers in rubber agroforestry systems [38]).

4.2. Plant Water Use Efficiency and Soil Water Conditions

Commonly, differences in leaf C isotope discrimination between C_3 and C_4 plants are great because of the great differences in their dark reactions in photosynthesis [39]. Therefore, the leaf $\delta^{13}C$ values of C_3 plants typically range between −20‰ and −37‰, and the leaf $\delta^{13}C$ values of C_4 plants mainly range between −12‰ and −16‰ [40,41]. That is, rubber trees and sharp-leaf galangals both belong to C_3 plants (Figure 4a).

For C_3 plants, their leaf $\delta^{13}C$ values could represent their long-term WUEs and reflect the plant response to soil drought [40]. Commonly, high $\delta^{13}C$ values always correspond to high WUEs [42,43]. Therefore, the similar leaf $\delta^{13}C$ values of the rubber trees in the RM and the RS-AFS (Figure 4a; Figure 6a) suggest that the WUE of rubber trees remained the same with intercropping with sharp-leaf galangal. It was discovered that the SWC in the RS-AFS did not differ from the RM in the extremely dry season, and the SWC in the RS-AFS was even higher than in the RM in the pronounced rainy season (Figure 2). The results above imply that the competition for water in the RS-AFS was not too intense to reduce soil water availability. In addition, the WUE of the rubber trees and sharp-leaf galangals in the RS-AFS displayed no apparent difference in different seasons (Table 2). This phenomenon thereby implies that the internal microclimatic environment in the RS-AFS was stable, because environmental factors (e.g., moisture and temperature) also affect plant leaf $\delta^{13}C$ greatly [43].

The soil water δ^2H and $\delta^{18}O$ values were higher in the extremely dry season than in the pronounced rainy season because the strong soil evaporation and low rainwater supply in the dry season generally result in the enrichment of the stable hydrogen and oxygen isotopes of soil water [19]. Similarly, the seasonal variations in rainfall also led to low SWC in the extremely dry season but high SWC in the pronounced rainy season. Theoretically speaking, the soil in rubber agroforestry systems could contain much more water because intercropping could increase interception, decrease runoff, and improve soil water infiltration and soil water-holding capacity [36]. Therefore, a significantly higher SWC was found in the RS-AFS in the pronounced rainy season. However, the SWC of the RS-AFS was as low as that of the RM in the extremely dry season. Excluding the causes of intense soil evaporation and low rainfall in this season (Figure 3a), these results might have been due to the extra consumption of the sharp-leaf galangal, which showed relatively lower WUEs compared to the rubber trees in the RS-AFS. In addition, rubber trees may also aggravate soil water consumption because they must access enough reserve water for leaf expansion during this season [25]. Anyway, the soil water in the RS-AFS was no less than that in the RM. However, the water in the middle soil layer of the RS-AFS consumed more relative to the RM in the extremely dry season (Figure 2a). This was probably because the rubber trees in the RS-AFS preferred to absorb water from the middle soil layers (Figure 3).

In short, intercropping sharp-leaf galangal could improve soil water conditions distinctly in the pronounced rainy season, but these benefits were not apparent in the extremely dry season. This finding was consistent with a previous similar study, which demonstrated that soil water shortages still existed in this kind of intercropping system in the dry season, and a soil drought would limit the normal growth and photosynthesis of sharp-leaf galangals [15]. Therefore, it seems essential to conduct appropriate irrigation in an RS-AFS for sharp-leaf galangals in the dry season, although such a shortage of soil water has limited impacts on rubber trees.

4.3. Plant Leaf Carbon, Nitrogen, and Phosphorus Concentrations and Ratios

Commonly, C, N, and P are the most essential nutrients in the composition of various proteins and genetic materials in plants [44]. In addition, C/N/P ratios are closely associated with many eco-physiological functions of plants [45,46], and thus some well-known hypotheses and concepts have been put forward, such as the growth rate hypothesis [47]. Therefore, the similar leaf nutrient concentrations and ratios between the rubber trees in the RM and the RS-AFS (Figure 4c–g; Figure 6c–g) suggest that the interspecific competition in the RS-AFS did not affect the leaf physiological functions of the rubber trees. This was probably because the water-absorbing patterns of the rubber trees in the RS-AFS were relatively fixed, and the main water-absorbing soil layers of plants are also the main soil layers where plants absorb soil N and P [48]. Therefore, the nutrient absorption of the rubber trees in the RS-AFS may not have been affected by the interspecific competition.

In addition, the significant and negative correlation between the leaf P of the rubber trees and the sharp-leaf galangals in the RS-AFS (Figure 5d) suggested their opposite P use strategies, and such P use was possibly affected by their water use behaviors, because their leaf $\delta^{13}C$ was significantly and positively correlated with their leaf P (Figure 5b,c). Because intercropping can increase soil P mobility in a rubber-based agroforestry system, a large proportion of P compounds accumulate in the deep soil layers [38]. Therefore, P acquisition by rubber trees in the RS-AFS benefited from water-absorbing patterns (Figure 3). However, such phenomena might not be consistent with the P requirement of intercropped sharp-leaf galangal, which has very shallow absorption areas. Therefore, interspecific competition in an RS-AFS may have a great impact on the growth of sharp-leaf galangal [15].

As indicated by the growth rate hypothesis, fast-growing organisms generally exhibit low C/P and N/P ratios in their tissues [47]. Therefore, the significantly lower leaf C/P and N/P ratios of the rubber trees in the RS-AFS relative to their intercrops (Figure 4) seemed to indicate that their growth was relatively more vigorous than that of the sharp-leaf galangals. Additionally, the opposite variation tendency in the leaf C/P and N/P ratios of the rubber trees and the sharp-leaf galangals in the RS-AFS (Figure 6f,g) demonstrated that their life and competition strategies were quite different. Generally, slow-growing species exhibit lower rates of resource acquisition and longer leaf life than fast-growing species do [49,50], and plants with long-lived leaves can reduce their nutrient requirements to maintain their leaf areas in resource-limited environments [23]. These characteristics are essentially consistent with those of sharp-leaf galangals [15]. Therefore, as evergreen perennials, the sharp-leaf galangals in the RS-AFS seemed to have chosen a persistence competition strategy, which is regarded as a sit-and-wait approach for available resources due to their characteristics of long leaf lives, fixed water-absorbing patterns, and slow growth rates [51]. However, the rubber trees showed a foraging competition strategy, which reflected rapid plant growth toward available resource patches; the isolation of resources; and sustained, rapid, and controlled growth toward more available resources [51].

5. Conclusions

Rubber trees in the RM absorbed more water from the surface soil layers, but they would take up more water from the shallow and middle soil layers after six years of intercropping with the sharp-leaf galangals. This was because the intercropped sharp-leaf galangals mainly depended on surface and shallow soil water, and the root niches of the rubber trees and the sharp-leaf galangals in the RS-AFS seemed to overlap greatly at the beginning of the intercropping. Due to the competitive foraging strategy of the rubber trees in the RS-AFS, the feeder roots of the rubber tree would enter deeper soil layers. Then, a hydrologic niche separation occurred, and the stable and complementary water use patterns of the rubber trees and the sharp-leaf galangal were formed in the RS-AFS. Therefore, the competition for water in this agroforestry system was weak, and such competition did not change the WUE of the rubber trees. It is worth mentioning that the soil in the RS-AFS could contain more water during the pronounced rainy season, but the SWCs of the RS-AFS were also as low as those in the RM during the extremely dry season. Therefore, irrigation during the dry season is necessary to maintain the healthy growth of sharp-leaf galangals. In addition, competition for soil P between the rubber trees and the sharp-leaf galangals in the RS-AFS was obvious, but there was no distinct influence on the P requirement for the growth of rubber trees. Therefore, P fertilization should also be applied properly in the planting rows of sharp-leaf galangal, especially in the surface soil layer. In brief, the interspecific competition below the ground did not affect the water use and nutrient uptake of the rubber trees.

Supplementary Materials: The following are available online at http://www.mdpi.com/1999-4907/10/10/924/s1, Figure S1: Schematic diagrams of (a) the rubber monoculture and (b) the rubber tree and sharp-leaf galangal agroforestry system (RS-AFS). Figure S2: Randomly selected blocks for sampling. Figure S3: (a) Information on the phenophases of rubber trees and (b) the monthly precipitation and monthly mean temperature during the study period (data were provided from the Xishuangbanna Station for Tropical Rainforest Ecosystem Studies). FS, fruit setting; FR, fruit ripening; DS, dormant stage; LS, leaf shedding; LF, leaf flushing; LE, leaf expansion; and FP, flowering phase.

Author Contributions: J.W. and W.L. designed the experiments; J.W., H.Z., and C.C. conducted the experiment, analyzed the data, and wrote the first draft; W.L. and X.J. revised and edited the manuscript.

Funding: This study was supported by the National Natural Science Foundation of China (31570622, 41701029, and 31800356), the National Postdoctoral Program for Innovative Talents (BX201700278), the China Postdoctoral Science Foundation (2018M633437), the Fourth Oriented Training Foundation for Postdoctors of Yunnan Province, the Scientific Research Foundation for Postdoctors of Yunnan Province, the "PhD Plus" Talents Training Program Foundation of XTBG, the CAS 135 Program (2017XTBG-F01), and the Foundation for the CAS Key Laboratory of Tropical Forest Ecology (09KF001B04).

Acknowledgments: We thank Fu Y., Deng Y., Liu M.N., and the Central Laboratory of XTBG for the help provided.

Conflicts of Interest: The authors declare no conflict of interest.

References

1. Cornish, K. Alternative natural rubber crops: Why should we care? *Technol. Innov.* **2017**, *18*, 244–255. [CrossRef]

2. Mann, C.C. Addicted to rubber. *Science* **2009**, *325*, 564–566. [CrossRef] [PubMed]

3. Fox, J.; Castella, J.C. Expansion of rubber (*Hevea brasiliensis*) in Mainland Southeast Asia: What are the prospects for smallholders? *J. Peasant Stud.* **2013**, *40*, 155–170. [CrossRef]

4. Wigboldus, S.; Hammond, J.; Xu, J.; Yi, Z.F.; He, J.; Klerkx, L.; Leeuwis, C. Scaling green rubber cultivation in Southwest China–An integrative analysis of stakeholder perspectives. *Sci. Total Environ.* **2017**, *580*, 1475–1482. [CrossRef] [PubMed]

5. Zhang, L.; Kono, Y.; Kobayashi, S. The process of expansion in commercial banana cropping in tropical China: A case study at a Dai village, Mengla County. *Agric. Syst.* **2014**, *124*, 32–38. [CrossRef]

6. Zhang, J.Q.; Corlett, R.T.; Zhai, D. After the rubber boom: Good news and bad news for biodiversity in Xishuangbanna, Yunnan, China. *Reg. Environ. Chang.* **2019**, *19*, 1–12. [CrossRef]

7. Li, H.; Ma, Y.; Aide, T.M.; Liu, W. Past, present and future land-use in Xishuangbanna, China and the implications for carbon dynamics. *For. Ecol. Manag.* **2008**, *255*, 16–24. [CrossRef]

8. Mann, C.C. Why We (Still) Can't Live Without Rubber. Available online: https://www.nationalgeographic.com/magazine/2016/01/southeast-asia-rubber-boom/ (accessed on 20 September 2019).

9. Song, Q.H.; Tan, Z.H.; Zhang, Y.P.; Sha, L.Q.; Deng, X.B.; Deng, Y.; Zhao, W. Do the rubber plantations in tropical China act as large carbon sinks? *iForest* **2013**, *7*, 42–47. [CrossRef]

10. Blagodatsky, S.; Xu, J.; Cadisch, G. Carbon balance of rubber (*Hevea brasiliensis*) plantations: A review of uncertainties at plot, landscape and production level. *Agric. Ecosyst. Environ.* **2016**, *221*, 8–19. [CrossRef]

11. Zhang, M.; Chang, C.; Quan, R. Natural forest at landscape scale is most important for bird conservation in rubber plantation. *Biol. Conserv.* **2017**, *210*, 243–252. [CrossRef]

12. Langenberger, G.; Cadisch, G.; Martin, K.; Min, S.; Waibel, H. Rubber intercropping: A viable concept for the 21st century? *Agrofor. Syst.* **2017**, *91*, 577–596. [CrossRef]

13. John, J.; Nair, A.M. Prospects of allelopathic research in multi-storey cropping systems. In *Allelopathy in Ecological Agriculture and Forestry*; Narwal, S.S., Hoagland, R.E., Dilday, R.H., Reigosa Roger, M.J., Eds.; Springer: Dordrecht, The Netherlands, 2000; pp. 159–179.

14. He, Z.H.; Ge, W.; Yue, G.G.L.; Bik-San Lau, C.; He, M.F.; But, P.P.H. Anti-angiogenic effects of the fruit of *Alpinia oxyphylla*. *J. Ethnopharmacol.* **2010**, *132*, 443–449. [CrossRef] [PubMed]

15. Cheng, H.; Liu, Q.F.; Yan, J.K.; Zhang, T.L.; Wang, Q.Y. Seasonal changes of photosynthetic characteristics of *Alpinia oxyphylla* growing under *Hevea brasiliensis*. *Chin. J. Plant Ecol.* **2018**, *42*, 585–594.

16. FAO and IAEA. Management of Agroforestry Systems for Enhancing Resource use Efficiency and Crop Productivity. Available online: https://www.iaea.org/publications/8181/management-of-agroforestry-systems-for-enhancing-resource-use-efficiency-and-crop-productivity (accessed on 29 September 2019).

17. Craine, J.M.; Dybzinski, R. Mechanisms of plant competition for nutrients, water and light. *Funct. Ecol.* **2013**, *27*, 833–840. [CrossRef]

18. Zechmeister-Boltenstern, S.; Keiblinger, K.M.; Mooshammer, M.; Peñuelas, J.; Richter, A.; Sardans, J.; Wanek, W. The application of ecological stoichiometry to plant-microbial-soil organic matter transformations. *Ecol. Monogr.* **2015**, *85*, 133–155. [CrossRef]

19. Fry, B. *Stable Isotope Ecology*; Springer: New York, NY, USA, 2006.

20. White, J.W.; Cook, E.R.; Lawrence, J.R. The DH ratios of sap in trees: Implications for water sources and tree ring DH ratios. *Geochim. Cosmochim. Acta* **1985**, *49*, 237–246. [CrossRef]

21. Silvertown, J.; Araya, Y.; Gowing, D. Hydrological niches in terrestrial plant communities: A review. *J. Ecol.* **2015**, *103*, 93–108. [CrossRef]

22. Youri, R.; Mathieu, J. Reviews and syntheses: Isotopic approaches to quantify root water uptake: A review and comparison of methods. *Biogeosciences* **2017**, *14*, 2199–2224.

23. Chapin, F.S., III; Matson, P.A.; Vitousek, P. *Principles of Terrestrial Ecosystem Ecology*; Springer: New York, NY, USA, 2011.

24. Everard, K.; Seabloom, E.W.; Harpole, W.S.; de Mazancourt, C. Plant water use affects competition for nitrogen: Why drought favors invasive species in California. *Am. Nat.* **2009**, *1*, 85–97. [CrossRef]

25. Guardiola-Claramonte, M.; Troch, P.A.; Ziegler, A.D.; Giambelluca, T.W.; Durcik, M.; Vogler, J.B.; Nullet, M.A. Hydrologic effects of the expansion of rubber (*Hevea brasiliensis*) in a tropical catchment. *Ecohydrol.* **2010**, *3*, 306–314. [CrossRef]

26. Wu, J.; Liu, W.; Chen, C. Can intercropping with the world's three major beverage plants help improve the water use of rubber trees? *J. Appl. Ecol.* **2016**, *53*, 1787–1799. [CrossRef]

27. Parnell, A.C.; Phillips, D.L.; Bearhop, S.; Semmens, B.X.; Ward, E.J.; Moore, J.W.; Inger, R. Bayesian stable isotope mixing models. *Environmetrics* **2013**, *24*, 387–399. [CrossRef]

28. R Core Team. *R: A Language and Environment for Statistical Computing Computer Software*; R Foundation for Statistical Computing: Vienna, Austria, 2014.

29. George, S.; Suresh, P.R.; Wahid, P.A.; Nair, R.B.; Punnoose, K.I. Active root distribution pattern of *Hevea brasiliensis* determined by *radioassay* of latex serum. *Agrofor. Syst.* **2009**, *76*, 275–281. [CrossRef]

30. Priyadarshan, P. *Biology of Hevea Rubber*; CABI: Wallingford, UK, 2011.

31. Kobayashi, N.; Kumagai, T.O.; Miyazawa, Y.; Matsumoto, K.; Tateishi, M.; Lim, T.K.; Yin, S. Transpiration characteristics of a rubber plantation in central Cambodia. *Tree Physiol.* **2014**, *34*, 285–301. [CrossRef]

32. Giambelluca, T.W.; Mudd, R.G.; Liu, W.; Ziegler, A.D.; Kobayashi, N.; Kumagai, T.O.; Yin, S. Evapotranspiration of rubber (*Hevea brasiliensis*) cultivated at two plantation sites in Southeast Asia. *Water Resour. Res.* **2016**, *52*, 660–679. [CrossRef]

33. Wu, J.; Liu, W.; Chen, C. How do plants share water sources in a rubber-tea agroforestry system during the pronounced dry season. *Agric. Ecosyst. Environ.* **2017**, *236*, 69–77. [CrossRef]

34. Bleby, T.M.; Mcelrone, A.J.; Jackson, R.B. Water uptake and hydraulic redistribution across large woody root systems to 20 m depth. *Plant Cell Environ.* **2010**, *33*, 2132–2148. [CrossRef]

35. Callaway, R.M. The detection of neighbors by plants. *Trends Ecol. Evol.* **2002**, *17*, 104–105. [CrossRef]

36. Jiang, X.J.; Liu, W.; Wu, J.; Wang, P.; Liu, C.; Yuan, Z.Q. Land Degradation Controlled and Mitigated by Rubber-based Agroforestry Systems through Optimizing Soil Physical Conditions and Water Supply Mechanisms: A Case Study in Xishuangbanna, China. *Land Degrad. Dev.* **2017**, *28*, 2277–2289. [CrossRef]

37. Pierret, A.; Maeght, J.L.; Clément, C.; Montoroi, J.P.; Hartmann, C.; Gonkhamdee, S. Understanding deep roots and their functions in ecosystems: An advocacy for more unconventional research. *Ann. Bot.* **2016**, *118*, 621–635. [CrossRef]

38. Liu, C.; Jin, Y.; Liu, C.; Tang, J.; Wang, Q.; Xu, M. Phosphorous fractions in soils of rubber-based agroforestry systems: Influence of season, management and stand age. *Sci. Total Environ.* **2018**, *616*, 1576–1588. [CrossRef] [PubMed]

39. O'Leary, M.H. Carbon isotopes in photosynthesis. *BioScience* **1988**, *38*, 328–336. [CrossRef]

40. Farquhar, G.D.; Ehleringer, J.R.; Hubick, K.T. Carbon isotope discrimination and photosynthesis. *Annu. Rev. Plant Biol.* **1989**, *40*, 503–537. [CrossRef]

41. Kohn, M.J. Carbon isotope compositions of terrestrial C3 plants as indicators of (paleo) ecology and (paleo) climate. *Proc. Natl. Acad. Sci. USA* **2010**, *107*, 19691–19695. [CrossRef] [PubMed]

42. Ehleringer, J.R.; Roden, J.; Dawson, T.E. Assessing ecosystem-level water relations through stable isotope ratio analyses. In *Methods in Ecosystem Science*; Sala, O.E., Jackson, R.B., Mooney, H.A., Howarth, R.W., Eds.; Springer: New York, NY, USA, 2000; pp. 181–198.

43. Marshall, J.D.; Brooks, J.R.; Lajtha, K. Sources of variation in the stable isotopic composition of plants. In *Stable Isotopes in Ecology and Environmental Science*; Michener, R., Lajtha, K., Eds.; John Wiley Sons: Chichester, UK, 2007; pp. 22–60.

44. Sardans, J.; Rivas-Ubach, A.; Penuelas, J. The elemental stoichiometry of aquatic and terrestrial ecosystems and its relationships with organismic lifestyle and ecosystem structure and function: A review and perspectives. *Biogeochemistry* **2012**, *111*, 1–39. [CrossRef]

45. Sasaki, T.; Yoshihara, Y.; Jamsran, U.; Ohkuro, T. Ecological stoichiometry explains larger-scale facilitation processes by shrubs on species coexistence among understory plants. *Ecol. Eng.* **2010**, *36*, 1070–1075. [CrossRef]

46. Sterner, R.W.; Elser, J.J. *Ecological Stoichiometry: The Biology of Elements from Molecules to the Biosphere*; Princeton University Press: Oxford, UK, 2002.

47. Elser, J.J.; Sterner, R.W.; Gorokhova, E.A.; Fagan, W.F.; Markow, T.A.; Cotner, J.B.; Weider, L.W. Biological stoichiometry from genes to ecosystems. *Ecol. Lett.* **2000**, *3*, 540–550. [CrossRef]

48. Barber, S.A. *Soil Nutrient Bioavailability: A Mechanistic Approach*; John Wiley Sons: New York, NY, USA, 1995.

49. Craine, J.M.; Fargione, J.; Sugita, S. Supply pre-emption, not concentration reduction, is the mechanism of competition for nutrients. *New Phytol.* **2005**, *166*, 933–940. [CrossRef]

50. Craine, J.M. Competition for nutrients and optimal root allocation. *Plant Soil* **2006**, *285*, 171–185. [CrossRef]

51. Keddy, P.; Fraser, L.H.; Wisheu, I.C. A comparative approach to examine competitive response of 48 wetland plant species. *J. Veg. Sci.* **1998**, *9*, 777–786. [CrossRef]

 *forests*

Article

Ecosystem Services and Importance of Common Tree Species in Coffee-Agroforestry Systems: Local Knowledge of Small-Scale Farmers at Mt. Kilimanjaro, Tanzania

Sigrun Wagner [1,*], Clement Rigal [2,3], Theresa Liebig [4], Rudolf Mremi [5], Andreas Hemp [6], Martin Jones [1], Elizabeth Price [1] and Richard Preziosi [1]

[1] Ecology and Environment Research Centre, Manchester Metropolitan University, Manchester M1 5GD, UK; m.jones@mmu.ac.uk (M.J.); e.price@mmu.ac.uk (E.P.); r.preziosi@mmu.ac.uk (R.P.)

[2] CIRAD, UMR SYSTEM, F-34398 Montpellier, France; clement.rigal@cirad.fr

[3] SYSTEM, University of Montpellier, CIHEAM-IAMM, CIRAD, INRA, Montpellier SupAgro, F-34060 Montpellier, France

[4] IITA Uganda, Plot 15 Naguru East Road, PO Box 7878, Kampala 00256, Uganda; theresa.liebig@gmail.com

[5] College of African Wildlife Management, Mweka P.O. Box 3031 Moshi, Tanzania; rmremi@mwekawildlife.ac.tz

[6] Department of Plant Systematics, Universität Bayreuth, Universitätsstr. 30, 95440 Bayreuth, Germany; andreas.hemp@uni-bayreuth.de

* Correspondence: sigrun.k.wagner@stu.mmu.ac.uk

Received: 7 October 2019; Accepted: 25 October 2019; Published: 1 November 2019

Abstract: *Research Highlights:* Global coffee production, especially in smallholder farming systems, is vulnerable and must adapt in the face of climate change. To this end, shaded agroforestry systems are a promising strategy. *Background and Objectives:* Understanding local contexts is a prerequisite for designing locally tailored systems; this can be achieved by utilizing farmers' knowledge. Our objective is to explore ecosystem services (ESs) provided by different shade tree species as perceived by farmers and possible factors (elevation, gender, and membership in local farmers groups) influencing these perceptions. We related these factors, as well as farmers' ESs preferences, to planting densities of tree species. *Materials and Methods:* During interviews with 263 small-scale coffee farmers on the southern slope of Mt. Kilimanjaro, they ranked the most common shade tree species according to perceived provision of the locally most important ESs for coffee farmers. We asked them to estimate the population of each tree species on their coffee fields and to identify the three ESs most important for their household. *Results:* Food, fodder, and fuelwood emerged as the most important ESs, with 37.8% of the respondents mentioning all three as priorities. Density of tree species perceived to provide these three ESs were significantly higher for farmers prioritizing these services compared to farmers that did not consider all three ESs in their top three. *Albizia schimperiana* scored the highest for all rankings of regulatory ESs such as coffee yield improvement, quality shade provision, and soil fertility improvement. Influence of elevation, gender, and farmer group affiliation was negligible for all rankings. *Conclusions:* This study shows the need to understand factors underlying farmers' management decisions before recommending shade tree species. Our results led to the upgrade of the online tool (shadetreeadvice.org) which generates lists of potential common shade tree species tailored to local ecological context considering individual farmers' needs.

Keywords: shade tree species; farmers' knowledge; East Africa

---

1. Introduction

Agroforestry is a promising agricultural production system due to its potential for climate change mitigation and adaptation [1–3]. Besides carbon sequestration [4], shade trees also improve local climatic conditions and reduce variability in microclimate and soil moisture [1]. Agroforestry is particularly important for coffee (*Coffea arabica* L.) production as climate change is expected to reduce the suitable production area for crops such as coffee [5–7]. In addition to regulatory services, the associated shade tree species can provide various direct ecosystem services (ESs) such as food, fodder, or fuelwood [8]. Furthermore, due to their diversity, agroforestry systems have the potential to provide diverse income sources which may act as social safety nets, increasing farmers' economic resilience in the face of coffee price volatility in global markets and possible crop failures [8–11]. However, ecological conditions, competition among associated species in the system, and farmers' individual objectives need to be considered in designing agroforestry systems to maximize the benefits and minimize the shortcomings of these systems [12,13].

Farmers can be very knowledgeable on factors that influence coffee productivity. From experience, they are aware of interaction(s) between shade tree species and coffee, as well as many direct ESs provided by specific tree species [14,15]. In some areas, however (e.g., the impact of individual tree species on pests and diseases), their knowledge might be limited [16–18]. Nevertheless, exploring farmers' knowledge might provide novel insights into interactions between shade tree species and coffee productivity. This local knowledge is vital in tailoring recommendations to local conditions.

Although several studies have investigated how the Chagga people living on Mt. Kilimanjaro in Tanzania use their natural environment [19–22], research has so far not identified which tree species are considered superior in providing relevant ESs for the local coffee farmers. Our aim is to assess indigenous knowledge of local farmers on Mt. Kilimanjaro regarding selection of shade trees that enhance coffee production and provide other ESs. A participatory approach based on van der Wolf et al. [23] allowed us to collect and study farmers' knowledge regarding shade tree species' provision of ESs. Following this approach, we identified shade tree species with high potential for coffee agroforestry systems on the southern slopes of Mt. Kilimanjaro. Similar studies have been conducted in other East African countries and the differences in findings [15,16,24,25] demonstrate the importance of locally specific investigations.

In this study, we explore the ESs provided by different shade tree species as perceived by farmers. We further examine if elevation, gender, and membership in local farmers' groups influence the perceived ESs provided, and the planting density of different shade tree species. Farm management may be based on farmers' knowledge and preference for specific shade tree species [15]. Therefore, we expect that the planting density of tree species will depend on the perceived ESs the tree species provides, as well as farmers' ESs preference.

The study aims to contribute to expanding the database of the online decision-support tool for tree selection in smallholder farming systems (shadetreeadvice.org) [23] and help tailor recommendations for important shade tree species for coffee farmers on Mt. Kilimanjaro.

2. Materials and Methods

2.1. Study Site

The research took place on the southern slopes of Mt. Kilimanjaro in Tanzania (Latitude 3°13′9″ S–3°17′41″ S; Longitude 37°9′24″ E–37°25′19″ E) (Figure 1). Here, the Chagga people have cultivated and converted the former forest into an agroforestry system (the so-called Chagga homegardens) over several centuries [21]. The main cultivation zone of *C. arabica* in these traditional coffee-banana plantations is located between 1000 m and 1800 m asl [26] and covers an area of nearly 80,000 ha [27]. Kilimanjaro National Park prevents expansion of coffee cultivation into higher elevations [19]. For our study area, Hemp [26] reports annual rainfall in the lower slopes (800 m to 1300 m asl) of between

900 mm to 1580 mm, and 1580 mm to 2200 mm at the higher slopes (1300 m to 1800 m asl) and mean annual temperatures of 23.4 to 18.8 °C and 18.8 to 16.1 °C, respectively.

Coffee farmers from eight communities participated in this study. Six communities work with the non-government organization Hanns R. Neumann Stiftung (HRNS) (Isuki, Lemira Mroma, Masama Mula, Mudio, Kiwakabo, and Mbokomu); two communities have no connection to the HRNS, are located in the center of the southern slope (Narum and Mweka), and were included in the study to have a wider representation of the study area (Figure 1).

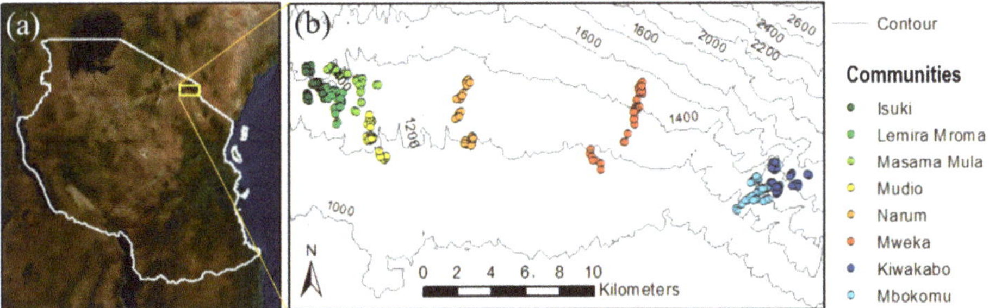

Figure 1. Location of the research area within Tanzania (**a**) and the distribution of the respondents along the southern slope of Mt. Kilimanjaro, divided into the eight communities (**b**).

2.2. Identification of Common Tree Species and Important Ecosystem Services

To identify the most common tree species and most important ESs for coffee farmers of the study area for subsequent data collection, we conducted focus group discussions (FGDs) [28]. In March 2019, we conducted three FGDs in the west (Isuki, Masama Mula and Mudio) and two FGDs in the east (Kiwakabo and Mbokomo). Participants were coffee farmers representing farmers' groups and independent farmers from the same community. Each FGD had between 9 and 15 attendees, leading to a total of 56 attendees.

The list of tree species that participants could choose from was based on Hemp [26], and the results of an investigation of 40 plots of Chagga homegardens in Narum and Mweka (unpublished data). As small-scale farmers in this region commonly intercropped coffee and banana (*Musa* spp.), we included the latter as a shade species. This resulted in an initial list of 58 shade tree species. We prepared technical sheets for each shade tree species showing pictures of the plant or plant parts, as well as the local name. We presented these sheets to the focus groups and asked them to rank the shade tree species according to their frequency in coffee fields in their area. Tree species shown to each FGD were then added or removed from the presented list of trees based on their rankings during previous FDGs. In subsequent data collection interviews with individual farmers, we only used the most common shade tree species. This list was composed of 22 tree species that were either ranked in the top 20 in at least three FGDs or in the top 10 of any single FGD (see the species listed in Table S1).

For identification of the locally most important ESs for farmers, we presented 25 ESs to the focus groups and asked the participants to add any additional ESs they considered important. The groups then ranked the services according to their perceived importance, giving us the final list of the 12 locally most important ESs for subsequent data collection interviews. The list included nine ESs that were ranked in the top 12 of at least four FGDs (food provision, shade provision, protection against wind, protection from heat, fodder supply, mulch provision, increased coffee yield, soil fertility improvement, and weed suppression); one ES ranked number four in one FGD and in the top 10 of two other FGDs (increasing coffee quality); one ES ranked as number two in two FGDs, but below 10 in all other FGDs (firewood supply); and one ES ranked as number one in one FGD, but only in the top 10 in one other FGD (soil moisture enhancement) (see all ESs listed in Table S2).

2.3. Shade Tree Species Ranking for Ecosystem Services

Ranking the most common tree species according to the most important ESs was the focus of data collection interviews with 263 small-scale coffee farmers along the southern slope of Mt. Kilimanjaro (Figure 1). In March and April 2019, we conducted at least 30 individual interviews in each community (Figure 1). We began with farmers that participated in the FGDs or whose farm we investigated; we then interviewed occupants of the fifth house away in each direction along the road. If the person in the fifth house was not a coffee farmer, declined participation, or was absent at the time, we asked at the next house(s) until a respondent was identified.

The respondents were asked to select the 10 tree species that they knew best out of the list of 22 most common shade tree species [23]. They were then asked to rank the 10 chosen shade tree species for each of the 12 locally most important ESs from the best (high provision of this ES) to the least performing (low provision of this ES) [16,23]. We also recorded gender, membership of the HRNS, and elevation of the respondents' home (using a GARMIN GPSMAP 64). We asked the respondents to name the three most important ESs for their household and the estimated number of individuals of the different tree species they had on their coffee fields, as well as the size of their coffee farm.

2.4. Data Analysis

We noted the number of tree species each farmer had of the 22 most common shade tree species, the percentage of farmers having each tree species in their fields, and the planting density of each tree species, using farmers' estimated number of trees and farm size. To identify the influence of elevation on the tree planting density of different tree species, we did linear regressions for each common shade tree species in R 3.5.0 [29]. We summarized the ESs respondents considered most important and tested gender differences with a chi-square test in R 3.5.0 [29].

Based on the method of van der Wolf, et al. [23], we used the BradleyTerry2 package in R 3.5.0 [29] to identify shade tree species best at providing specific ESs, as perceived by the farmers. We excluded interview respondents with less than five years of experience as well as tree species that were ranked less than 10 times for a particular ESs [18]; this left 20 of the 22 tree species in the analysis. As explained by Rigal el al. [18], the ranks for each ESs need to be converted into pairwise comparisons to fit the Bradley-Terry model. For each tree species and ES, this model calculates scores, which are comparative values representing the likelihood that one tree species performs better than another tree species in providing an ES [18]. We normalized the scores between 0 and 1 to be able to compare them [18]. Besides the scores, quasi-standard errors were calculated to indicate how frequently a species was included in the ranking and how consistently the respondents ranked this species [18]. We compared the scores pairwise using a Wald test. The more pairs that are significantly different, the more it reflects an agreement of farmers upon the ranking of the tree species. Therefore, large numbers of pairs that are significantly different indicate that the analysis is robust [18]. In our results, the lowest percent of pairs that were different was 67%, and the highest was 89%.

To assess the influence of farmers' objectives on their management practices, we compared shade tree densities on coffee farms between groups of farmers with different sets of priorities. More specifically, we split respondents into two groups: those who had selected the combination of the three most important ESs for small-scale coffee farmer households at Mt. Kilimanjaro as their top three priorities and those who had not. We then compared the shade tree density of species perceived to perform high for the combination of these ESs and tested differences between the two farmer groups using t-tests in R 3.5.0 [29].

To identify if gender, affiliation to a farmers' group, or elevation influenced perceived provision of ESs by shade tree species, we split the data sets by gender, membership of the HRNS, and into two elevation groups (threshold was the median 1336 m asl). We ran the BradleyTerry analysis for these subgroups of respondents and compared the resulting scores [15,16].

2.5. Ethical Approval

Ethical approval for this study was obtained from the Faculty Research Ethics and Governance Committees of the Manchester Metropolitan University, Faculty of Science and Engineering, on 26 May 2017, with application code SE1617108C.

3. Results

3.1. Main Characteristics of Respondent

Of the total respondents, 96 were women (36.5%) and 167 men (63.5%). The farm size ranged between 0.1 and 8 ha, the average farm size was 0.7 ha with 83.3% (219 respondents) having less than 1 ha. The elevation ranged from 1148 m to 1748 m asl, with an average elevation of 1343 m asl and a median elevation of 1336 m asl (Figure 2). Ninety-seven respondents (37%) were members of the HRNS, while 166 were non-members (63%).

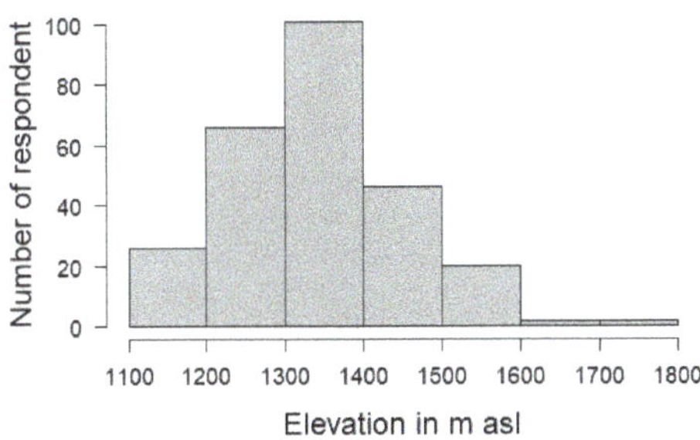

Figure 2. Distribution of respondents along elevation.

3.2. Tree Species Distribution

Sixty percent of the farmers reported that they had 10 or more of the 22 most common shade tree species on their coffee farms. The most common shade tree species is *Musa* spp. grown by all respondents, followed by *Grevillea robusta* A. Cunn. ex R. Br., *Albizia schimperiana* Oliv., and *Persea americana* Miller with 94.3%, 90.9%, and 83.7%, respectively (Figure 3). *Musa* spp. is by far the shade species with the highest density (1089 $\pm$ 106 tree ha^{-1}) (Figure 3). *Grevillea robusta* is second densest (39.1 $\pm$ 3.29 tree ha^{-1}), closely followed by *Markhamia lutea* (Benth.) K. Schum. (26.6 $\pm$ 4.90 tree ha^{-1}). Just a few farmers (about 20 percent) grow *M. lutea*. However, those that do grow it have a high density of the species on their fields. Despite their presence on more than 83% of coffee farms, the densities of *A. schimperiana* (7.9 $\pm$ 0.46 tree ha^{-1}) and *P. americana* (12.3$\pm$0.98 tree ha^{-1}) are significantly lower than those of the above species (Figure 3).

Neither the total shade tree density nor the *Musa* spp. density are significantly influenced by elevation. However, we observed differences in density of some tree species. Linear regressions show significant reduction in densities of *Cordia africana* Lam. ($F_{(1,261)} = 12.91$, $p < 0.001$), *Mangifera indica* L. ($F_{(1,261)} = 5.03$, $p = 0.026$) and *Senna siamea* (Lam.) H. S. Irwin & Barneby ($F_{(1,261)} = 3.96$, $p = 0.048$) with increasing elevation—the densities are reduced by 1.1 tree ha^{-1}, 0.4 tree ha^{-1} and 0.3 tree ha^{-1} respectively for every 100 m increase. We detected a significant increase in density towards higher elevations for *Margaritaria discoidea* (Baill.) G. L. Webster ($F_{(1,261)} = 30.88$, $p < 0.001$) and *P. americana* ($F_{(1,261)} = 11.77$, $p < 0.001$) (with an increase of 100 m, the density increases by 2.4 tree ha^{-1} and 2.7 tree ha^{-1} respectively).

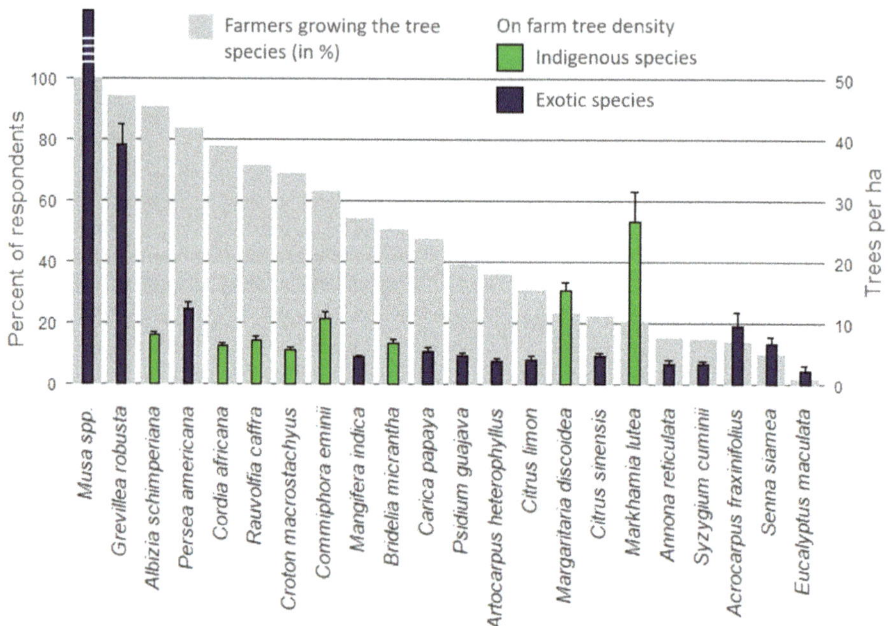

Figure 3. Percent of respondents reporting each species on their farm and average (+SE) tree density on those farms. The density of *Musa* spp. is much higher than the scale shown in the graph (1089 trees per ha on average).

3.3. Important Ecosystem Services

On the southern slope of Mt. Kilimanjaro, food provisioning is by far the most essential ES for coffee farmers. Of the 263 respondents, more than 75% selected food provision as the most important ES for their household, and more than 95% ranked it in the top three (Table 1). The second locally most important ESs is fuelwood supply, which was ranked first by 10% of the respondents and within the top three ESs by nearly 60% of the respondents (Table 1). More than 50% of the respondents also ranked fodder supply among the top three ESs, followed by shade provision, soil fertility improvement and increased coffee yield (Table 1). A chi-square test showed no significant differences in ES preference between genders. The only significant difference between the two elevation groups was that respondents at a higher elevation included soil moisture enhancement more often in the top three ESs than respondents at lower elevation (4.4% and 1.0% respectively, $X^2 = 7.2$, $p < 0.01$).

3.4. Pairwise Comparison

The analysis shows that the ranking of shade tree species is consistent for most ESs. We observed the clearest discrimination between tree species in the ranking for mulch provision and protection from heat with 89.1% and 87.9% of the pairwise comparisons of tree species' scores being significantly different ($p < 0.05$), followed by increase in coffee yield and quality (Table 2). Most difficult to rank were weed suppression, food provision, and protection against wind with 66.7%, 71.1%, and 72.5% of all pairs being significantly different ($p < 0.05$) (Table 2).

Table 1. Ranking of Ecosystem services on a household level.

Ecosystem Services (ESs)	Selected as First ES	Among the First 3 ESs
Food provision	76.4%	95.4%
Firewood supply	10.3%	59.5%
Fodder supply	2.7%	55.3%
Shade provision	3.4%	31.7%
Soil fertility improvement	2.7%	17.2%
Increased coffee yield	0.0%	15.6%
Soil moisture enhancement	1.5%	8.0%
Increased coffee quality	0.8%	5.7%
Protection against wind	1.1%	5.3%
Mulch provision	0.8%	3.4%
Protection from heat	0.0%	1.5%
Weed suppression	0.4%	1.1%

Table 2. Percent of significantly different pairwise comparisons of species' scores.

Ecosystem Service	Number of Tree Species Included in the Ranking	Percent of Significant Differences between Pairs ($p < 0.05$)
Mulch provision	11	89.1
Protection from heat	12	87.9
Increased coffee yield	11	87.3
Increased coffee quality	10	86.7
Soil moisture enhancement	11	85.5
Shade provision	12	84.8
Fodder supply	11	83.6
Firewood supply	13	82.1
Soil fertility improvement	13	76.9
Protection against wind	16	72.5
Food provision	10	71.1
Weed suppression	13	66.7

3.5. Tree Ranking

Scores of the tree species ranked according to the three most important ESs for small-scale coffee farmers at the southern slopes of Mt. Kilimanjaro show major differences (Figure 4). Of the most common tree species providing food, all are exotic. Firewood is mainly obtained from indigenous species, except for *G. robusta*, which is exotic.

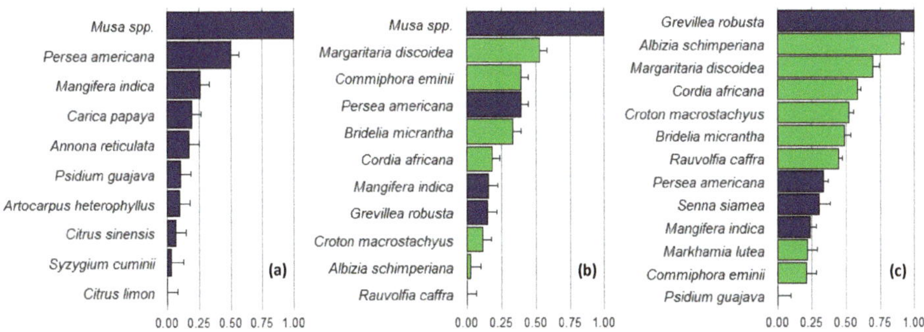

Figure 4. Scores and quasi-standard errors of tree species for (**a**) food provision, (**b**) fodder supply and (**c**) firewood supply. Dark blue bars represent exotic, and green bars indigenous tree species.

For a better recommendation of tree species regarding multiple regulatory ESs associated with coffee production, the following ESs were combined into three categories: (**a**) coffee production enhancement (combining increase in coffee yield and quality), (**b**) protection from climatic hazards (combining protection from heat, wind and shade provision), (**c**) soil quality enhancement (combining mulch provision, soil fertility and soil moisture enhancement). For each ES category, the scores of shade tree species were averaged over the set of combined ESs (Figure 5).

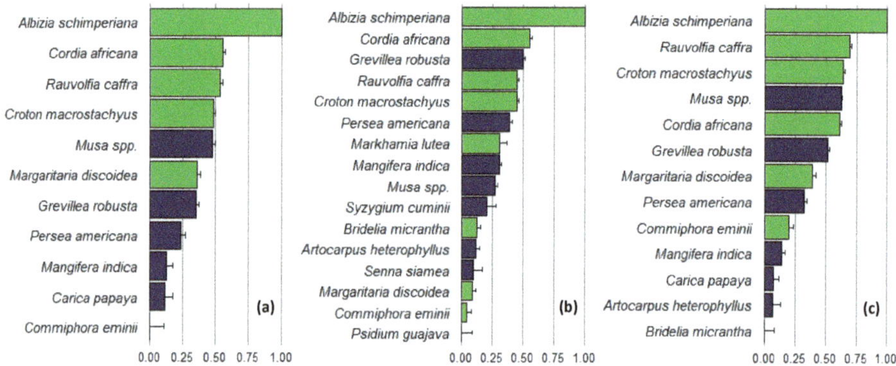

Figure 5. Scores and quasi-standard errors of the tree species ranked according to: (**a**) increase of coffee yield and quality, (**b**) protection from heat, wind and shade provision and (**c**) mulch provision, soil fertility and soil moisture enhancement. Green bars represent indigenous, and dark blue bars exotic tree species.

Albizia schimperiana is the highest ranked tree species for all three ES categories. Also within the top five for all three ES categories are *C. africana*, *Croton macrostachyus* Hochst. ex Delile, and *Rauvolfia caffra* Sond.. All of these tree species are indigenous. *Musa* spp. is important for coffee yield and quality, as well as for soil enhancement, while *G. robusta* contributes to protection from climatic hazards and soil quality enhancement.

3.6. Effect of Priorites on Shade Tree Density

Food, fodder, and firewood are the most important ESs for small-scale coffee farmer households (99 respondents (37.8%) selected this combination). To assess the influence ES priorities have on tree species selection, we averaged the scores of each shade tree species for the combination of these three ESs and compared the density of the five best performing tree species between two groups: respondents that selected these three ESs as most important versus those that did not. The planting density of three of these five tree species is significantly higher (*t*-test, $p < 0.05$) for respondents that selected these ESs as most important compared to those that did not (Figure 6). We found differences for exotic, but not indigenous species.

3.7. Elevation, Gender, and Farmer Group Affiliation

The ranking in higher elevations was not significantly different from the lower elevations. The only deviation we observed is that *M. indica* was included in the ranking of shade tree species providing fodder for the lower elevations, while *R. caffra* was included in the higher elevations (Figure S1). Farmers did, however, rank both of these tree species low for this ES.

There were slight differences considering gender and affiliation to a farmers group of the HRNS. Women included *Annona reticulata* L. in their ranking for food provision, and they ranked *Musa* spp. significantly higher for shade provision and soil fertility compared to men. Men ranked *G. robusta* significantly higher for shade provision, while *R. caffra* was ranked higher for soil fertility (Figure S2). Respondents that were members of HRNS ranked *M. discoidea* significantly higher than non-members for protection from heat (Figure S3). For coffee quality, members of HRNS ranked *C. macrostachyus* significantly higher and *Musa* spp. significantly lower than non-members (Figure S3).

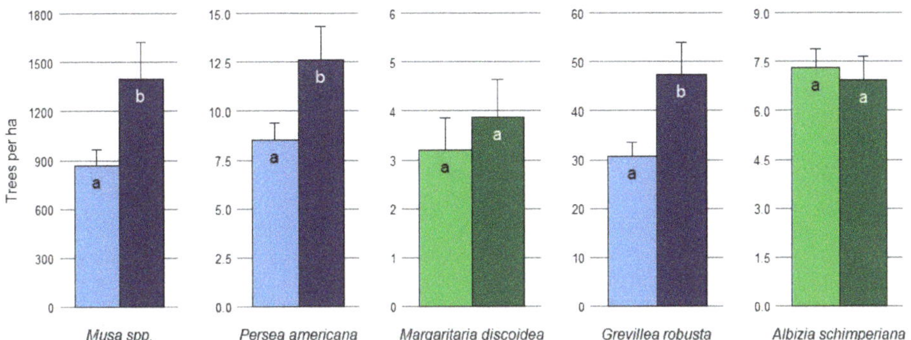

Figure 6. Average (+SE) plant density of the five highest ranked tree species providing the services food, fodder and firewood. The dark bars show respondents who selected the three ESs as the three most important for their family, while the light bars indicate the plant density for respondents who did not select all three. Significant difference between plant density is shown by difference in letters (a and b) ($p < 0.05$). Blue bars represent exotic, and green bars indigenous species.

4. Discussion

4.1. Improtant Ecosystem Services

Nearly all the farmers in our study area considered food provision by shade trees their top priority, followed by the provisioning services fodder and fuelwood supply (Table 1). These ESs were considered more important than regulatory services for coffee production such as shade provision, soil fertility improvement, and increased coffee yield. ES priorities were not significantly different between genders. Respondents of other studies have various priorities. At Mt. Elgon (Uganda), farmers rather prioritized ESs such as mulch provision, erosion control and temperature regulation [16], coffee yield, soil moisture enhancement, and quick leaf decomposition in Central Uganda [24]. Differences in the importance of ESs may reflect differences in environmental conditions, as well as market access [16,30]. One reason participants of the FGDs mentioned for excluding timber from the list of important ESs was the political limitation of tree harvesting [31]. The complexity of agroforestry systems and diverse interactions between different pests and predators on a land-scale level [32–35] might be the reason coffee farmers at Mt. Kilimanjaro did not observe any effect of shade tree species on pests and did not consider it an important ESs provided by shade trees.

4.2. Highly Ranked Tree Species

Albizia schimperiana is the most important shade tree species for coffee production on the southern slope of Mt. Kilimanjaro for the provision of most ESs included in this study. Several studies also report the importance of *A. schimperiana* for coffee production in Ethiopia [36–38]. As a leguminous plant, *A. schimperiana* can form a symbiotic relationship with rhizobia bacteria to fix atmospheric nitrogen, resulting in increased soil fertility. Other studies also show that, with an open wide-spreading crown, *A. schimperiana* provides good shade cover for coffee production, leading to improved microclimate and coffee yield [26,39]. Another advantage of *A. schimperiana* is that its leaves emerge in the dry season [36]. This means it can provide shade when there is a lot of sun and prevents coffee from being too densely shaded during the rainy season.

Despite the benefits of *A. schimperiana* and the perceived provision of multiple regulatory ESs by coffee farmers at Mt. Kilimanjaro, other tree species occurred at a much higher density (Figure 3). From general observations, it seems that the farms have mostly mature trees and that a new generation of *A. schimperiana* is missing. More research on the population structure of *A. schimperiana* in this area is required to get a better understanding of future development and challenges. Belay, et al. [36] found for their study region in Ethiopia that the population structure of *A. schimperiana* had a U-shape with more stems in lower and higher diameter classes, showing selective cutting or extraction of medium sized individuals. Potential explanations could be the increased exploitation of *A. schimperiana* for timber and fuelwood, and its low growth rate [37,40]. Other important aspects to examine are natural regeneration and the success of propagation.

A closely related species that commercial coffee plantations commonly include in their fields is *Albizia gummifera* (J. F. Gmel.) C. A. Sm. This species is favored by farmers in Ethiopia [41,42]. Other *Albizia* species are considered important for providing multiple ESs in East Africa, such as, but not limited to, mulch, shade, improvement of microclimate, coffee yield and soil moisture in Uganda [16,24]. Unfortunately some *Albizia* species are also alternative hosts for black coffee twig borer in the closely related Robusta coffee (*Coffea canephora* Pierre ex Froehn.) posing a potential risk [23].

The other three shade tree species associated with improvements of conditions for coffee production are also indigenous (*C. africana*, *C. macrostachyus* and *R. caffra*). *Rauvolfia caffra* ranked as an important shade tree species which is in line with other findings from Mt. Kilimanjaro [22,26]. Fernandes, et al. [19] report the potential of *R. caffra* to suppress various coffee pests. Its contribution to the production of traditional banana beer underlines its importance in traditional Chagga homegardens [19,22,40]. *Croton macrostachyus* provides good litter, preserves soil moisture, facilitating high coffee yield and high bean weight [43]. *Cordia africana* is also reported as an essential shade tree in coffee production in Kenya [15], Uganda [16] and Ethiopia [42,44]. Even though Kufa, et al. [39] found high coffee yields under *C. africana*, they also reported the highest yield variations under this tree species. As *C. africana* is a high quality timber tree, the economic value might be a major reason for farmers to grow it [42,43].

Even though the four tree species discussed above (*A. schimperiana*, *C. africana*, *C. macrostachyus*, and *R. caffra*) are multipurpose tree species and considered the best to enhance soil quality, create a suitable environment for coffee production and benefit coffee yield and quality, their densities in coffee farms were lower than that of the exotic fast growing *G. robusta* (Figure 3). In Kenya, native tree species were also considered to provide a healthy environment, but their abundance was low due to their slow growth rate [15]. In Tanzania, another reason for the preference of *G. robusta* might be that the wood can be utilized more easily than for native tree species, as native tree species require a permit to be cut down [31,45].

Farmers' management decisions to plant or remove a certain tree species in their plantations is usually based on their knowledge or tree preference [14,15,46]. We therefore need to look at their ESs priorities as socio-economic factors might influence farmers' choices for on-field composition. Our results confirm that the density of exotic tree species perceived to provide food, fodder and fuelwood are higher when farmers prioritize these ESs (Figure 6). This is especially the case when the tree species are exotic, as their presence in the field is usually due to management rather than natural occurrences. Some other studies also show the importance of shade tree species for coffee production is matched with their planting densities [36,42]. Bukomenko, et al. [24], however, found a mismatch between ESs that are important for respondents and the trees they have on their fields. Graefe, et al. [30] used a similar methodology for cocoa production in Ghana and also reported disparities between higher ranked tree species suitable for cocoa intercropping and their abundance in the northern part of the cocoa belt with marginal conditions for cocoa production. This might be an adaptation strategy to diversify income, since they confirmed our findings for farms in the wetter southern region with optimal cocoa production conditions [30]. The match between shade tree density and prioritized ESs appears to be more consistent with direct short- or mid-term benefits for farmers, such as food, fodder, and firewood supply, rather than regulatory service provisioning, such as climate modification and soil fertility improvement [15]. This becomes evident from the lower density of *A. schimperiana* (the most highly ranked tree species for regulatory services) in comparison to *G. robusta* and *P. americana*, which provide direct outputs like fuelwood and food (Figure 3). Our findings stress the importance of understanding the socio-economic component when investigating tree species distribution.

4.3. Factors Influencing Tree Species Ranking and Distribution

The first factor influencing tree species distribution, as just discussed, is farmers' preference and the ESs they consider important for their household.

In general, there is agreement on the ranking of tree species among small-scale coffee farmers at Mt. Kilimanjaro. We can confirm the finding by Gram, et al. [16] that local knowledge regarding ranking of tree species is gender blind, as there were negligible differences in our study. Besides gender, participation in a farmer group did not influence the ranking. Farmers participating in farmer groups meet regularly, receive trainings and exchange knowledge and therefore might have better access to different information sources. This could have led to ranking differences, as other researchers have shown the influence of promotion activities of certain tree species on the perception and distribution of those species [18,46]. In Kenya, *G. robusta* was considered suitable for intercropping with coffee, despite having similar traits to those of tree species believed to negatively affect coffee production [15]. For example, the root system for both *Eucalyptus* spp. and *G. robusta* are perceived to be wide spreading [15]. This discrepancy in perception of different tree species could be due to promotion activities from extension services [15]. Such biases need to be considered when using local knowledge to inform recommendations.

Some studies report an influence of elevation on the distribution and ranking of tree species in East Africa [16,26]. However, we found that neither the presence nor density of tree species in the coffee fields of small-scale farmers at Mt. Kilimanjaro varied much across elevations for most investigated species. The reason could be that our focus was on the most common tree species that are well known by many farmers rather than the whole natural flora as reported by Hemp [26]. *Mangifera indica* and *S. siamea* are known to grow better at lower elevations [40]; it is therefore not surprising that their density is higher at lower elevations. Rather unexpected is the reduced presence and density of *C. africana* with increased elevation as the suitable range for this tree species in the Kilimanjaro regions is reported to be between 1200 m and 2000 m asl. [40]. We therefore conclude that the density of *C. africana* is influenced by socio-economic factors rather than by environmental factors. In Uganda, *C. africana* was perceived to perform well for all ESs [16] and is therefore found in farms at all elevations. Even though this tree species is also perceived to perform highly for regulatory services at Mt. Kilimanjaro, this was not a priority for most farmers. Food being the highest priority explains the presence of either *M. indica* or *P. americana* in 90% of the coffee farms, and the climatic needs for these two species explain the decrease in density of *M. indica* and the increase of density of *P. americana* along the elevation gradient [40]. Another influencing aspect might be increased distance to markets at higher elevations and therefore increased importance of self-sufficiency to farmers. The increased density of *M. discoidea* with increased elevation might be due to its importance for fodder supply. At lower elevations, farmers might have better access to other sources of fodder.

Rankings were not influenced by elevation, despite the relationship between shade tree distribution and elevation. This confirms the findings of Lamond, et al. [15] that farmers' knowledge of tree attributes affecting field interactions between shade tree and coffee is consistent along an elevation gradient.

4.4. Tree Species Ranking in East Africa

Shade tree species in coffee agroforestry systems vary greatly among regions and even within East Africa; hence, it is not possible to generalize findings for the region. Studies with a similar approach to farmers' knowledge of tree species have been carried out in Rwanda [25], Uganda [16,24], and Kenya [15]. Figure 7 combines the tree species included in the rankings and shows the overlaps in the species reported. Only seven shade tree species were commonly recorded in all countries and most of them are exotic. There are also very important tree species that only appear in the ranking of one country, for example *A. schimperiana*, the most important shade tree species for the Mt. Kilimanjaro region.

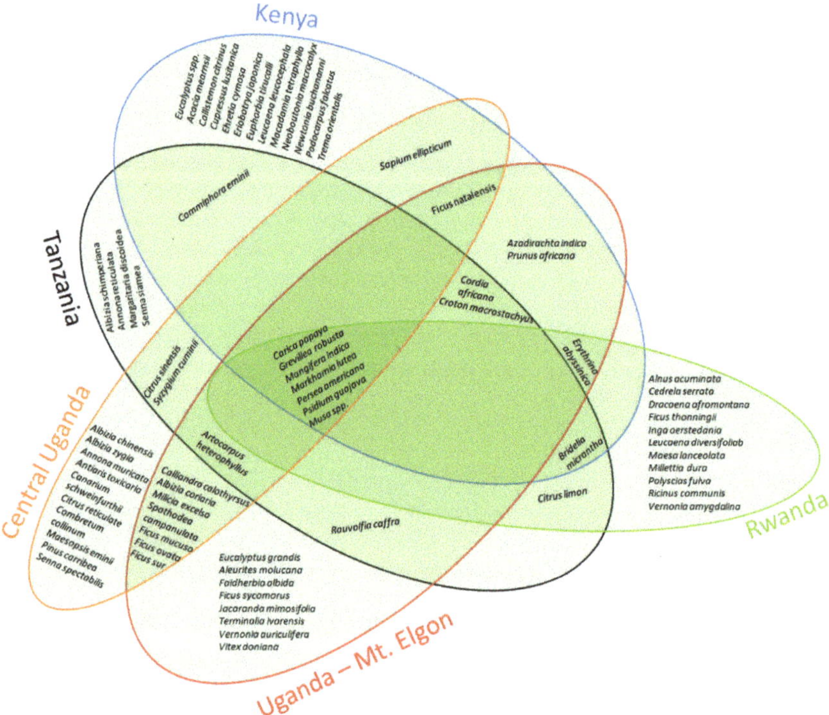

Figure 7. Shade species included in different rankings in East Africa [15,16,24,25].

The comparison of our results with those from similar research done in other East African countries shows the importance of considering the local context in this type of research based on local knowledge. The results linked to specific shade tree species are specific to locations and cannot be generalized. Including an approach based on functional ecology, linking the scores of shade tree species with their attributes can help generalizing results in future studies. This may lead to recommendations of characteristics of tree species that are generally more acceptable. Lamond, et al. [15] already focused more on ranking attributes rather than tree species and Albertin, et al. [47] investigated which tree characteristics are important for coffee farmers. Focusing on tree characteristics might also help to recommend shade tree species that are not very common, but might be a good fit for farmers since ranking is limited to the most common shade tree species, neglecting the importance of rather rare species.

4.5. Resilience of Coffee Agroforestry Systems

Some of the ESs shade tree species can provide, such as temperature regulation or soil moisture enhancement, might help mitigate the impacts of climate change [1,2]. However, not all shade tree species are perceived as similarly effective in providing said services as the scores show (Figure 5). Tree species not influenced by elevation are considered more climate change resilient [16]. This is the case for all investigated shade tree species for our study area within the investigated elevation range, besides *C. africana*, *M. indica*, *S. siamea*, *P. americana*, and *M. discoidea*.

A challenge of the present methodology is the small number of shade tree species that can be included in ES rankings. In order to enhance the ability of the agroforestry system to recover from external pressure and be more resilient in the face of climate change, it is important to protect the biodiversity of the farming systems [48]. It is therefore important to not only recommend the common species mentioned in this paper, but to retain a variety of tree species in the local ecosystem.

A better understanding of the optimal shade intensity and therefore the optimal shade tree density is also essential to tailor advice to farmers. Even though *A. schimperiana* can provide several services, knowledge of the optimal planting density and best management practices in terms of pruning will still be required to achieve optimal shade levels and utilize all potential benefits for coffee production.

For socio-economic resilience, it is important to consult with farmers about their preferences prior to recommending a list of shade tree species, to ensure that the advice fits with their objectives and their constraints. The online decision-support tool for tree selection (shadetreeadvice.org) [23] is a first step in tailoring recommendations for important shade tree species to farmers' preferences. It also needs to be considered that with recurring low coffee prices, focusing primarily on shade tree species that optimize coffee production might not be economically sustainable. The emphasis might shift towards other ESs and shade tree species that increase the economic resilience of coffee farms.

5. Conclusions

This study demonstrated the link between farmers' preference for certain ESs and the planting density of shade tree species that provide these ESs. This shows the importance of understanding the factors underlying farmers' management decisions before recommending shade tree species. Despite being aware of negative crop-tree interactions, farmers might include tree species that are not necessarily beneficial to coffee production in order to acquire other services such as food, fodder, and firewood provisioning, all considered priorities for farmers on the southern slopes of Mt. Kilimanjaro.

Local knowledge of tree species' benefits can be very valuable to local producers; however, it needs to be complemented with expert knowledge to identify biases and fill in knowledge gaps. Even though our study has confirmed that local knowledge of tree species is gender blind, it could still be influenced by other factors such as differences in access to information, access to markets, and/or other socio-economic factors. Contrary to other studies, we did not observe an influence of elevation on the perceptions of tree species' provisioning of ESs. Nevertheless, it will be important to consider environmental aspects in future studies. Another limitation of this methodology is that it only includes the most common shade tree species, leaving aside rare indigenous species with high potential for agroforestry systems. This underestimates the importance of less common species, which might even be superior in providing certain ESs. It may also give the impression that the common tree species alone are enough to support a resilient coffee production system. One approach to improving recommendations might be focusing on the traits shown by highly ranked tree species. This will not only help comparing results from different regions and generalizing recommendations, but also ensure that a wider range of species are included.

Supplementary Materials: The following are available online at http://www.mdpi.com/1999-4907/10/11/963/s1, Table S1: Tree species used for the ranking, Table S2: Ecosystem services used for the ranking, Figure S1: Scores and quasi-standard errors of tree species for fodder supply at (a) lower elevations (1148–1335 m asl) and (b) higher elevations (1336–1748 m asl). Red bars show tree species with significantly different scores between the two groups, Figure S2: Scores and quasi-standard errors of tree species for food provision (a,b), shade provision (c,d), and soil fertility (e,f) divided by gender (women are presented in a, c and e; men are presented in b, d and f). Red bars show tree species with significantly different scores between the two groups, Figure S3: Scores and quasi-standard errors of tree species for protection from heat (a,b), and increasing coffee quality (c,d) divided by affiliation to a farmers group (non-members are presented in a and c; members are presented in b and d). Red bars show tree species with significantly different scores between the two groups.

Author Contributions: Conceptualization, S.W.; Data curation, S.W. and T.L.; Formal analysis, S.W. and C.R.; Funding acquisition, S.W. and R.P.; Investigation, S.W.; Methodology, S.W.; Project administration, S.W., M.J., E.P. and R.P.; Software, S.W., C.R. and T.L.; Supervision, M.J., E.P. and R.P.; Visualization, S.W.; Writing—Original Draft, S.W.; Writing—Review & Editing, S.W., C.R., T.L., R.M., A.H., M.J., E.P. and R.P.

Funding: This research received no external funding.

Acknowledgments: This research would not have been possible without the participation of coffee farmers at Mt. Kilimanjaro. We thank all respondents for their time and for sharing their experience. Special thanks to Laurence Jassogne for her support during the research design and Happyness Thadei Mwenda, Thomas Terosi and Emmanuel Ruheta for their assistance in communicating with the farmers. We are grateful for the logistical support from the Hanns R. Neumann Stiftung. We thank the Manchester Metropolitan University for their financial support, the College for African wildlife management (MWEKA), the Kilimanjaro Research Group and the International Institute of Tropical Agriculture (IITA) for their collaboration and the Tanzania Commission for Science and Technology (COSTECH) for granting us the research permit.

Conflicts of Interest: The authors declare no conflict of interest.

References

1. Lin, B.B. Agroforestry management as an adaptive strategy against potential microclimate extremes in coffee agriculture. *Agric. For. Meteorol.* **2007**, *144*, 85–94. [CrossRef]
2. Mbow, C.; Smith, P.; Skole, D.; Duguma, L.; Bustamante, M. Achieving mitigation and adaptation to climate change through sustainable agroforestry practices in Africa. *Curr. Opin. Environ. Sustain.* **2014**, *6*, 8–14. [CrossRef]
3. Rahn, E.; Läderach, P.; Baca, M.; Cressy, C.; Schroth, G.; Malin, D.; van Rikxoort, H.; Shriver, J. Climate change adaptation, mitigation and livelihood benefits in coffee production: Where are the synergies? *Mitig. Adapt. Strateg. Glob. Chang.* **2014**, *19*, 1119–1137. [CrossRef]
4. Jose, S.; Bardhan, S. Agroforestry for biomass production and carbon sequestration: An overview. *Agrofor. Syst.* **2012**, *86*, 105–111. [CrossRef]
5. Adhikari, U.; Nejadhashemi, A.P.; Woznicki, S.A. Climate change and eastern Africa: A review of impact on major crops. *Food Energy Secur.* **2015**, *4*, 110–132. [CrossRef]
6. Bunn, C.; Läderach, P.; Rivera, O.O.; Kirschke, D. A bitter cup: Climate change profile of global production of Arabica and Robusta coffee. *Clim. Chang.* **2015**, *129*, 89–101. [CrossRef]
7. Craparo, A.C.W.; Van Asten, P.J.A.; Läderach, P.; Jassogne, L.T.P.; Grab, S.W. Coffea arabica yields decline in Tanzania due to climate change: Global implications. *Agric. For. Meteorol.* **2015**, *207*, 1–10. [CrossRef]
8. Reed, J.; van Vianen, J.; Foli, S.; Clendenning, J.; Yang, K.; MacDonald, M.; Petrokofsky, G.; Padoch, C.; Sunderland, T. Trees for life: The ecosystem service contribution of trees to food production and livelihoods in the tropics. *For. Policy Econ.* **2017**, *84*, 62–71. [CrossRef]
9. Bacon, C. Confronting the Coffee Crisis: Can Fair Trade, Organic, and Specialty Coffees Reduce Small-Scale Farmer Vulnerability in Northern Nicaragua? *World Dev.* **2005**, *33*, 497–511. [CrossRef]
10. Tscharntke, T.; Clough, Y.; Bhagwat, S.A.; Buchori, D.; Faust, H.; Hertel, D.; Hoelscher, D.; Juhrbandt, J.; Kessler, M.; Perfecto, I.; et al. Multifunctional shade-tree management in tropical agroforestry landscapes—A review. *J. Appl. Ecol.* **2011**, *48*, 619–629. [CrossRef]
11. Charles, R.; Munishi, P.; Nzunda, E. Agroforestry as Adaptation Strategy under Climate Change in Mwanga District, Kilimanjaro, Tanzania. *Int. J. Environ. Prot.* **2013**, *3*, 29–38.
12. Beer, J.; Muschler, R.; Kass, D.; Somarriba, E. Shade management in coffee and cacao plantations. *Agrofor. Syst.* **1998**, *38*, 139–164. [CrossRef]
13. Schroth, G.; Lehmann, J.; Rodrigues, M.R.L.; Barros, E.; Macêdo, J.L. V Plant-soil interactions in multistrata agroforestry in the humid tropics. *Agrofor. Syst.* **2001**, *53*, 85–102. [CrossRef]
14. Cerdán, C.R.; Rebolledo, M.C.; Soto, G.; Rapidel, B.; Sinclair, F.L. Local knowledge of impacts of tree cover on ecosystem services in smallholder coffee production systems. *Agric. Syst.* **2012**, *110*, 119–130. [CrossRef]
15. Lamond, G.; Sandbrook, L.; Gassner, A.; Sinclair, F.L. Local knowledge of tree attributes underpins species selection on coffee farms. *Exp. Agric.* **2016**, 1–15. [CrossRef]
16. Gram, G.; Vaast, P.; Van Der Wolf, J.; Jassogne, L. Local tree knowledge can fast-track agroforestry recommendations for coffee smallholders along a climate gradient in Mount Elgon, Uganda. *Agrofor. Syst.* **2018**, *92*, 1625–1638. [CrossRef]
17. Liebig, T.; Jassogne, L.; Rahn, E.; Läderach, P.; Poehling, H.M.; Kucel, P.; Van Asten, P.; Avelino, J. Towards a collaborative research: A case study on linking science to farmers' perceptions and knowledge on Arabica Coffee Pests and Diseases and Its Management. *PLoS ONE* **2016**, *11*, e0159392. [CrossRef]

18. Rigal, C.; Vaast, P.; Xu, J. Using farmers' local knowledge of tree provision of ecosystem services to strengthen the emergence of coffee-agroforestry landscapes in southwest China. *PLoS ONE* **2018**, *13*, e0204046. [CrossRef]

19. Fernandes, E.C.M.; Oktingati, A.; Maghembe, J. The Chagga home gardens: A multi-storeyed agro-forestry cropping system on Mt. Kilimanjaro, northern Tanzania. *Food Nutr. Bull.* **1985**, *7*, 29–36. [CrossRef]

20. Hemp, A. An ethnobotanical study on Mt. Kilimanjaro. *Ecotropica* **1999**, *5*, 147–166.

21. Hemp, A.; Hemp, C. *Environment and Worldview: The Chaga Homegardens Part I: Ethnobotany and Ethnozoology*; Archaeopress: Oxford, UK, 2009; ISBN 978-1-4073-0449-6.

22. Mollel, N.P.; Fischer, M.; Hemp, A. Usable wild plant species in relation to elevation and land use at Mount Kilimanjaro, Tanzania. *Alp. Bot.* **2017**, *127*, 145–154. [CrossRef]

23. van der Wolf, J.; Jassogne, L.; Gram, G.; Vaast, P. Turning local knowledge on Agroforestry into an online decision-support tool for tree selection in smallholders' Farms. *Exp. Agric.* **2016**, *55*, 50–66. [CrossRef]

24. Bukomeko, H.; Jassogne, L.; Tumwebaze, S.B.; Eilu, G.; Vaast, P. Integrating local knowledge with tree diversity analyses to optimize on-farm tree species composition for ecosystem service delivery in coffee agroforestry systems of Uganda. *Agrofor. Syst.* **2019**, *93*, 755–770. [CrossRef]

25. Smith Dumont, E.; Gassner, A.; Agaba, G.; Nansamba, R. The utility of farmer ranking of tree attributes for selecting companion trees in coffee production systems. *Agrofor. Syst.* **2019**, *93*, 1469–1483. [CrossRef]

26. Hemp, A. The banana forests of Kilimanjaro: Biodiversity and conservation of the Chagga homegardens. *Biodivers. Conserv.* **2006**, *15*, 1193–1217. [CrossRef]

27. Hemp, A.; Oleson, E.; Buchroithner, M.F. Kilimanjaro. Physiographic map with land use and vegetation, scale 1:100,000. In *Arbeitsgemeinschaft für Vergleichende Hochgebirgsforschung*; TU Dresden: Munich, Germany, 2017; ISBN 978-3-9816552-3-0.

28. Smithson, J. Chapter 21: Focus Groups. In *The SAGE Handbook of Social Research Methods*; SAGE Publications: London, UK, 2008; pp. 356–369.

29. R Core Team. *R: A Language and Environment for Statistical Computing*; R Foundation for Statistical Computing: Vienna, Austria, 2018.

30. Graefe, S.; Meyer-Sand, L.F.; Chauvette, K.; Abdulai, I.; Jassogne, L.; Vaast, P.; Asare, R. Evaluating Farmers' Knowledge of Shade Trees in Different Cocoa Agro-Ecological Zones in Ghana. *Hum. Ecol.* **2017**, *45*, 321–332. [CrossRef]

31. Parliament of the United Republic of Tanzania. The Forest Act No. 14. 2002. Available online: http://parliament.go.tz/polis/uploads/bills/acts/1454074677-ActNo-14-2002.pdf (accessed on 1 October 2019).

32. Staver, C.; Guharay, F.; Monterroso, D.; Muschler, R.G. Designing pest-suppressive multistrata perennial crop systems: Shade-Grown coffee in Central America. *Agrofor. Syst.* **2001**, *53*, 151–170. [CrossRef]

33. Teodoro, A.; Klein, A.M.; Reis, P.R.; Tscharntke, T. Agroforestry management affects coffee pests contingent on season and developmental stage. *Agric. For. Entomol.* **2009**, *11*, 295–300. [CrossRef]

34. De la Mora, A.; García-Ballinas, J.A.; Philpott, S.M. Local, landscape, and diversity drivers of predation services provided by ants in a coffee landscape in Chiapas, Mexico. *Agric. Ecosyst. Environ.* **2015**, *201*, 83–91. [CrossRef]

35. Martínez-Salinas, A.; DeClerck, F.; Vierling, K.; Vierling, L.; Legal, L.; Vílchez-Mendoza, S.; Avelino, J. Bird functional diversity supports pest control services in a Costa Rican coffee farm. *Agric. Ecosyst. Environ.* **2016**, *235*, 277–288. [CrossRef]

36. Belay, B.; Zewdie, S.; Mekuria, W.; Abiyu, A.; Amare, D.; Woldemariam, T. Woody species diversity and coffee production in remnant semi-natural dry Afromontane Forest in Zegie Peninsula, Ethiopia. *Agrofor. Syst.* **2019**, *93*, 1793–1806. [CrossRef]

37. Tadesse, G.; Zavaleta, E.; Shennan, C. Effects of land-use changes on woody species distribution and above-ground carbon storage of forest-coffee systems. *Agric. Ecosyst. Environ.* **2014**, *197*, 21–30. [CrossRef]

38. De Beenhouwer, M.; Geeraert, L.; Mertens, J.; Van Geel, M.; Aerts, R.; Vanderhaegen, K.; Honnay, O. Biodiversity and carbon storage co-benefits of coffee agroforestry across a gradient of increasing management intensity in the SW Ethiopian highlands. *Agric. Ecosyst. Environ.* **2016**, *222*, 193–199. [CrossRef]

39. Kufa, T.; Yilma, A.; Shimber, T.; Nestere, A.; Taye, E. Yield performance of Coffea arabica cultivars under different shade trees at Jimma Research Center, southwest Ethiopia. In Proceedings of the The Second International Symposuim on Multi-Strata Agroforestry Systems with Perennial Crops, CATIE, Turrialba, Costa Rica, 17–21 September 2007.

40. Mbuya, L.P.; Hsanga, H.P.; Ruffo, C.K.; Birnie, A.; Tengnas, B. *Useful Trees and Shrubs for Tanzania—Identification, Propagation and Management for Agricultural and Pastoral Communities*; Regional Soil Conservation Unit (RSCU), Swedish International Development Authority (SIDA): Nairobi, Kenya, 1994; ISBN 9966-896-16-3.

41. Ango, T.G.; Börjeson, L.; Senbeta, F.; Hylander, K. Balancing ecosystem services and disservices: Smallholder farmers' use and management of forest and trees in an agricultural landscape in southwestern Ethiopia. *Ecol. Soc.* **2014**, *19*, 30. [CrossRef]

42. Denu, D.; Platts, P.J.; Kelbessa, E.; Gole, T.W.; Marchant, R. The role of traditional coffee management in forest conservation and carbon storage in the Jimma Highlands, Ethiopia. *For. Trees Livelihoods* **2016**, *25*, 226–238. [CrossRef]

43. Ebisa, L. Effect of Dominant Shade Trees on Coffee Production in Manasibu District, West Oromia, Ethiopia. *Sci. Technol. Arts Res. J.* **2014**, *3*, 18–22. [CrossRef]

44. Teketay, D.; Tegineh, A. Traditional tree crop based agroforestry in coffee producing areas of Harerge, Eastern Ethiopia. *Agrofor. Syst.* **1991**, *16*, 257–267. [CrossRef]

45. Nath, C.D.; Schroth, G.; Burslem, D.F.R.P. Why do farmers plant more exotic than native trees? A case study from the Western Ghats, India. *Agric. Ecosyst. Environ.* **2016**, *230*, 315–328. [CrossRef]

46. Valencia, V.; West, P.; Sterling, E.J.; Garcia-Barrios, L.; Naeem, S. The use of farmers' knowledge in coffee agroforestry management: Implications for the conservation of tree biodiversity. *Ecosphere* **2015**, *6*, 1–17. [CrossRef]

47. Albertin, A.; Nair, P.K.R. Farmers' Perspectives on the Role of Shade Trees in Coffee Production Systems: An Assessment from the Nicoya Peninsula, Costa Rica. *Hum. Ecol.* **2004**, *32*, 443–463. [CrossRef]

48. Fischer, J.; Lindenmayer, D.B.; Manning, A.D. Biodiversity, ecosystem function, and resilience: Ten guiding principles for commodity production landscapes. *Front. Ecol. Environ.* **2006**, *4*, 80–86. [CrossRef]

Article

Assessment of the Diverse Roles of Home Gardens and Their Sustainable Management for Livelihood Improvement in West Java, Indonesia

Jeong Ho Park [1], Su Young Woo [1,*], Myeong Ja Kwak [1], Jong Kyu Lee [1], Sundawati Leti [2] and Trison Soni [2]

[1]   Department of Environmental Horticulture, University of Seoul, Seoul 02504, Korea;
      parkjeongho82@gmail.com (J.H.P.); 016na8349@hanmail.net (M.J.K.); gpl90@naver.com (J.K.L.)
[2]   Faculty of Forestry, Bogor Agricultural University, Jakarta 16680, Indonesia; lsundawati@gmail.com (S.L.);
      trisonsoni@yahoo.com (T.S.)
*   Correspondence: wsy@uos.ac.kr; Tel.: +82-10-3802-5242

Received: 7 October 2019; Accepted: 31 October 2019; Published: 2 November 2019

Abstract: Home garden is a traditional agroforestry system, which is an ecologically and socio-economically sustainable land use system in West Java, Indonesia. It plays a fundamental role in providing subsistence food and income to local people through a multi-strata structure. Despite the importance of the home garden, which is strongly linked with quality of living, there is still a lack of quantitative data and information. Therefore, we quantified the economic and ecological characteristics of home gardens in the present study to evaluate their diverse roles. In addition, general strategies that are applicable to home gardens in West Java were developed for sustainable management. The results of this study indicated that: (1) large landholding size showed a significantly higher Net Present Value (NPV) than small landholding size when the home gardens were dominated by fruit tree species, (2) species richness, species diversity, and carbon stock did not differ significantly among the different types and sizes of home gardens in West Java, and (3) multi-layered and diverse species composition is considerable for sustainable management of home gardens in terms of income generation and against urbanization and commercialization in West Java, Indonesia. Further studies should be considered for developing a standardized and generalized model that is able to evaluate and quantify the various ecosystem values that are generally acceptable and applicable in rural areas.

Keywords: home garden; margalef index; ahannon-wiener index; sustainable management; West Java; Indonesia

---

1. Introduction

Pezer [1] reported that 767 million people are estimated to be suffering from hunger and approximately one billion people live in extreme poverty. At the same time, 80% of poor people are living in rural areas and 64% of them are engaged in agricultural activities including crop cultivation, animal husbandry, forestry, fisheries, and aquaculture as their main income source and for food. In order to achieve poverty eradication and sustainable social, economic, and environmental development, the international community adopted Sustainable Development Goals (SDGs), which are a set of 17 objectives with 169 targets to be achieved from 2015 to 2030. Most of the poorest people who are heavily dependent on natural resources for their living are concentrated in the Southeast Asian region. In this region, more than 47.3% of the people are living in rural areas and most of them are likely to rely on agricultural activities, which provide local people with daily food and income sources. Moreover, they are particularly vulnerable to livelihood risks caused by climate change and other anthropogenic impacts, such as shifting cultivation mainly related to agricultural activities [2,3].

Agroforestry is a bridge between agriculture and forestry, and has been considered for several decades as a series of land management approaches combining trees, agricultural crops, livestock production, and other activities [4,5]. It is also considered to be a potential way to improve socio-economic conditions, environmental sustainability, and food security [4,6–10]. Agroforestry as an integrated land use system is widely practiced by more than 1.2 billion people in the world due to its unique characteristics of small land and low labor requirements, less input costs, and a location close to home [4,8]. Home gardens, which are one of the traditional agroforestry systems, are defined by a variety of characteristics in accordance with the local physical environment, ecological and economic situation, and cultural characteristics [11,12], but they are generally defined as multi-species, multi-storied, and multi-purpose gardens located close to a home [12,13]. The multi-layered structure of home gardens, which is created by the combination of various cultivated plants and wild plants, is responsible for several benefits and services of home garden systems [14].

Pekarangan is defined as a traditional home garden in Indonesia and is widely used in scientific research as one of the agroforestry systems related to its interactions with livelihood and the environment [15]. Many researchers have described home gardens as traditional agroforestry systems that are ecologically and socially sustainable land use systems. There have been numerous studies focused on the diverse structure, and socio-cultural and ecological functions of home gardens [14,16–19].

Home gardens play a fundamental role in providing subsistence food and income to indigenous people, and in serving as an important habitat for wild flora and fauna through a multi-strata structure in the area. Despite the importance of home gardens, which are strongly linked with quality of living, not much quantitative data and information on home gardens is available. In order to compare the economic value of the home gardens between the different types and sizes, we investigated the Net Present Value (NPV) and the B/C ratio, which is widely used for calculating a value of cost and benefits of the home gardens. In addition, we compared ecological value of the home gardens between the different types and sizes. We adopted the Margalef index and the Shannon-Wiener index's commonly used indicator for the ecological condition to investigate the species richness and species diversity, respectively.

In addition, there has been an increase in population while industrial structures have been changed mainly from agricultural to non-agricultural activities in areas with a growth of cities in West Java. However, there is still a lack of information on how to develop home gardens in a sustainable manner based on their ecological characteristics. Therefore, the research questions were set up as: (1) Do ecological and economic values vary according to the dominant species and the landholding size of home gardens? (2) Are species composition and land holding size important factors for sustainable management of home gardens?

The objectives of the study were:

(1) To quantify the economic and ecological characteristics using several indices that represent the specific characteristics of home gardens to evaluate their diverse roles, and

(2) To develop generally applicable strategies for the sustainable management of home gardens in West Java, Indonesia.

One study found a home garden with fruit tree species are more profitable than wood tree species regardless of the size of home gardens, but there was no difference in ecological characteristics between different types and sizes of home gardens in the Sukabumi region. However, further studies should be considered for developing a standardized and generalized model by collecting more information such as increased numbers of sample plots and respondents in the Sukabumi region.

2. Materials and Methods

2.1. Study Sites

This study was conducted in the Hegarmanah and Cicantayan villages, which administratively belong to the Cicantayan sub-district, Sukabumi Regency, in West Java, Indonesia. These villages lie between 6°57′ and 7°25′ north latitudes, and between 106°49′ and 107°00′ east longitudes (Figure 1). Both the Hegarmanah and Cicantayan villages have an altitude between 100 m and 1000 m (average of 500 m in the Cicantayan village and 600 m in the Hegarmanah village), and also have an average of approximately 3000 mm/year of precipitation and an average temperature of 32 °C [20]. The population density in the Cicantayan village is bigger than that of the Hegarmanah village due to the small area and large population in the Cicantayan village. Home gardens, as one of the typical farming systems commonly used in those villages, are strongly related to the livelihood of local people since it generates the main income source as rice, horticulture products, crops, and fruits for those villages.

Figure 1. Location of the villages of Hegarmanah and Cicantayan as study sites in the Cicantayan District, Sukabumi Regency, in West Java, Indonesia. (source: https://www.mapsland.com/asia/indonesia).

The climatic conditions of average precipitation and temperature of both of the villages are similar to each other. General information for the Hegarmanah and Cicantayan villages is summarized in Table 1.

Table 1. General information of the Hegarmanah village and the Cicantayan village.

	Hegarmanah Village	Cicantayan Village
Area (ha)	1573	600
Population	6006	8483
Population density (people/ha)	3.86	14.14
No. of household	1826	1698
Average precipitation (mm/year)	3000	3000
Average temperature (°C)	32	32

The land use types of these villages can be distinguished into the following three categories: rice paddy, dry land, which includes agriculture and forest areas, and buildings/gardens, which includes residential areas and home gardens. Hegarmanah village only has 81 ha of rice paddy. Most of the land in Hegarmanah is classified into buildings/gardens with 844 ha, which comprises 53.6% of the total land area. On the other hand, the Cicantayan village has 298.5 ha of rice paddy, which is 49.7% of the total land area, and 221.5 ha of buildings/gardens (Table 2).

Table 2. Land use type and size of the Hegarmanah village and Cicantayan village.

Village	Land Use Type (ha)			Total (ha)
	Rice Paddy	Dry Land	Building/Garden	
Hegarmanah	81	648	844	1573
Cicantayan	299	80	222	600

2.2. Data Collection

The diverse roles of home gardens have been reviewed. It is important to quantify the tangible and intangible benefits for a better understanding of their different roles. In this study, economic and ecological characteristics were identified and quantified in order to evaluate the diverse roles of home gardens using different indices. Primary data was collected through a household survey and in-depth interviews in the villages of Hegarmanah and Cicantayan in Sukabumi Regency, Indonesia. The household surveys were conducted using a structured questionnaire.

In order to conduct financial analysis, 27 households (total of 139 house members, which is approximately 1% of the total population in both villages) were selected and surveyed using a structured questionnaire for the type and size of home gardens. The questionnaire included information on "general information of households and home gardens", "household land ownership and land use systems", "household's main income source and planation activities", and "general cost for plantation activities and expected benefits". During the analysis process, not all of the data was used due to some missing data and the analysis was based on cost and revenue of farm-based activities [21].

For the analysis of ecological characteristics, a total of 29 home gardens (some of the households had two home gardens) were selected to implement the plant inventory survey. The results of the preliminary discussions with local people at the study sites showed that there were two different main income sources: fruit and wood materials. There were two different types of home gardens in this study: wood-dominated home gardens (WDH) and fruit-dominated home gardens (FDH). WDH were dominated by wood tree species, such as mahoni (*Swietenia mahagoni* (L.) Jacq.), gmelina (*Gmelina arborea* Roxb.), suren (*Toona sureni* (Blume) Merr.), jabon (*Anthocephalus cadamba* (Roxb.) Miq.), and teak (*Tectona grandis* L.f.). FDH were dominated by fruit tree species, such as manggis (*Garcinia mangostana* L.), rambutan (*Nephelium lappaceum* L.), and avocado (*Persea americana* Mill.). Among the 29 home gardens, 14 were classified as WDH and 15 were classified as FDH. There were also herbaceous plants, such as vegetables, ornamental plants, and crops, in both the WDH and FDH, which were primarily consumed by family members and not for sale.

Selected home gardens were further categorized into two different landholding size classes: small and large, based on the median value of land size in each type of home garden. Both WDH and FDH were additionally categorized as "WDHS" and "FDHS," which means landholding size smaller than 0.08 ha, and "WDHL" and "FDHL," which means landholding size bigger than or the same as 0.08 ha.

Vegetation characteristics of each home garden were determined within 400 m^2 plots (20 m × 20 m) for trees that were more than 10 cm in diameter at breast height (DBH), 100 m^2 subplots (10 m × 10m) for trees that had DBH values of 5 cm–10 cm and shrubs, and 1 m^2 sub-subplots (1 m × 1 m) for herbaceous plants, such as vegetables, ornamentals, weeds, and spices. The total sample area for this study was 1.16 ha for trees, and 0.29 ha for small trees and shrubs. The DBH (unit: cm) and height (unit: m) were measured for individual tree species with DBH values ≥1.5 cm. The information on the uses and local names were collected from households through questionnaires and interviews. All of the collected species were recorded with local names and scientific names.

2.3. Financial Analysis

Different criteria can be used to evaluate and quantify the economic value of home gardens, and the most widely used methods are the net present value (NPV), the benefit-cost ratio (B/C ratio), and an internal rate of return [22]. The net present value is an absolute measure that estimates the net worth of trees planted in the home gardens by calculating the value of costs and benefits of the home garden system as a whole [23–25]. The benefit-cost ratio is a relative measure and is calculated by dividing the sum of discounted revenues by the sum of discounted cost [24,25]. Normally, when the B/C ratio is bigger than 1 and the NPV is a positive value, then a home garden is considered profitable or feasible [26]. The economic values presented below were compared with the dominant type of home garden, such as WDH and FDH, and the size of the home gardens, which was divided based on the median value of 0.8 ha.

2.3.1. Net Present Value (NPV)

The net present value was calculated using Equation (1) [24,26].

$$\sum\nolimits_{n=1}^{n} (Bn - Cn) / (1 + i)^{n} \tag{1}$$

where *Bn* is the benefit each year, n is the number of years, *Cn* is the cost each year, and *i* is the interest rate, which was assumed to be 12% following the interest rate of loan from the bank in Indonesia. Detailed input and output data were collected using the questionnaire.

2.3.2. Benefit-Cost Ratio (B/C Ratio)

Input for home gardens included costs for fertilizer, pesticides, tools, and materials, as well as labor costs for land clearing, planting, and harvesting. On the other hand, output from home gardens included benefits from selling fruits and timber products. Intangible benefits, such as aesthetics, ornamentation, and shading effects, were not considered in this study. The benefit-cost ratio was determined using the following equation [24,26].

$$\sum\nolimits_{n=1}^{n} \frac{Bn}{(1+i)^{n}} / \frac{Cn}{(1+i)^{n}} \tag{2}$$

where *Bn* is the benefit each year, n is the number of years, *Cn* is the cost each year, and *i* is the interest rate, which was assumed to be 12%.

2.3.3. Sensitivity Analysis

The sensitivity analysis was conducted by adding a 10% increment to the price of fertilizer, pesticide, and labor, and a reduction of 10% in the market price of timber and fruit. Moreover, the interest rates were controlled by four types and were assumed to be: 10%, 12%, 14%, and 16%.

2.4. Stand Structure and Ecological Characteristics Analysis

Standard ecological references use many different indices to estimate the diversity of a site, and the Shannon-Weiner index is the most commonly used diversity indicator [27]. Furthermore, Nagendra [28] noted that the most commonly used indices are a combination of richness and evenness. Richness refers to the number of different land cover types within a site and evenness refers to the relative percentage of land distributed among these different cover types. In this study, complete inventories were included to calculate species diversity (Shannon-Wiener index), species richness (Margalef index), and species evenness (Simpson index). However, herbaceous plant species, such as ornamental plants, vegetables, weeds, and spices, were not included when assessing the stand structure characteristics (relative density, frequency, and relative frequency) and the aboveground biomass calculation.

2.4.1. Quantitative Structure

The quantitative characteristics of the stand structure were analyzed using relative density (RD), frequency (F), and relative frequency (RF). These were calculated using the following Formulas (3)–(5).

$$RD = \text{(Total number of individuals of a species)/(Total number of individuals of all of the species)} \times 100 \,(\%), \quad (3)$$

$$F = \text{(Total number of samples in which the species occur)/(Total number of samples enumerated)} \times 100 \,(\%), \quad (4)$$

$$RF = \text{(Frequency of the species in the stand)/(Sum of the frequencies of all of the species in the stand)} \times 100 \,(\%). \quad (5)$$

2.4.2. Species Richness (Margalef Index)

The Margalef index is a species diversity index divided into two types of species richness: how many types exist in the area and an assessment of species evenness or dominance, which means how individual species are distributed among the community [15]. The Margalef index can be used to provide an understanding of the species richness of the WDH and FDH. This index adjusts the number of species sampled in an area by the log of the total number of individuals sampled and summed over the species as follows [27].

$$\text{Margalef index} = \frac{(S-1)}{\ln(N)} \quad (6)$$

where S is the total number of species and N is the total number of individuals in the sample plots.

2.4.3. Species diversity (Shannon-Wiener Index)

The Shannon-Wiener index is the most commonly used diversity indicator in plant communities. It has a value of zero when there is only one species in a community and a maximum value when all of the species are present in equal abundance [27]. The index is calculated using Equation (7).

$$\text{Shannon-Wiener index} = -\sum_{i=1}^{S}\left(pi(\ln pi)\right) \quad (7)$$

where S is the total number of species and pi is the frequency of the ith species.

2.4.4. Species Evenness (Simpson Index)

The Simpson index is used to emphasize the evenness of the species [28]. Producing values from 0 to 1, the Simpson index defines the probability that two equal sized and randomly selected home gardens belong to the different home garden areas. Thus, the index is calculated using Equation (8) below.

$$\text{Simpson index} = 1 - \frac{\sum n(n-1)}{N(N-1)} \quad (8)$$

where n is number of individuals of each species and N is total number of individuals of all species.

2.5. Statistical Analysis

Data were analyzed using SPSS Statistics 25 (SPSS Inc., Chicago, IL, USA). The various statistical procedures utilized in this study included analysis of variance (ANOVA) to compare the characteristics of different types and size categories of home gardens.

3. Results and Discussions

The 27 households and 29 home gardens were randomly selected to analyze the economic and ecological characteristics of each household in both the Hegarmanah village and the Cicantayan village. A total of 13 households were classified as WDH and 14 households were classified as FDH. Selected home gardens were further categorized into two different landholding size classes indicated as: "WDHS", "WDHL", "FDHS", and "FDHL".

Each household had three to seven total family members (4.5 members on average). A total of one to six members from each household were dependent on the living and income sources from the home gardens. WDH had mean areas of 0.03 ha and 1.17 ha in small landholding size and in large landholding size, respectively. FDH had mean areas of 0.04 ha and 0.57 ha in small landholding size and in large landholding size, respectively. In general, the large landholding size in FDH had the furthest distance of 416.7 m from the household followed by distances for WDH with large landholding size and small landholding size of 271.4 m and 90 m, respectively (Table 3).

Table 3. Characteristics of selected home gardens in woody-dominated home gardens (WDH) and fruit-dominated home gardens (FDH) with its landholding size.

Types	N	Area (ha)	Distance (m)
WDHS	7	0.03 ± 0.02	90.0 ± 80.4
WDHL	7	1.17 ± 1.29	271.4 ± 205.9
FDHS	6	0.04 ± 0.02	49.5 ± 75.9
FDHL	9	0.57 ± 0.93	416.7 ± 433.7

Values are means ± SD.

In the study sites, there were 28 species of woody and fruit tree species with 306 individuals, which had DBH values of more than 1.5 cm, and 12 species of herbaceous species including vegetables, ornamental plants, and weeds (Table 4). There were 13 woody tree species that are primarily considered to be long-term income sources and are used as raw materials for building houses or fences. These species include sengon (*Paraserianthes falcataria* (L.) I.C.Nielsen), suren (*T. sureni*), jabon (*A. cadamba*), and mahoni (*S. macrophylla*). There are also 15 fruit tree species that are used for annual income generation of households, but are also sometimes consumed by local people for nutritional support, such as durian (*Durio zibethinus* L.), manggis (*G. mangostana*), rambutan (*N. lappaceum*), and bacang (*Mangifera foetida* Lour.) found in the study sites. Among these species, suren (*T. sureni*) and sengon (*P. falcataria*) were the most favorable woody tree species for the households, which occupied 45% of the total tree species. In addition, manggis (*G. mangostana*) and durian (*D. zibethinus*) were the most favorable fruit tree species.

Table 4. List of tree species mainly distributed in woody-dominated home gardens (WDH) and fruit-dominated home gardens (FDH).

WDH	FDH
Scientific Name	**Scientific Name**
Paraserianthe falcataria	*Durio zibethinus*
Toona sureni	*Garcinia mangostana*
Anthocephalus cadamba	*Nephelium lappaceum*
Agathis alba	*Mangifera foetida*
Swietenia macrophylla	*Archidendron pauciflorum*
Manglietia glauca	*Artocarpus heterophyllus*
Vitex pinnata	*Myristica fragrans*
Maesopsos eminii	*Lansium domesticum var. Aqueum*
Tectona grandis	*Parkia speciosa*
Peronema canescens	*Ceiba pentandra*
Gmelina arborea	*Citrus* sp.
Schima wallichii	*Aleurites moluccanus*
Neofelis nebulosa	*Persea americana*
	Lansium domesticum
	Syzygium aqueum

3.1. Economic Values of the Home Gardens

A summary of the quantified economic values of the home gardens is presented in Table 5. This table shows the B/C ratio and NPV according to the types and sizes of home gardens. The B/C ratio of all of the different types and sizes was bigger than 1, and a positive NPV meant that they were profitable at a 12% interest rate.

Table 5. Quantified economic value of different types and sizes of the home gardens.

Types	B/C Ratio	NPV (year^{-1})	NPV (ha^{-1} year^{-1})
WDHS	3.7 ± 2.4 [a]	300.7 ± 228.8 [ab]	10,622.3 ± 9538.1 [a]
WDHL	4.2 ± 2.0 [a]	1105.8 ± 643.9 [ab]	4177.7 ± 6870.5 [a]
FDHS	2.5 ± 1.8 [a]	274.1 ± 310.1 [b]	3618.0 ± 3146.0 [a]
FDHL	5.2 ± 2.8 [a]	1311.5 ± 932.4 [a]	4080.6 ± 3980.6 [a]

Values are means ± SD. Superscript lower-case letters indicate significant differences among the types and sizes according to Scheffe's test at $p < 0.05$. NPV is in USD.

There was only a significant difference in NPV between FDHS with $274.10 USD per year and FDHL with $1311.50 USD per year. In the case of WDH, NPV was $300.70 USD per year and $1105.80 USD per year for WDHS and WDHL, respectively, but there was no significant difference.

In general, the B/C ratio and NPV per year showed a similar trend that large landholding size had higher values than small landholding size in both WDH and FDH. However, there were no significant differences due to the large standard deviation value. The results of the presented economic value indicated that home gardens have various income sources derived from their unique structure and different species composition of each home garden within the study sites. NPV per year per ha in WDHS was $10,622.30 USD, but there was not a significant difference from other types and sizes of home gardens due to the large standard deviation value. Mohan et al. [29] quantified the financial values of home gardens based on size (small, medium, and large) in Kerala, India. The results showed that the large size of home gardens had the largest mean financial values. However, small size home gardens had the largest mean profit generated per unit area (m^2). In addition, Alam [30] and Rahman et al. [25] studied the financial benefits of home gardens in Bangladesh and showed that financial benefits increased along with increased farm size.

The variable range of the B/C ratio and NPV may have been caused by different levels of productivity and species composition in the home gardens. Rohadi et al. [21] summarized different smallholder timber plantations including teak plantation home gardens, rubber plantations, and palm oil plantations in Indonesia. Commercial plantations, such as palm oil plantations and rubber plantations showed relatively higher values for their B/C ratio and NPV value. The tegalan system, mostly planted with food crops and teak, showed a B/C ratio range of 1.59–6.21 and palm oil plantations had a B/C ratio value of 10.22. In addition, Current and Scherr [23] summarized the financial analysis of different agroforestry systems and agricultural intercrops showed an average B/C ratio of 1.79, which is comparable with that of alley cropping (2.10), perennial intercrops (1.75), taungya (2.50), and woodlots (0.97).

3.2. Sensitivity Analysis

Financial sensitivity analysis was conducted to determine whether or not the land-use is vulnerable to changes in the cost and benefit components as well as the interest rate. The effects of four interest rates (10%, 12%, 14%, and 16%), and two percentage changes with a 10% increase in cost and 10% decrease in benefits to BCR and NPV were examined. The results are presented in Tables 6–8.

Table 6. Sensitivity of the B/C ratio and NPV on different home garden types and sizes to changes in cost and benefit components.

Types	Normal		10% Increase of Cost		10% Decrease of Benefits	
	B/C Ratio	NPV	B/C Ratio	NPV	B/C Ratio	NPV
WDHS	3.7 ± 2.4 [a]	300.7 ± 228.8 [a]	3.4 ± 2.2 [a]	287.8 ± 230.2 [a]	3.3 ± 2.2 [a]	259.0 ± 208.6 [a]
WDHL	4.2 ± 2.0 [a]	1105.8 ± 643.9 [a]	3.8 ± 1.8 [a]	1086.3 ± 604.3 [a]	3.8 ± 1.8 [a]	956.8 ± 568.3 [a]
FDHS	2.5 ± 1.8 [a]	274.1 ± 310.1 [a]	2.3 ± 1.7 [a]	251.8 ± 302.2 [a]	2.2 ± 1.6 [a]	230.2 ± 273.4 [a]
FDHL	5.2 ± 2.8 [a]	1311.5 ± 932.4 [a]	4.7 ± 2.6 [a]	1280.6 ± 1913.7 [a]	4.7 ± 2.5 [a]	1143.9 ± 827.3 [a]

Values are means ± SD. Superscript lower-case letters indicate significant differences among the values on normal, 10% increase of cost, and 10% decrease of benefits, according to Duncan's test at $p < 0.05$. NPV is in USD.

Table 7. Sensitivity of the B/C ratio on different home garden types and sizes to changes in different interest rates on 10%, 12%, 14%, and 16%.

Types	Interest Rate (%)			
	10	12	14	16
WDHS	3.8 ± 2.4 [a]	3.7 ± 2.4 [a]	3.6 ± 2.4 [a]	3.5 ± 2.3 [a]
WDHL	4.2 ± 2.0 [a]	4.2 ± 2.0 [a]	4.1 ± 2.0 [a]	4.1 ± 2.0 [a]
FDHS	2.5 ± 1.8 [a]	2.5 ± 1.8 [a]	2.5 ± 1.8 [a]	2.4 ± 1.8 [a]
FDHL	5.3 ± 2.9 [a]	5.2 ± 2.8 [a]	5.1 ± 2.7 [a]	5.0 ± 2.6 [a]

Values are means ± Standard Deviation. Superscript lower-case letters indicate significant differences among the interest rates, according to Duncan's test at $p < 0.05$.

Table 8. Sensitivity of NPV on different home garden types and sizes to changes in different interest rates on 10%, 12%, 14%, and 16%.

Types	Interest Rate (%)			
	10	12	14	16
WDHS	4.5 ± 3.4 [a]	300.7 ± 228.8 [a]	280.6 ± 215.8 [a]	266.2 ± 208.6 [a]
WDHL	16.2 ± 9.4 [a]	1105.8 ± 643.9 [a]	1050.4 ± 611.5 [a]	1000.0 ± 582.7 [a]
FDHS	4.0 ± 4.5 [a]	274.1 ± 310.1 [a]	259.0 ± 295.0 [a]	244.6 ± 280.6 [a]
FDHL	19.2 ± 13.6 [a]	1311.5 ± 932.4 [a]	1244.6 ± 892.1 [a]	1187.1 ± 848.9 [a]

Values are means ± SD. Superscript lower-case letters indicate significant differences among the interest rate according to Duncan's test at $p < 0.05$. NPV values are in USD.

Table 6 indicates the changes in the B/C ratio and NPV indicates the home gardens when the cost components were increased by 10% and the benefit components were decreased by 10%. When the cost components were increased by 10%, the B/C ratio decreased by almost 10% in all of the different types and landholding sizes. However, the NPV decreased from USD 15.4 to USD 13.3 (approximately 11.9% reduction) in WDHL and from USD 3.8 to USD 3.2 (approximately 7.0% reduction) in FDHS. When the benefit decreases by 10%, the B/C ratio decreased similarly by about 10% in all of the different types and land holding sizes, but the NPV decreased from USD 18.2 to USD 15.9 (approximately 12.5% reduction) in FDHL and from USD 3.8 to USD 3.2 (approximately 17.0% reduction) in FDHS (Table 6). In a previous study, Mohan et al. [29] resulted in a relatively smaller level of change in the annual profit range (0.24%–2.46%) compared to our study, since we consider price fluctuation of all input cost since Mohan et al. [29] consider only labor cost. Although we were not able to find any significant differences between the increase in cost and decrease in benefits in all of the sites, home gardens are more sensitive to a decrease in benefits, which is caused by fluctuations in market prices, than an increase in cost, since home gardens have less input costs, so the B/C ratio and NPV are more dependent on the benefits.

Tables 7 and 8 show the changes in the B/C ratio and NPV of the different types and sizes based on various interest rates of 10%, 12%, 14%, and 16%. The B/C ratio was 3.8, 3.7, 3.6, and 3.5 with interest rates of 10%, 12%, 14%, and 16%, respectively, in WDHS, but there were no significant differences among the interest rates in all of the different types and sizes of home gardens (Table 8). Changes in

NPV are shown in Table 9 based on the different interest rates and the NPV decreased constantly while the interest rate increased in all of the different types and sizes of home gardens, but there were no significant differences. FDHL is still preferable for households with the highest NPV, which is followed by WDHL.

Table 9. List of tree species planted in home gardens in the study sites and characteristics of a quantitative stand structure.

No.	Scientific Name	Usage	RD (%)	F	RF (%)
1	*Maesopsis eminii*	Wood	0.96	6.90	1.59
2	*Agathis alba*	Wood	1.93	10.34	2.38
3	*Persea americana*	Fruit	0.96	10.34	2.38
4	*Mangifera foetida*	Fruit	1.29	13.79	3.17
5	*Lansium domesticum*	Fruit	2.25	13.79	3.17
6	*Durio zibethinus*	Fruit	4.50	27.59	6.35
7	*Gmelina arborea*	Wood	0.32	3.45	0.79
8	*Anthocephalus cadamba*	Wood	4.50	13.79	3.17
9	*Syzygium aqueum*	Fruit	0.32	3.45	0.79
10	*Tectona grandis*	Wood	0.96	10.34	2.38
11	*Archidendron pauciflorum*	Fruit	1.93	17.24	3.97
12	*Citrus* sp.	Fruit	0.64	3.45	0.79
13	*Ceiba pentandra*	Fruit	0.32	3.45	0.79
14	*Aleurites moluccanus*	Fruit	0.64	6.90	1.59
15	*Lansium domesticum* var. *aqueum*	Fruit	1.93	13.79	3.17
16	*Vitex pinnata*	Wood	0.32	3.45	0.79
17	*Swietenia macrophylla*	Wood	0.64	6.90	1.59
18	*Swietenia mahagoni*	Wood	1.29	6.90	1.59
19	*Garcinia mangostana*	Fruit	15.11	58.62	13.49
20	*Manglietia glauca*	Wood	0.32	3.45	0.79
21	*Artocarpus heterophyllus*	Fruit	2.57	17.24	3.97
22	*Myristica fragrans*	Fruit	3.22	6.90	1.59
23	*Parkia speciosa*	Fruit	1.93	13.79	3.17
24	*Schima wallichii*	Wood	0.64	6.90	1.59
25	*Nephelium lappaceum*	Fruit	2.57	20.69	4.76
26	*Euodia roxburghiana*	Wood	0.64	6.90	1.59
27	*Paraserianthes falcataria*	Wood	20.26	44.83	10.32
28	*Peronema canescens*	Wood	0.64	6.90	1.59
29	*Toona sureni*	Wood	24.76	62.07	14.29
30	*Camellia sinensis*	Fruit	1.61	10.34	2.38

3.3. Ecological Values of Home Gardens

The results of our study showed a smaller number for plant diversity than those of a previous study on plant diversity of traditional home gardens in Indonesia due to the characteristics of the home gardens. Arifin et al. [31] analyzed vegetation structure dynamics of traditional home gardens in Indonesia, and summarized six major factors that influence the vegetation structure of home gardens as follows: (1) small open space area, (2) land fragmentation, (3) different owner, (4) changes in function of some part of the home gardens, (5) plant popularity trend, and (6) economic condition changes. Kehlenbeck et al. [14] also found that commercialization, fragmentation, and urbanization threatened the plant diversity of home gardens in Indonesia. Farmers living within the study sites have their own job for their major income source and it is not possible for them to spend all of their time cultivating diverse plant species in their home gardens. In addition, the number of family members tends to decrease, because the younger generation wants to stay in urban areas to generate their own income. As a result, farmers tried to plant wood trees and fruit trees, which needed less effort to manage and achieved longer term benefits than those of tangible crops, including vegetables and herbal plants.

Quantitative stand structure characteristics are shown in Table 10. The results showed that suren (*T. sureni*) constituted the highest percentage (RD 24.76%) of the relative density followed by sengon (*P. falcataria*, RD 20.26%), manggis (*G. mangostana*, RD 15.11%), durian (*D. zibethinus*, RD 4.50%), and jabon (*A. cadamba*, RD 4.50%). Those five species occupied almost 70% of the home garden vegetation. Woody species (58.20%) occupied more than fruit species (41.80%). It was also revealed that suren is the most frequently occurring species with RF of 14.29%, followed by manggis, sengon, and durian with RF values of 13.49%, 10.32%, and 6.35%, respectively. In this study, we also showed that the RF of five frequently occurring woody species (suren, sengon, jabon, jati, and agathis) and five frequently occurring fruit species (manggis, durian, rambutan, nangka, and jengkol) were the same at 32.54%. In the case of woody species, suren and sengon occupied more than 55.37% of the total woody species (Table 9).

Table 10. Ecological values of different home garden types and sizes.

Types	N	Margalef Index	Shannon-Wiener Index	Simpson Index
WDHS	7	1.8 ± 0.4 [a]	1.4 ± 0.4 [a]	0.6 ± 0.1 [a]
WDHL	7	1.8 ± 0.5 [a]	1.3 ± 0.4 [a]	0.6 ± 0.2 [a]
FDHS	6	2.0 ± 0.5 [a]	1.6 ± 0.2 [a]	0.8 ± 0.1 [a]
FDHL	9	1.9 ± 0.5 [a]	1.5 ± 0.4 [a]	0.7 ± 0.1 [a]

Values are means $\pm$ SD. Superscript lower-case letters indicate significant differences among the types and sizes, according to Scheffe's test at $p < 0.05$.

Table 10 shows the ecological features of the 29 households surveyed in this study, according to the dominant species and the landholding size classes. Species richness values were determined to be 2.0 and 1.9 in FDHS and FDHL, respectively. As for WDHS and WDHL, the species' richness value was determined to be 1.8. The species' richness values were not significantly different among the different types and sizes of home gardens (Table 10). Saha et al. [32] also found that smaller home gardens had higher richness values in Kerala, India. Mohan et al. [27] assessed the ecological diversity in home gardens and compared the species' richness, according to the landholding size as small, medium, large, and commercial purpose home gardens. The commercial purpose of home gardens showed the lowest species' richness value (Margalef index: 5.43) and the small home gardens showed the highest species' richness value (Margalef index: 6.42). Rahman et al. [25] tried to explore the species composition with ecological features in homestead agroforestry systems in Northern Bangladesh and found that the Margalef index values ranged from 4.93 to 5.76.

Considering the species' diversity values through the Shannon-Wiener index and Simpson index, the values were determined to be 1.4 and 0.6 in WDHS and 1.3 and 0.6 in WDHL, respectively (Table 10). It was also determined that the values were 1.6 and 0.8 in FDHS and 1.5 and 0.7 in FDHL, respectively. However, we could not find a significant difference in ecological values among the different types and sizes of home gardens (Table 10).

In previous research, it has been determined that mean Shannon indices vary widely in tropical home gardens and those have been reported to range from 3.0 in West Java, Indonesia [33], 2.03 in West Java, Indonesia [34], 3.02–3.28 in Ethiopia [35], and 1.9–2.7 in Thailand [36]. In addition, mean Shannon indices were determined to be 2.0 in the dry zone in Sri Lanka [37], 1.71 in Cuba [38], 1.15–1.42 in the state of Kerala in India [27], and 3.36 in Bangladesh [25]. The Shannon-Wiener index values ranged from 1.5–3.5 and were seldom more than 4.5 [37]. The Shannon-Wiener index characterizes the proportion of species abundance in the population, being at a maximum when all of the species are equally abundant and being at the lowest when the sample only contains one species [37]. Mohan et al. [27] assumed that home gardens that were more than 1 ha in size were more likely to look like agricultural fields or plantations. Thus, this could cause lower species richness and diversity. Commercial home gardens have larger areas, a smaller number of species, a higher number of species, and a lower Shannon-Wiener diversity index value than non-commercial home gardens [34].

3.4. Sustainable Management of Home Gardens in West Java, Indonesia

Due to the rapid sprawl of the cities in West Java, starting with Jakarta, the population has increased, and industrial structures have changed mainly from agricultural to nonagricultural activities. Home gardens in the area of West Java have been faced with population pressure on the land and fragmentation, an increase in land cost, a growing market economy, and large-scale land conversion to non-agricultural activities that are primarily related to urbanization and commercialization in the area. In most of the changes presently observed, home gardens have lost their original characteristics in terms of rich biodiversity and multiple dimensions of the household economy [39]. Gangopadhyay and Balooni [40] noted that the sizes of the home gardens were negatively affected by urbanization, and Arifin et al. [41] found that home gardens were in critical trouble due to ecological and financial aspects when the size of home gardens fell to below 100 m^2.

The productivity of home gardens is related to a number of factors including multi-layered species composition, diversity, climatic parameters, and management intensity [25,30]. Well-developed home gardens provide households with high-nutrient food items through annuals and perennials [42], and multi-strata systems are more sustainable and profitable for the households by promoting income generation throughout the year as well as pest and disease prevention [43]. Species composition used to be decided by farmers' preferences for household consumption and market value for income generation. Fruit-growing is presently considered to be a proper activity related to the production economy throughout the year and it is well adapted to garden conversion, which has a higher productivity margin under the lower availability of land on Java Island [39,44]. Herbaceous plants, such as vegetables, starchy crops, and spice plants, are seasonal plants that are highly affected by climatic parameters, such as rainfall and temperature, and they are easily accessed through the market for subsistence consumption. On the other hand, when considering economic aspects, ornamental species are more attractive for households than vegetables due to the market economy in urbanized areas in particular [14,19,34,45–48].

One of the most important functions of home gardens is providing a sustainable income source over the short-term as well as over the long-term. It has been reported that the growing of fruit trees is a proper agricultural activity related to the production economy [39,44]. Fruit tree species, such as manggis (*G. mangostana*), bacang (*M. foetida*), durian (*D. zibethinus*), and rambutan (*N. lappaceum*), are important sources for generating income by selling to the market a highly marketable product. In the study sites, manggis and durian showed the highest frequency and these are one of the products that had the largest production values in the Sukabumi Regency. However, households are not able to generate income by fruit selling throughout the year, since fruit production is seasonal, and dependent on the climate and healthy tree conditions. In this case, ornamental plants are able to alternatively provide sustainable income for households as well as aesthetic function in response to urbanization [45]. Ornamental plants are used to plant in the space between houses as a fence to emphasize aesthetic function with medicinal plants and clove trees [39]. Arifin et al. [31] listed 103 species of ornamental plants that are planted in the home gardens on Java Island. Coffee (*Coffea canephora* var. robusta) was also cultivated by some households to replace the ornamental plants with higher income-generating plants. Considering further long-term benefits, sengon (*P. falcataria*) is the most preferable tree species for households in the study sites due to its short harvesting rotation of 7–10 years and high market demand [49]. Jati (*T. grandis*) and mahoni (*S. mahagoni*) are high value trees in the domestic market as well as in the international market with 20 to 30 years of harvesting rotation. Therefore, these trees act as saving accounts and safety nets just in case households urgently need money for special cases, such as a wedding ceremony or when building a new house [21].

4. Conclusions

In the present study, we quantified economic and ecological characteristics using several indices that represent the specific characteristics of home gardens to evaluate the diverse roles of home gardens. The results of this study indicated that the financial status of the households and large landholding sizes had an NPV of 18.23 per year, which was significantly higher than that of small landholding sizes of 3.81 per year when the home gardens were dominated by fruit tree species (FDH). On the other hand, home gardens dominated by woody-tree species (WDH) did not show a difference in NPV values between small landholding sizes and large landholding sizes. In addition, there was no significant difference in the B/C ratio within the study sites. Although we were not able to find any significant differences between an increase in cost and a decrease in benefits at all of the sites, home gardens are more sensitive to a decrease in benefits, which is caused by a fluctuation in the market price rather than an increase in cost. This is due to home gardens having a lower input cost. Therefore, the B/C ratio and NPV are more vulnerable to the benefits.

The diversity status values of the home gardens were very similar to each other even though they had different dominant tree species, types, and land sizes. There were 29 tree species with 311 individuals and 12 species of herbaceous plants including ornamental, vegetables, and tea. Suren (*T. sureni*) and sengon (*A. chinensis*) comprised 45% of the total number of trees, and manggis (*G. mangostana*) and durian (*D. zibethinus*) comprised 20% of the total number of trees within the study sites. Those four species also showed the highest relative density (RD) and relative frequency (RF) among the species. Their species composition was also fairly similar, which indicates that home gardens retain some specific species that are considered to be important for farmers' consumption as well as income generation. Considering the ecological characteristics of the study sites, there was no difference in species richness, species diversity, and evenness among the different types and sizes of home gardens.

Under the limited availability of land area on the entire island of Java especially with regard to the urbanization and commercialization pressure of the study sites, sustainable management of home gardens needs to consider livelihood improvement through multi-layered and diverse species composition. Combination of fruit tree species and ornamental plants are able to generate continuous income throughout a year due to higher market demand. Additional wood tree species act as a savings account in the home gardens.

A further study should be considered for developing a standardized and generalized model that is able to evaluate and quantify the various ecosystem values generally acceptable and applicable in rural areas, particularly for the Sukabumi region in West Java. To this end, we recommend selecting more classified local people based on their income, income source, and labor type in order to make a more exact calculation of the B/C ration and NPV. In addition, it is recommended to survey not only trees but also herbaceous plants including ornamental plants, which are increasing demand from the market. Therefore, the financial status can be analyzed more precisely.

Author Contributions: J.H.P., S.Y.W., and S.L. conceived and designed the experiments. J.H.P. and T.S. performed the experiments. J.H.P., M.J.K., and J.K.L. analyzed the data. J.H.P. and S.Y.W. wrote the paper. S.L. and S.Y.W. reviewed and edited the paper.

Funding: The R&D Program for Forest Science Technology through the Korea Forest Service (Korea Forestry Promotion Institute), grant number No. 2018120B10-1920-AB01, funded this research.

Acknowledgments: This study was carried out with the support of 'R&D Program for Forest Science Technology (Project No. 2018120B10-1920-AB01) provided by the Korea Forest Service (Korea Forestry Promotion Institute).

References

1. Pezer, M. *Poverty and Shared Prosperity 2016: Taking on Inequality*; The World Bank: Washington, DC, USA, 2016. [CrossRef]
2. FAO. *The State of Food and Agriculture–Social Protection and Agriculture: Breaking the Cycle of Rural Poverty*; Food and Agriculture Organization of the United Nations: Rome, Italy, 2015. [CrossRef]
3. FAO. *Food and Agriculture: Key to Achieving the 2030 Agenda for Sustainable Development*; Food and Agriculture Organization of the United States: Rome, Italy, 2017.

4.   Jamnadass, R.; Place, F.; Torquebiau, E.; Malézieux, E.; Iiyama, M.; Sileshi, G.W.; Kehlenbeck, K.; Masters, E.; McMullin, S.; Weber, J.C. *Agroforestry, Food and Nutritional Security*; ICRAF Working Paper, No. 170; World Agroforestry Centre: Nairobi, Kenya, 2013. [CrossRef]

5.   Van Noordwijk, M.; Lasco, R. *Agroforestry in Southeast Asia: Bridging the Forestry–Agriculture Divide for Sustainable Development*; Policy Brief No. 67, Agroforestry Options for ASEAN Series No. 1, Bogor, Indonesia, ASEAN-Swiss Partnership on Social Forestry and Climate Change; World Agroforestry Centre (ICRAF) Southeast Asia Regional Program: Jakarta, Indonesia, 2016.

6.   Alavalapati, J.R.; Shrestha, R.K.; Stainback, G.A.; Matta, J.R. Agroforestry development: An environmental economic perspective. *Agrofor. Syst.* **2004**, *61*, 299–310. [CrossRef]

7.   Nair, P.R. Classification of agroforestry systems. *Agrofor. Syst.* **1985**, *3*, 97–128. [CrossRef]

8.   Kiptot, E.; Franzel, S.; Degrande, A. Gender, agroforestry and food security in Africa. *Curr. Opin. Environ. Sustain.* **2014**, *6*, 104–109. [CrossRef]

9.   Torquebiau, E.F. A renewed perspective on agroforestry concepts and classification. *Comptes Rendus Acad. Sci.* **2000**, *323*, 1009–1017. [CrossRef]

10. Wiersum, K.F. Tree gardening and taungya on Java: Examples of agroforestry techniques in the humid tropics. *Agrofor. Syst.* **1982**, *1*, 3–70. [CrossRef]

11. Kumar, B.M.; Nair, P.R. The enigma of tropical homegardens. *Agrofor. Syst.* **2004**, *61*, 135–152. [CrossRef]

12. Kumar, V.; Tripathi, A.M. Vegetation Composition and Functional Changes of Tropical Homegardens: Prospects and Challenges. In *Agroforestry for Increased Production and Livelihood Security*; Gupta, S.K., Panwar, P., Kaushal, R., Eds.; New India Publishing Agency: New Delhi, India, 2017; pp. 475–505. ISBN 9789385516764.

13. Hodgkin, T. Home gardens and the maintenance of genetic diversity. In *Proceedings of the Second International Home Garden Workshop*; Watson, J.W., Eyzaguirre, P.B., Eds.; Biodiversity International: Rome, Italy, 2001; pp. 14–18. ISBN 9290435178.

14. Kehlenbeck, K.; Arifin, H.S.; Maass, B.L. Plant diversity in homegardens in a socio-economic and agro-ecological context. In *Stability of Tropical Rainforest Margins: Linking Ecological, Economic and Social Constraints*; Tscharntke, T., Leuschner, C., Zeller, M., Guhardja, E., Bidin, A., Eds.; Springer: Berlin, Germany, 2007; pp. 295–317. ISBN 9783540302896.

15. Kaswanto, R.L.; Nakagoshi, N. Landscape ecology-based approach for assessing pekarangan condition to preserve protected area in West Java. In *Designing Low Carbon Societies in Landscapes*; Nakagoshi, N., Mabuhay, J.A., Eds.; Springer: Tokyo, Japan, 2014; pp. 289–311. ISBN 9784431548188.

16. Galhena, D.H.; Freed, R.; Maredia, K.M. Home gardens: A promising approach to enhance household food security and wellbeing. *Agric. Food Secur.* **2013**, *2*, 48–62. [CrossRef]

17. Huai, H.; Hamilton, A. Characteristics and functions of traditional homegardens: A review. *Front. Biol. Chin.* **2009**, *4*, 151–157. [CrossRef]

18. Kehlenbeck, K.; Maass, B.L. Crop diversity and classification of homegardens in Central Sulawesi, Indonesia. *Agrofor. Syst.* **2004**, *63*, 53–62. [CrossRef]

19. Soemarwoto, O. Homegardens: A traditional agroforestry system with a promising future. In *Agroforestry: A Decade of Development*; Steppler, H.A., Nair, P.R., Eds.; International Centre for Research in Agroforestry: Nairobi, Kenya, 1987; pp. 157–170. ISBN 929059036X.

20. Badan Pusat Statistik. *Kecamatan Cicantayan Dalam Angka 2017*; BPS Kabupaten Sukabumi: Sukabumi, Indonesia, 2017; 72p.

21. Rohadi, D.; Kallio, M.; Krisnawati, H.; Manalu, P. Economic incentives and household perceptions on smallholder timber plantations: Lessons from case studies in Indonesia. In Proceedings of the International Community Forestry Conference, Montpellier, France, 24–26 March 2010.

22. Guo, Z.; Zhang, Y.; Deegen, P.; Uibrig, H. Economic analyses of rubber and tea plantations and rubber-tea intercropping in Hainan, China. *Agrofor. Syst.* **2006**, *66*, 117–127. [CrossRef]

23. Current, D.; Scherr, S.J. Farmer costs and benefits from agroforestry and farm forestry projects in Central America and the Caribbean: Implications for policy. *Agrofor. Syst.* **1995**, *30*, 87–103. [CrossRef]

24. Momen, R.U.; Huda, S.S.; Hossain, M.K.; Khan, B.M. Economics of the plant species used in homestead agroforestry on an off-shore Sandwip Island of Chittagong District, Bangladesh. *J. For. Res.* **2006**, *17*, 285–288. [CrossRef]

25. Rahman, S.A.; Baldauf, C.; Mollee, E.M.; Pavel, M.A.A.; Mamun, M.A.A.; Toy, M.M.; Sunderland, T. Cultivated plants in the diversified homegardens of local communities in Ganges Valley, Bangladesh. *Sci. J. Agric. Res. Manag.* **2013**, *2013*, 1–6. [CrossRef]

26. Singh, A.K.; Gohain, I.; Datta, M. Upscaling of agroforestry homestead gardens for economic and livelihood security in mid–tropical plain zone of India. *Agrofor. Syst.* **2016**, *90*, 1103–1112. [CrossRef]

27. Mohan, S.; Nair, P.R.; Long, A.J. An assessment of ecological diversity in homegardens: A case study from Kerala State, India. *J. Sustain. Agric.* **2007**, *29*, 135–153. [CrossRef]

28. Nagendra, H. Opposite trends in response for the Shannon and Simpson indices of landscape diversity. *Appl. Geogr.* **2002**, *22*, 175–186. [CrossRef]

29. Mohan, S.; Alavalapati, J.R.; Nair, P.R. Financial analysis of homegardens: A case study from Kerala state, India. In *Tropical Homegardens: A Time-Tested Example of Sustainable Agroforestry*; Kumer, B.M., Nair, P.K.R., Eds.; Springer: Dordrecht, The Netherlands, 2006; pp. 283–296. ISBN 9781402049477.

30. Alam, M. Valuation of tangible benefits of a homestead agroforestry system: A case study from Bangladesh. *Hum. Ecol.* **2012**, *40*, 639–645. [CrossRef]

31. Arifin, H.S.; Kaswanto, R.L.; Nakagoshi, N. Low Carbon Society Through Pekarangan, Traditional Agroforestry Practices in Java, Indonesia. In *Designing Low Carbon Societies in Landscapes*; Nakagoshi, N., Mabuhay, J.A., Eds.; Springer: Tokyo, Japan, 2014; pp. 129–143. ISBN 9784431548188.

32. Saha, S.K.; Nair, P.R.; Nair, V.D.; Kumar, B.M. Soil carbon stock in relation to plant diversity of homegardens in Kerala, India. *Agrofor. Syst.* **2009**, *76*, 53–65. [CrossRef]

33. Karyono, I. Homegarden in Java. Their structure and function. In *Tropical Home Gardens*; Landauer, K., Brazil, M., Eds.; The United Nations University: Tokyo, Japan, 1990; pp. 138–146.

34. Abdoellah, O.S.; Hadikusumah, H.Y.; Takeuchi, K.; Okubo, S. Commercialization of homegardens in an Indonesian village: Vegetation composition and functional changes. In *Tropical Homegardens: A Time-Tested Example of Sustainable Agroforestry*; Kumer, B.M., Nair, P.K.R., Eds.; Springer: Dordrecht, The Netherlands, 2006; pp. 233–250. ISBN 9781402049477.

35. Amberber, M.; Argaw, M.; Asfaw, Z. The role of homegardens for in situ conservation of plant biodiversity in Holeta Town, Oromia National Regional State, Ethiopia. *Int. J. Biodivers. Conserv.* **2014**, *6*, 8–16. [CrossRef]

36. Gajaseni, J.; Gajaseni, N. Ecological rationalities of the traditional homegarden system in the Chao Phraya Basin, Thailand. *Agrofor. Syst.* **1999**, *46*, 3–23. [CrossRef]

37. Mattsson, E.; Ostwald, M.; Nissanka, S.P.; Pushpakumara, D.K.N.G. Quantification of carbon stock and tree diversity of homegardens in a dry zone area of Moneragala district, Sri Lanka. *Agrofor. Syst.* **2015**, *89*, 435–445. [CrossRef]

38. Wezel, A.; Bender, S. Plant species diversity of homegardens of Cuba and its significance for household food supply. *Agrofor. Syst.* **2003**, *57*, 9–49. [CrossRef]

39. Michon, G.; Mary, F. Conversion of traditional village gardens and new economic strategies of rural households in the area of Bogor, Indonesia. *Agrofor. Syst.* **1994**, *25*, 31–58. [CrossRef]

40. Gangopadhyay, K.; Balooni, K. Technological infusion and the change in private, urban green spaces. *Urban For. Urban Green.* **2012**, *11*, 205–210. [CrossRef]

41. Arifin, H.S.; Sakamoto, K.; Chiba, K. Effects of the fragmentation and change of the social and economical aspects on the vegetation structure in the rural homegardens of West Java. *J. Jpn. Inst. Landsc. Architect.* **1997**, *60*, 489–494. [CrossRef]

42. Krishnal, S.; Weerahewa, J.; Gunaratne, L.H.P. Role of Homegardens in Achieving Food Security in Batticaloa District, Sri Lanka. In Proceedings of the International Conference on Economics and Finance Research IPEDR, Chennai, India, 10 March 2012; IACSIT Press: Singapore, 2012.

43. Jensen, M. Soil conditions, vegetation structure and biomass of a Javanese homegarden. *Agrofor. Syst.* **1993**, *24*, 171–186. [CrossRef]

44. Kubota, N.; Hadikusumah, H.Y.; Abdoellah, O.S.; Sugiyama, N. Changes in the Performance of the Homegardens in West Java for Twenty Years. *Jpn. J. Trop. Agric.* **2002**, *46*, 143–151. [CrossRef]

45. Arifin, H.S.; Sakamoto, K.; Chiba, K. Effects of urbanization on the vegetation structure of home gardens in West Java, Indonesia. *Jpn. J. Trop. Agric.* **1998**, *42*, 94–102. [CrossRef]

46. Clarke, L.W.; Li, L.; Jenerette, G.D.; Yu, Z. Drivers of plant biodiversity and ecosystem service production in home gardens across the Beijing Municipality of China. *Urban Ecosyst.* **2014**, *17*, 741–760. [CrossRef]

47. Panyadee, P.; Sutjaritjai, N.; Inta, A. The effects of distance from the urban center on plant diversity and composition in homegardens of Shan communities in Thailand. *Thai J. Bot.* **2012**, *4*, 83–94.
48. Wiersum, K.F. Diversity and change in homegarden cultivation in Indonesia. In *Tropical Homegardens: A Time-Tested Example of Sustainable Agroforestry*; Kumer, B.M., Nair, P.K.R., Eds.; Springer: Dordrecht, The Netherlands, 2006; pp. 13–24. ISBN 9781402049477.
49. Roshetko, J.M.; Delaney, M.; Hairiah, K.; Purnomosidhi, P. Carbon stocks in Indonesian homegarden systems: Can smallholder systems be targeted for increased carbon storage? *Am. J. Altern. Agric.* **2002**, *17*, 138–148. [CrossRef]

Review

A Natural Capital Approach to Agroforestry Decision-Making at the Farm Scale

Zara E. Marais [1,*], Thomas P. Baker [1], Anthony P. O'Grady [2], Jacqueline R. England [3], Dugald Tinch [4] and Mark A. Hunt [1]

[1] School of Natural Sciences, University of Tasmania, ARC Industrial Training Centre for Forest Value, Sandy Bay, TAS 7001, Australia; Thomas.Baker@utas.edu.au (T.P.B.); M.Hunt@utas.edu.au (M.A.H.)
[2] CSIRO Land and Water, Sandy Bay, TAS 7001, Australia; Anthony.O'Grady@csiro.au
[3] CSIRO Land and Water, Clayton South, VIC 3169, Australia; Jacqui.England@csiro.au
[4] Tasmanian School of Business and Economics, University of Tasmania, Sandy Bay, TAS 7001, Australia; Dugald.Tinch@utas.edu.au
* Correspondence: Zara.Marais@utas.edu.au; Tel.: +61-423-776-474

Received: 23 September 2019; Accepted: 1 November 2019; Published: 5 November 2019

Abstract: Background: Agroforestry systems can improve the provision of ecosystem services at the farm scale whilst improving agricultural productivity, thereby playing an important role in the sustainable intensification of agriculture. Natural capital accounting offers a framework for demonstrating the capacity of agroforestry systems to deliver sustained private benefits to farming enterprises, but traditionally is applied at larger scales than those at which farmers make decisions. Methods: Here we review the current state of knowledge on natural capital accounting and analyse how such an approach may be effectively applied to demonstrate the farm-scale value of agroforestry assets. We also discuss the merits of applying a natural capital approach to agroforestry decision-making and present an example of a conceptual model for valuation of agroforestry assets at the farm scale. Results: Our findings suggest that with further development of conceptual models to support existing tools and frameworks, a natural capital approach could be usefully applied to improve decision-making in agroforestry at the farm scale. Using this approach to demonstrate the private benefits of agroforestry systems could also encourage adoption of agroforestry, increasing public benefits such as biodiversity conservation and climate change mitigation. However, to apply this approach, improvements must be made in our ability to predict the types and amounts of services that agroforestry assets of varying condition provide at the farm or paddock scale.

Keywords: review; ecosystem services; agroforestry; natural capital; economic benefits

---

1. Introduction

1.1. Background

The projected increase in global demand for agricultural commodities is expected to be met mainly through the continued intensification of agricultural production [1]. Production gains to-date have placed pressure on stocks of natural capital and the ecosystem services that they provide [2–4]. Future strategies for intensification must balance the need to increase yields with objectives such as climate change mitigation and adaptation, improved soil and water management, and the protection of ecosystem services that support production [5]. Agroforestry is one land management strategy that farmers could employ to meet this challenge. Agroforestry describes any land-use system, practice, or technology, where woody perennials are integrated with agricultural crops and/or animals in the same land management unit (e.g., shelterbelts, alley cropping, integrated remnant vegetation) [6]. Proponents of agroforestry describe it as a 'win-win' approach, as carefully designed systems can balance the production of food, fibre, and fuel while restoring natural capital and thereby enhancing the provision of ecosystem services (e.g., erosion control, microclimate regulation) [7]. Increasing forest cover is also the cheapest and most direct method to reduce atmospheric concentration of greenhouse gases [8], and while most of this is likely to occur on land unsuitable for agriculture, agroforestry has been recognised as an important component of this reforestation effort [9].

Although the benefits of agroforestry systems are well-researched, adoption of agroforestry in temperate developed agricultural systems, particularly in Australia, remains constrained [10,11]. While technical, social and policy impediments exist [12], studies have shown that the perceived economic value of trees is often an important determinant of a farmer's decision to adopt agroforestry [13,14]. Clear demonstration of the capacity of agroforestry systems to deliver long-term economic benefits to the farm enterprise may therefore improve levels of uptake [15], which could increase delivery of public benefits such as biodiversity conservation and climate change mitigation. Concepts that capture both commercial and non-commercial benefits, such as the valuation of ecosystem services as part of a broader natural capital accounting approach, may be useful tools in this regard. These concepts may also be useful for developing tools that improve agroforestry-related decision-making at the farm scale (i.e., deciding what type of agroforestry system best suits the objectives of the enterprise). This review considers how a natural capital approach, which has traditionally been applied at national or regional scales, may be practically applied to demonstrate the value of agroforestry systems and improve agroforestry decision-making at the farm or paddock scale.

1.2. Natural Capital and Agriculture

Natural capital is the stock of renewable and non-renewable resources (e.g., plants, animals, air, water, soils, and minerals) that combine to yield a flow of ecosystem services, which in turn provide a variety of benefits to people [16,17]. All industries depend to some extent on natural capital and its benefits, and most businesses also impact on natural capital through their operations or use of products. Primary industries are particularly reliant on stocks of natural capital. In the case of agriculture, producers manage stocks of natural capital to deliver provisioning services in the form of food and fibre. At the same time, management activities may affect the capacity of the same natural capital to provide services into the future. Because interactions between agricultural businesses and natural capital may not immediately affect market values, cash flows, or prices, impacts and dependencies on natural capital are typically considered externalities and are often under-valued or not considered at all in valuation. Intensified production coupled with a failure to account for impacts on natural capital has led to the depletion of natural capital stocks (e.g., soil, biodiversity, water, vegetation) across many of the world's agricultural landscapes [18–20].

To address this, approaches that account for impacts and dependencies on natural capital have recently been developed [21–23]. Building on several decades of environmental economics research [24], natural capital accounting provides information on the stocks and flows of natural resources in a given ecosystem, region, or indeed enterprise, in physical or monetary terms. This information facilitates measurement and tracking of natural capital and an examination of how actions inhibit or improve its capacity to generate goods and services on an ongoing basis. Most natural capital accounting work that has been undertaken to-date focuses on valuing natural capital stocks for the purpose of conserving biodiversity at global, national, and regional scales [22,25,26]. While interest in the application of natural capital accounting to agriculture is increasing, particularly with the recent release of The Economics of Ecosystems and Biodiversity (TEEB) AgriFood report [27], the System of Environmental-Economic Accounting for Agriculture, Forestry and Fisheries [28], and the Natural Capital Finance Alliance Agriculture Sector Guide [29], the concept is rarely applied in the context of farm-scale decision-making.

When applied to agriculture at the farm scale, natural capital accounting can be used to determine the nature and magnitude of a farming operation's impacts and dependencies on natural capital and the associated business risks and opportunities [21,29,30]. This can help farmers and investors identify the specific types and levels of farming activity that pose material risks in terms of impacts or dependencies on natural capital. Conversely, the same approach can be used to identify management interventions that reduce these risks. In the case of agroforestry, there may be unexploited potential to increase adoption by using natural capital accounting to demonstrate farm-scale benefits or avenues for risk mitigation. Where sufficient information is available, these concepts can also be applied to compare the benefits of alternative agroforestry scenarios at the paddock or farm scale (Section 3).

1.3. Approach

In Section 2, we consider how a natural capital accounting framework could be applied to demonstrate the economic benefits of agroforestry at the farm scale and whether existing methods for quantifying and valuing ecosystem services are suitable in this context, based on a review of:

1. The conceptual framework for natural capital accounting (Section 2.1);
2. Methods for quantifying ecosystem services at the farm scale (Section 2.2);
3. Methods for valuing ecosystem services at the farm scale (Section 2.3).

In Section 3 we discuss how natural capital accounting may be usefully and practically applied to improve farm-scale agroforestry decision-making (Sections 3.1 and 3.2). We present an example of a conceptual model that could be used to this effect (Section 3.3). This conceptual model is based on the findings of existing reviews on ecosystem services in agroforestry systems, as well as direct references from farmers. We also highlight the challenges and opportunities presented by this decision-making approach and suggest areas for further research (Section 3.3).

2. Natural Capital at the Farm Scale

2.1. Applying the Natural Capital Accounting Framework to Agroforestry

The conceptual framework underpinning natural capital accounting (Figure 1) consists of natural capital *assets* which, depending on their condition, provide a flow of ecosystem *services* from which we derive value in the form of *benefits* to business and society. In the context of agroforestry systems, the asset is the integrated 'woody' component, e.g., shelterbelts, woodlots, or integrated remnant vegetation. Ecosystem services and benefits provided by these assets are likely to be numerous and diverse and will depend on the condition of the vegetation (e.g., composition, structure, configuration) [31].

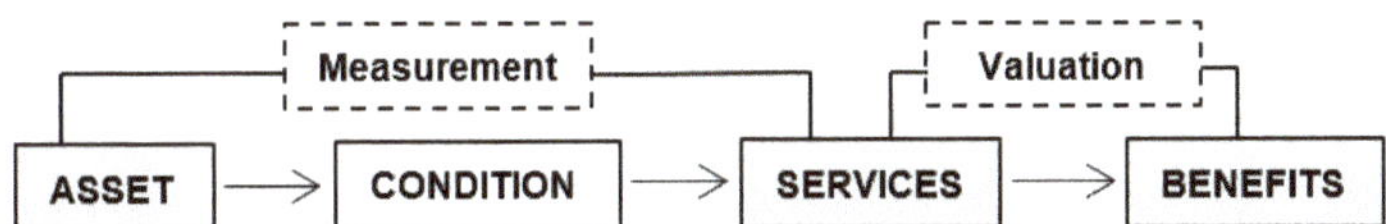

Figure 1. Flowchart adapted from the Natural Capital Protocol [21] illustrating the relationship between a natural capital *asset*, the *condition* of that asset, the ecosystem *services* that flow from the asset, and the *benefits* that those services provide to people.

Identification of ecosystem services and the benefits that they yield is central to the natural capital accounting framework. To reduce inconsistencies in measurement and valuation of services due to omission and/or double counting, the concept of 'final ecosystem services' [32] has been developed within ecosystem accounting frameworks. 'Final services' are directly obtained by specific human beneficiaries and are distinct from ecosystem functions/processes, or 'supporting services' (e.g., photosynthesis) [32–39]. Although the term 'final services' has been retained there is growing consensus among experts that, to reflect the role that they play in producing final services, 'intermediate' services (e.g., pollination) must also be considered in ecosystem accounting [40]. This is an important development in the context of agroforestry systems, as most of the services provided by agroforestry assets are considered intermediate. Although the debate on ecosystem accounting approaches and ecosystem service classification is ongoing [41], coverage of this debate is beyond the scope of this review. Rather, current classification concepts are used in this review to identify relevant ecosystem services for the purpose of discussing the merit of valuing these services to aid in farm-scale decision-making. The classification system currently used in the System of Environmental-Economic Accounting–Experimental Ecosystem Accounting (SEEA-EEA), Common International Classification of Ecosystem Services (CICES) [42], applies a suitably broad interpretation of final ecosystem services, which includes several intermediate services and is therefore well-suited to agroforestry systems. An example of the application of CICES (V5.1) classification is provided below (Table 1) for a list of services compiled from several reviews on agroforestry ecosystem services [7,43,44]. The CICES system provides an efficient means of identifying and classifying ecosystem services in an agroforestry context, reduces double-counting, and allows for inclusion of the full range of services described in the cited reviews.

Table 1. Farm-scale ecosystem services provided by agroforestry assets (adapted from CICES V5.1).

Section	Group	Service
Provisioning	Cultivated terrestrial plants for nutrition, materials, or energy	Cultivated trees or shrubs grown for nutritional purposes (food), fibres and other materials (timber), or energy (fuel)
Regulation & Maintenance	Mediation of wastes/toxic substances by living processes	Sequestration of atmospheric carbon
	Mediation of nuisances of anthropogenic origin	Noise attenuation Visual screening
	Regulation of baseline flows and extreme events	Control of erosion rates Hydrological cycle and water flow regulation (including flood control) Wind protection
	Lifecycle maintenance, habitat, and gene pool protection	Pollination (habitat for pollinators)
	Pest and disease control	Pest control (habitat for pest-predators)
	Regulation of soil quality	Decomposition and fixing processes and their effect on soil quality
	Water conditions	Regulation of the chemical condition of freshwaters through run-off control and nutrient uptake by trees and shrubs
	Atmospheric composition and conditions	Regulation of temperature and humidity, including ventilation and transpiration
Cultural	Physical and experiential interactions with natural environment	Characteristics of agroforestry systems that enable activities promoting health, recuperation, or enjoyment through active or immersive interactions or passive or observational interactions
	Intellectual and representative interactions with natural environment	Characteristics of agroforestry systems that are resonant in terms of culture or heritage or enable aesthetic experiences
	Other biotic characteristics that have a non-use value	Characteristics of agroforestry systems that have an existence value or an option or bequest value

While there is a good understanding of the services that can be provided by agroforestry systems (Table 1), measurement or valuation of these services at the farm or paddock scale has been more limited. However, research in this area is developing rapidly, and there have been several recent studies that value a combination of private and public ecosystem services at the farm scale [45–47]. In Sections 2.2 and 2.3 we consider the current methodologies for both measurement and valuation to determine their application to agroforestry at the farm scale.

2.2. Measuring Ecosystem Services at the Farm Scale

Measurement of ecosystem services (Figure 1) is often a pre-requisite to their valuation [48]. High demand for information to support decision-making in resource management has stimulated rapid progress in the development of approaches to measuring ecosystem services [49,50]. Here we provide an overview of the leading methods and tools for measuring ecosystem services and their suitability in the context of farm-scale measurement of services provided by agroforestry assets (see Table 1).

Availability and quality of primary data varies between different ecosystem services, but for many services, a lack of data is the most significant constraint to their quantification [51,52]. As a result, most quantitative estimates of ecosystem service provision at the landscape scale are based on secondary data or spatial proxies, which tend to be derived from either topographical data or land use land cover (LULC) datasets [53,54]. While estimates based on LULC proxies are useful for broad or rapid assessments over large areas [55,56], they are generally unsuitable for fine-scale (e.g., farm-scale) assessments as the coarse resolution of LULC data may not account for actual spatial variability in biophysical measurements of ecosystem services [52]. Importantly for farm-scale agroforestry assessments, readily available remotely sensed LULC data often fail to capture fine-scale landscape features such as shelterbelts and individual trees, which provide important ecosystem services at smaller scales. Use of LULC proxies also requires well-established links between land cover and ecosystem service provision. At a fine scale, ecosystem service provision is highly dependent on the condition of the natural capital (e.g., vegetation structure and composition). Although resolution of LULC data is improving, many aspects of condition remain difficult to establish from remotely sensed land cover data. This makes proxy-based techniques particularly unsuitable for farm-scale agroforestry assessments, where the condition of the asset (e.g., the configuration and height of a shelterbelt) has a significant influence on provision of key services (e.g., wind speed reduction).

One alternative to proxy-based measurement is the use of models that can capture processes at finer scales [57]. Models consider a wider set of local ecological variables as inputs and are therefore more reliable for fine-scale assessments, compared to LULC proxy-based measurement. One widely applied fine-scale modelling tool is InVEST: Integrated Valuation of Ecosystem Services and Trade-offs [58]. InVEST estimates levels of ecosystem services and their economic value using a suite of models ranging in complexity from proxy-based mapping e.g., carbon sequestration, to complex site-specific process models, e.g., pollination services [59]. Its ability to capture relatively fine-scale processes makes InVEST a potentially useful tool for measuring agroforestry ecosystem services at the farm-scale, although to our knowledge, it has yet to be used for such purposes. Several other advanced models exist that cater specifically for agroforestry systems, although they focus primarily on provisioning services and typically require a high degree of technical competency, e.g., CABALA, Farm Forestry Toolbox, for predicting quantities of timber/fibre; Yield-SAFE, SCUAF, APSIM, for predicting crop growth with tree interactions; and SPIF, for timber and environmental outcomes [60–66].

To improve the breadth and usefulness of fine-scale models, we first need to improve our understanding of how different natural capital assets influence ecosystem service inflows to agricultural systems and how the condition of these assets affects the types and amounts of services provided. Simple field measurements could then be used as either direct indicators, or model inputs, to accurately quantify multiple ecosystem services. For example, the USDA Forest Service's online toolkit 'i-Tree' contains a series of models that estimate ecosystem services provided by trees based on their physical properties [67]. Using simple input requirements, e.g., diameter at breast height, species, total height, alongside environmental and location variables, i-Tree employs a suite of models to forecast the provision of a range of services such as pollution reduction, public health benefits, carbon sequestration, and avoided run-off. While services important in an agroforestry context such as crop/livestock shelter, erosion control, indirect pollination and biological control are not yet included, an approach similar to i-Tree could be taken to quantify ecosystem services provided by agroforestry systems at the farm scale. However, we first need to address gaps in our understanding of how the condition of agroforestry assets (e.g., species composition, height, root depth, and configuration in relation to crops, livestock, and other landscape features) affects the services that they provide. Using these physical characteristics as inputs alongside environmental data, existing models may be able to predict quantities for several services including provisioning services (e.g., timber/fibre, food, and fuel), and regulating services (carbon sequestration, erosion control, and microclimate regulation).

One approach to improve accuracy of existing models is to conduct 'bottom-up' assessments where services are measured directly at the farm or paddock level, providing fine-resolution site-specific information that is directly relevant to the farmer. Sandhu et al. [68] and Porter et al. [69] measured biophysical indicators of multiple ecosystem services in order to compare land management techniques based on the value of services that they provide. These studies provide examples of how a wide range of services may be quantified at the farm or paddock scale based on observational data. In the case of cultural services where supply is more closely related to user appreciation than to ecosystem condition, measurement can also be achieved through incorporation of qualitative techniques such as interviews and surveys [45,70,71]. Participatory methods could be used in conjunction with biophysical measurement to ensure cultural ecosystem services are adequately represented in farm-scale assessments of agroforestry systems. Although broad uptake of bottom-up approaches is limited by the practical constraints and costs of data collection, they are likely to play an important role in improving the accuracy and relevance of existing models.

For measurement of ecosystem services provided by agroforestry systems at the farm scale, the key is striking an appropriate balance between practicality and the suitability of outputs for decision making. While rough estimates of ecosystem service supply can be derived relatively easily from an LULC proxy, farmers are generally faced with decisions at finer scales (i.e., the paddock or farm scale) that require more detailed site-specific information. In these cases, use of fine-scale models supported by quantitative and qualitative primary data appears to be the most appropriate approach to measuring a wide range of ecosystem services at the farm scale. While there are many promising techniques and packages that could be applied to agroforestry systems, there are still key gaps that need to be addressed, e.g., quantifying the impact of condition.

2.3. Valuing Ecosystem Services at the Farm Scale

Once ecosystem services have been quantified, the next step is to determine the extent to which these services are valued by relevant beneficiaries. Ecosystem service valuation may also be conceptualized as the measurement of the dividends or 'ecosystem income' yielded by natural capital [72–74]. As described by Fenichel et al. [74], marginal valuation of natural capital for the purpose of constructing accounts requires an understanding of the links between natural capital, human behaviour, and valued service flows. They identify the importance of political and social institutions in driving the management of ecosystem assets which impact upon the ecosystem income, or flow of value from ecosystem services. They further relate the values of ecosystem income and ecosystem stocks to sustainability at a country level, in essence as a measure of genuine savings [75]. However, accounting for the value of stocks of natural assets at this macro level is beyond the scope of this review, which focuses instead on the valuation of ecosystem service flows from agroforestry assets to inform decision-making at the farm or paddock scale. Here we describe methods for economic valuation of relevant ecosystem services (Table 2) and discuss different approaches to valuation of agroforestry systems at the farm scale. For the purposes of this review, the value of an asset refers to its Total Economic Value (TEV) (Figure 2), which encompasses both 'use' and 'non-use' values [76]. In this context, TEV is defined as the aggregation of the values of all service flows generated by natural capital both now and in the future [76].

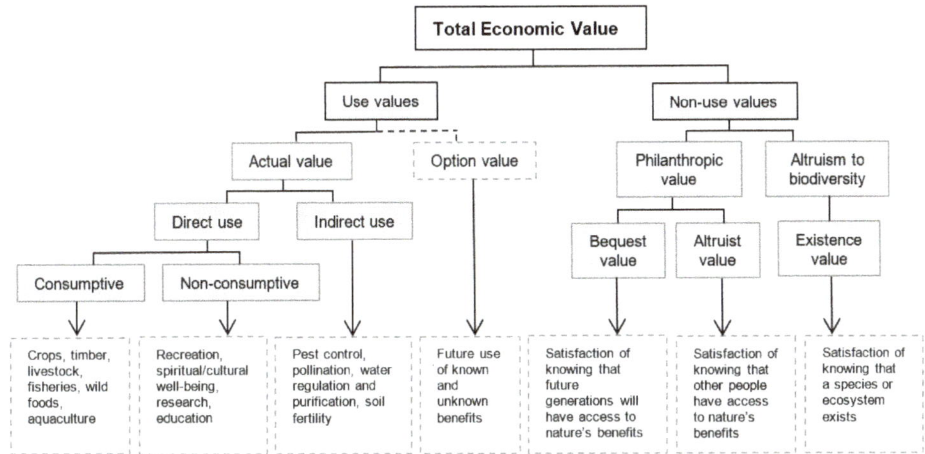

Figure 2. TEV typology adapted from [77], which classifies values associated with direct use, indirect use, and non-use of service flows generated by natural capital.

An important consideration when valuing ecosystem services is defining the beneficiary. Ecosystem services provided by agroforestry assets can be valued based on the benefits that they provide to the public (e.g., erosion control for improved downstream water quality), to the farmer (e.g., erosion control for retention of soil), or a combination of these approaches. As the purpose of valuation in the context of this review is to demonstrate the long-term benefits of agroforestry to farmers, we reviewed valuation strategies focusing on the farmer as the beneficiary.

It is important to note that valuation pathways of ecosystem services provided directly and indirectly by agroforestry assets vary in complexity. While agroforestry provides provisioning services that are directly harvested from the trees/shrubs themselves, e.g., food, fibre, or fuel, agroforestry assets also provide regulating services that indirectly influence other flows of provisioning services on the farm (e.g., increasing lamb survival through regulation of microclimate). In addition, agroforestry assets can also influence stocks of other forms of natural capital (e.g., by providing habitat for insects) which can indirectly influence flows of regulating services such as pollination. Therefore, some valuation pathways lead to monetary values (e.g., market value for provisioning services), whereas others lead to less-tangible forms of value (e.g., farmer well-being). In many cases, particularly where the intention is to justify an investment in agroforestry assets, valuation pathways that lead to a marketable product will form a compelling case. However, non-market values such as amenity, cultural value, and bequest value can also be important drivers for decision-making on farms. Monetary values alone will often fail to capture the full value of an agroforestry asset, which is why it is important to consider a range of ecosystem services that provide a broader perspective of value.

Economic valuation of agroforestry as a land-use system usually takes one of two forms: either a financial analysis of revenues received by the landowner at the enterprise or farm-scale, or an expanded analysis that includes 'externalities' or impacts beyond the farm boundaries [78]. Although some farm-scale financial analyses include hypothetical payments for regulating ecosystem services or taxes for disservices (e.g., pollution) [46,79], non-provisioning ecosystem services are generally not included in traditional farm-scale profitability studies. In studies where regulating and cultural services such as soil protection, carbon sequestration, air quality, and amenity are included, these services tend to be valued with public beneficiaries in mind, rather than as 'inflows' to the agricultural enterprise [80]. Exceptions do exist, including work by Ovando et al. [45], in which the private amenity of Mediterranean agroforestry farms is considered to ultimately be 'consumed' by the farmer through its effect on land prices [47,81]. Despite increasing demand for information in this space, there are still a limited number of studies assessing farm-scale economic benefits of agroforestry systems based on a broad range of use and non-use values, and fewer still that focus on the value of regulating services from a productivity perspective.

Table 2. Methods for valuing ecosystem services provided by agroforestry assets at the farm scale with examples of how they might be applied if the farmer is considered the primary beneficiary.

Valuation Method	Description	Services (See Table 1) that Could Be Valued Using This Method
Direct market valuation	Where commercial markets exist for services, market prices can be used to represent their value.	Food, fibre, timber, or fuel from cultivated trees or shrubs. Sequestration of atmospheric carbon
Production function	Where a service plays an intermediate role in the production of a marketable good, production functions can be used to estimate the contribution of that service as a proportion of the market price.	Pollination (habitat for pollinators), e.g., Morse and Calderone [82]. Regulation of temperature and humidity, including ventilation and transpiration. Wind protection.
Averted expenditure	Service is valued based on costs associated with declining benefits due to the loss of that service.	Control of erosion rates, e.g., [83]. Regulation of the chemical condition of freshwaters through run-off control and nutrient uptake by trees and shrubs. Hydrological cycle and water flow regulation (including flood control).
Replacement cost	Service is valued based on the cost of replacing that service entirely with an artificial or technical solution. This method is often employed to value regulating services in agriculture.	Pollination (habitat for pollinators), e.g., Winfree et al. [84]. Pest control (habitat for pest-predators). Decomposition and fixing processes and their effect on soil quality, e.g., Sandhu et al. [68], Alam et al. [85]. Control of erosion rates. Regulation of temperature and humidity, including ventilation and transpiration.
Revealed preference: hedonic pricing	Estimates the value of people's preferences for characteristics of a place based on their contribution to property prices.	Various (potentially difficult to isolate value of individual services) e.g., Polyakov et al. [86].
Stated preference: contingent valuation or choice experiment	These methods use questionnaires about hypothetical scenarios of environmental change to estimate economic value.	Use and non-use values of a broad range of services including: amenity, cultural heritage, recreation, aesthetics, and existence or bequest value, e.g., Shrestha and Alavalapati [87].
Benefit transfer	Where resources do not allow for original economic valuation using one of the above methods, it is possible to use data from comparable studies to value services.	Any of the above

Most agroforestry valuation studies that incorporate a broad suite of ecosystem services employ an equally broad suite of valuation methods, e.g., Porter et al. [69]. This usually includes market valuation, avoided expenditure, replacement costs, and some form of stated preference. Benefit transfer is often used for some or all of these valuations, depending on the focus of the study and the resources available to the investigator.

Some agroforestry valuation studies consider multiple beneficiaries, combining private and public perspectives. For example, de Jalón et al. [88] and Kay et al. [46] use a range of valuation techniques to compare the productivity and profitability of different agroforestry landscapes against conventional agricultural and forestry systems. Across these studies, the value of sequestered carbon is based on a carbon price, disservices of soil erosion and nitrogen/phosphorus surplus are valued based on the cost of removing these materials from public watercourses, and pollination is valued according to a production function [46,88]. Services such as windspeed reduction and noise reduction are excluded despite their potential to deliver significant private benefits to farmers. Valuation studies that combine private and public benefits may be appropriate in some cases, for example when designing payments for ecosystem services. However, the objectives of the agroforestry venture must be clear to ensure that key ecosystem services are included and that the results of the valuation are relevant to the decision-maker.

If the purpose of the valuation is to encourage private investment in agroforestry, it makes sense to focus on ecosystem services that deliver private benefits to farmers and value those services accordingly. Porter et al. [69] and Alam et al. [85] take this approach, borrowing techniques used by Sandhu et al. [68] to value field-scale ecosystem services in agroforestry systems. Production of food and raw materials is valued at market prices; nitrogen regulation, soil formation, groundwater recharge, and pollination are valued according to replacement costs; biological control of pests according to avoided cost of pesticides; and aesthetics through benefit transfer, derived from a contingent valuation study. The broad range of services included in these studies, and the focus on the farmer as the beneficiary in most valuation methods, ensures that the final estimate of each system's economic value reflects a range of values that are directly relevant to the farmer.

In natural capital accounting, valuation methods should be chosen to suit the purpose of the study and the types of services that are being valued. In the case of agroforestry systems, there is merit in recognising the role of farmers as decision-makers and ensuring that the information produced is directly relevant to them. Strategies for achieving this could include incorporating a broad range of use and non-use values and valuing regulating services from a productivity perspective, rather than as externalities.

3. A Natural Capital Approach to Agroforestry Decision-Making at the Farm Scale

As farmers consider strategies to enhance the long-term productivity of their enterprise while protecting the natural capital base that supports it, they are likely to benefit from the availability of tools that support their decision-making. Here we draw on findings from Section 2 to discuss the usefulness and feasibility of applying a natural capital approach to farm-scale agroforestry decision-making.

3.1. Advantages of a Natural Capital Approach

As demonstrated in Section 2, a natural capital accounting framework can be applied to agroforestry systems to establish the value of agroforestry assets at the farm scale. The framework identifies links between stocks of natural capital, ecosystem service provision, and farm-scale benefits (value). Farmers who conceptualise their farm in this way and understand these links may be more inclined to adopt strategies that protect or enhance natural capital. Natural capital accounting can therefore be useful in justifying private investment in agroforestry. Farmers may also choose to communicate their awareness and management of natural capital impacts and dependencies to internal or external stakeholders to attract new investors or customers. Indeed, agribusiness lenders are showing increasing interest in using natural capital approaches to account for the value of natural capital stocks in farm valuations and credit risk assessments [89].

The natural capital approach also highlights the flexibility of agroforestry systems, i.e., that they can be designed to deliver a range of benefits depending on the objectives of the farm enterprise. Farmers who are looking to adopt agroforestry will be faced with decisions about the type, extent, location, and configuration of agroforestry assets. Natural capital approaches can be used to compare the benefits of different agroforestry options, in terms of the value of the ecosystem services that each might provide. In this way, there is potential for the natural capital framework to be used as the basis for the development of tools that assist farmers in choosing between alternative agroforestry scenarios based on costs and benefits to the enterprise (Section 3.3).

3.2. Existing Frameworks for Natural Capital Accounting at the Farm Scale

While general awareness of the role of natural capital in agriculture is increasing [90], the concept is rarely applied in the context of farm-scale decision-making. There are still relatively few studies that attempt to value or account for stocks of natural capital at a scale that is useful for decision-making on farms. Although natural capital accounting is being used broadly to appeal for changes in agricultural practice that will protect the natural capital base [27], little practical guidance exists for farmers and other practitioners looking to construct accounts of their own. This may be due in part to a lack of consensus on the best approach for farm-scale natural capital assessment and accounting. Here we describe several tools and frameworks that may fill this gap and bring us a step closer to a standardised, practical natural capital approach to farm-scale decision-making.

At the outset, it is necessary to undertake some form of natural capital assessment to understand risks and dependencies relating to natural capital stocks and to gain an appreciation of the value of specific natural capital assets to the farm. The Natural Capital Protocol provides a general approach for natural capital assessments [21]. Although the Natural Capital Protocol offers little guidance on how their approach may be implemented in practice, other projects have applied the framework to undertake natural capital assessments in agriculture, e.g., the FAO's report on Natural Capital Impacts in Agriculture, which highlights trade-offs between different farming practices (e.g., organic vs. conventional) based on costs to human health and ecosystems [30]. Although some case studies touch on internal benefits, most valuations are not considered from the perspective of the farmer, and this approach is therefore not useful as a template for assessments to support farm-scale decision-making. In a more transferable approach, Ascui and Cojoianu [29] provide a generic procedure for lenders to undertake farm-specific natural capital credit risk assessments (based on the Natural Capital Protocol). In their approach, biophysical indicators (such as percentage vegetation cover) are valued based on evaluation of risks to the lender, which informs whether credit should be extended to the farmer. Although their approach focuses on the value perspective of the lender, there is scope for this procedure to be used by farmers to prioritise management interventions based on assessment of key risks to their business.

Once natural capital risks, dependencies, and the value of natural capital assets have been established, farmers may wish to track the value or condition of natural capital assets through time to inform decisions around investment and operations. Three frameworks exist that provide a standardised approach to natural capital accounting at the farm scale. These are founded on the SEEA-EEA, which has not yet developed to cover farm-scale accounts but nonetheless provides a framework for tracking changes in the extent, condition, and monetary value of ecosystem assets over time across a given spatial area [22]. There is also potential for SEEA-EEA itself to be developed for use at the farm scale in the future. The Wentworth Group's 'Accounting for Nature' method is currently being adapted for use at the farm scale [23] and focuses on the construction of 'asset condition accounts' which provide information about changes to the condition of assets over time, based on measuring biophysical indicators. The second framework proposes an 'ecological balance sheet' (EBS) that enables the application of accrual accounting principles to ecological assets at the farm scale [91]. The advantage of the EBS is that it deliberately incorporates natural capital accounts into the farm's existing accounting system so that financial and environmental performance can be tracked simultaneously. Perhaps the most advanced of the existing frameworks is the 'Agroforestry Accounting System' (AAS) which estimates total income accrued from a range of market and non-market products delivered by agroforestry systems [92,93]. While application of the AAS to-date has focused on comparing the value of woodland agroforestry systems to other forest types [45], there is potential for this framework to be applied more broadly: at different scales and for different types of agroforestry systems. Each of these existing frameworks brings us closer to tracking the condition and value of natural capital assets through time at a scale that is useful for decision-making on farms.

Although these frameworks form a sound theoretical foundation for farm-scale natural capital accounting, it is important to recognise that they all rely on evidence-based conceptual models that demonstrate how agricultural systems function. In agriculture, key forms of natural capital may include soils, vegetation, fauna (including livestock and fisheries), and water [91]. Although it is conceptually easy to calculate stocks of the asset (woody vegetation) and determine its condition (i.e., age, structure, species composition, configuration, etc.), each form of natural capital yields multiple ecosystem services and disservices that may interact in additive, synergistic, or detractive ways. Many of these services are difficult to quantify, interactions between them are often poorly understood, and condition is rarely tracked. Additionally, there is a gap in our ability to predict the types and amounts of services that assets of varying condition provide at the farm scale, and how these services translate to benefits received by the farmer. While efforts are underway to improve our understanding of the value of some natural capital assets in complex agricultural systems [94], we do not yet have an adequate model for agroforestry assets. Conceptual models must also account for the impact that changes in asset condition have on value, particularly in agroforestry systems where the condition of the asset can significantly affect service provision. Such a model would greatly improve the applicability of existing natural capital accounting tools to farm-scale agroforestry decision-making.

3.3. A Conceptual Model for Agroforestry Decision-Making

A conceptual model for valuation of agroforestry assets may serve multiple purposes: firstly, to establish common understanding of causal pathways for the flow of benefits from agroforestry assets and, secondly, to facilitate rapid assessment of the benefits of various agroforestry options. Here we present an example of a conceptual model for farm-scale valuation of an agroforestry asset (Figure 3) and discuss how it may be used as the basis for farm-scale decision-making.

The model in Figure 3 illustrates how the framework in Figure 1 can be applied conceptually to an agroforestry system where the 'asset' is a shelterbelt and the farmer is considered the beneficiary. This conceptual model is based on studies describing the ecosystem services provided by agroforestry systems [7,43,44,95] and was developed in consultation with farmers and colleagues working in the field. This model (Figure 3) illustrates benefits in a temperate pasture/livestock system but could be adapted to suit other systems such as dairy or horticulture.

Although many of the services listed in Table 1 are featured in the model, some have been adapted or broken down into a series of biophysical processes to highlight interactions and trade-offs within the system. For example, the service of 'regulation of temperature' is captured in the provision of shade and the reduction in wind speed provided by the shelterbelt. Each pathway within the conceptual model linking the asset to a benefit involves a combination of measurement and valuation of one or more ecosystem services. For example, the extent of wind speed reduction caused by the shelterbelt can be measured, as can the resulting effects on evaporation and pasture growth on the leeward side of the shelterbelt [96,97]. Once the relationship between wind speed reduction and pasture yield has been quantified, this service can be valued based on the extent to which the increase in yield reduces costs associated with supplementary feeding and the positive effect that this has on gross profit margin. Depending on the situation, the effect of competition may also be measured, and the associated pasture yield decrease accounted for. Potential valuation pathways in the conceptual model will vary considerably in terms of methods and complexity.

From an accounting perspective, the development of conceptual models is an important first step in valuing and accounting for changes in natural capital assets on farms. Conceptual models are useful for establishing common understanding of key causal pathways amongst experts and stakeholders, [98]. In this case, it is useful for practitioners to build an understanding of the multiple ecosystem services that may flow from agroforestry assets, and the types of benefits that these services provide. This common understanding will enable more consistent valuation of agroforestry assets in accounting exercises at various scales (e.g., Accounting for Nature, AAS, SEEA-EEA). Conceptual models can be developed further to include a broader range of beneficiaries (e.g., the general public) and used as a 'blueprint' for valuation to suit a range of purposes. For example, government agencies may use an adapted version of the model in Figure 3 to determine the return on investment in agroforestry assets at the farm or landscape scale, considering both private and public benefits. Lenders and investors may also use similar models to conceptualise the value of agroforestry assets from a risk management perspective [29]. Conceptual models are an ideal tool for this purpose given their flexibility and capacity to clearly communicate relationships within complex systems such as agroforestry systems. These models can be more powerful if underpinned by an evidence-based review [99].

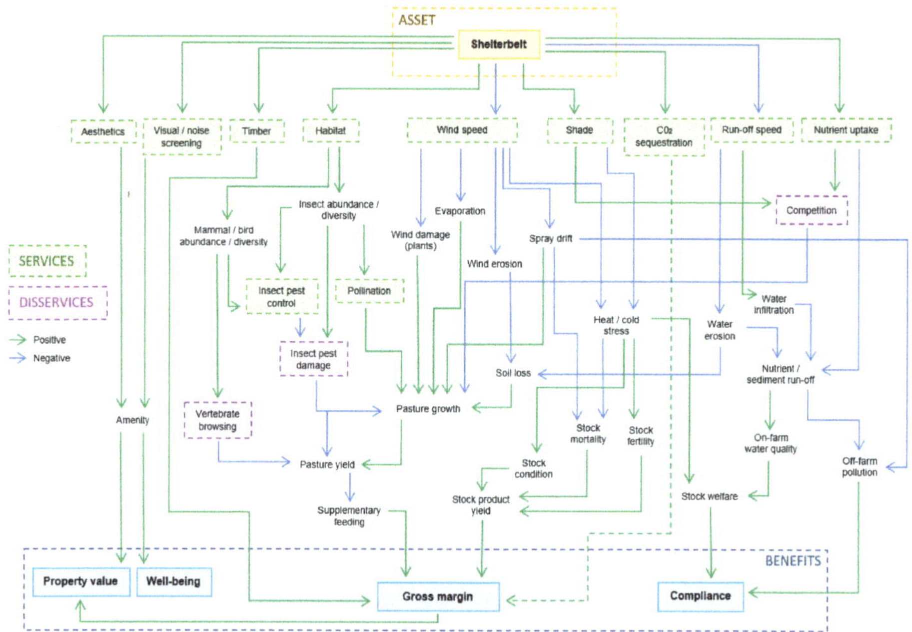

Figure 3. Conceptual model for ecosystem services and associated benefits provided by one common type of agroforestry asset (shelterbelt) in a temperate pasture/livestock system. Blue lines represent negative effects (i.e., reduction), and green lines positive effects.

The conceptual model also provides the basis for development of tools that can assist in agroforestry-related decision-making at the farm or paddock scale. Farmers are the primary decision-makers and creating tools that cater for them and the types of decisions that they face is crucial. The farm-scale value of services provided by agroforestry assets may be highly dependent on the location of the farm, the objectives of the farm enterprise, and the context of the asset within the farm [100]. Farmers require tools that enable them to make decisions about investing in agroforestry systems and designing them in such a way that maximises benefits to their particular enterprise. Conceptual models can enable them to make these decisions without having to undertake complex, expensive natural capital assessments that would require direct measurement and valuation of ecosystem services. For example, a farmer planning to invest in agroforestry would first need to decide what type of asset best suits the objectives of their enterprise. They may seek to maximise provision of services that improve productivity or reduce operational risk while waiting for longer-term returns from marketable wood products. If one of their priorities is to reduce lamb losses due to cold winds they may decide to invest in shelterbelts, based on the benefits demonstrated in a conceptual model of this system (Figure 3). The next phase will involve deciding how many shelterbelts to plant, the dimensions and orientation of each shelterbelt, and their location in relation to other elements of the farm. In making these decisions they may refer to other sections of the conceptual model to consider a wider range of potential benefits (e.g., amenity, reducing spray drift) and disbenefits (e.g., competition effects). Used in this way, conceptual models can provide a low-cost, rapid approach to agroforestry decision-making at the farm or paddock scale.

Although the evidence base that supports conceptual models for farm-scale valuation of agroforestry assets is growing [7], there are still gaps in our biophysical understanding of agroforestry systems [95]. While a lack of quantitative evidence may not necessarily restrict the usefulness of these models for farm-scale decision-making, it is helpful to have confidence in the direction of relationships (i.e., positive or negative) and the relative quantities of ecosystem services provided by different types of assets. Where conceptual models currently fall short is in demonstrating the impact of asset condition on the flow of services and benefits. Having chosen to plant shelterbelts, a farmer may eventually have to decide on the configuration and composition of the shelterbelts. They are also likely to be interested in changes to the flow of services and benefits over time, from planting to harvest/senescence. The effect of asset condition at fine scales is an important research gap that must be filled in order to improve the usefulness of these conceptual models.

Where sufficient quantitative evidence exists, conceptual models can also form the basis of more precise, predictive tools for decision-making. These tools may facilitate fine-scale, quantitative valuation of services that are of particular importance to farmers (e.g., shelter). Increasingly, valuation methods are being incorporated into ecosystem service models (e.g., InVEST, i-Tree Eco v6) and economic analysis tools, some of which are designed specifically for integrated farming systems (e.g., Imagine, Farm-SAFE) [101,102]. Conceptual models can guide the development of these tools by demonstrating the complexity of the system as a whole, ensuring that the tools account for interactions and trade-offs that might otherwise be missed. To improve useability, it may be advantageous to compile all relevant models into a single toolkit (similar in style to InVEST or i-Tree) or to incorporate ecosystem service models into an existing package (e.g., Farm Forestry Toolbox) or farm enterprise platform (e.g., DAS Rural Intelligence Platform, FarmMap4D) [62,103,104]. Data accessibility (including cost and usability) is an important consideration in the development of such a toolkit, as a collaborative approach is likely to greatly improve the scope and reliability of outputs.

Conceptual models can enhance the applicability of existing natural capital accounting tools to farm-scale agroforestry decision-making. They can improve consistency in the valuation of agroforestry assets for accounting purposes, guide rapid decision-making at the farm or paddock scale, and form the basis for development of quantitative decision-making tools. To improve the useability of conceptual models in this context, we need to expand the evidence base that supports them with particular focus on the impact of asset condition on ecosystem service provision.

4. Conclusions

The natural capital accounting framework provides a logical and increasingly consistent approach to the valuation of impacts and dependencies on natural capital. Findings from this review suggest that there is potential for this framework to be usefully applied to demonstrate the capacity of agroforestry systems to deliver sustained private benefits to farming enterprises.

Despite difficulties in obtaining information for many ecosystem services, tools and models for measuring services continue to advance and improve. In the case of measuring ecosystem services provided by agroforestry systems, the key is striking an appropriate balance between practicality and the relevance of outputs to decision-making. Use of fine-scale models supported by quantitative and qualitative primary data may be the most appropriate approach to measuring a wide range of ecosystem services at the farm scale. While promising advancements continue to be made in the development of tools to model service provision at these fine scales, there are still some key gaps that need to be addressed, e.g., quantifying the impact of condition.

As the evidence base for the value of natural capital in agriculture continues to grow, methods and tools for measuring this value are also improving. Methods for valuing ecosystem services should be chosen to suit the purpose of the valuation and the types of services that are being valued. In the context of demonstrating farm-scale benefits of agroforestry, valuations should be directed at farmers as key beneficiaries, incorporate a broad range of use and non-use values, and value regulating services from a productivity perspective rather than as externalities. Natural capital accounting can be applied to communicate the broad range of values that farmers can derive from agroforestry assets, thereby encouraging appropriate levels of investment.

A natural capital approach can also be applied to assist farmers in making decisions about agroforestry at the farm or paddock scale. While work is currently underway to develop a standardised natural capital approach to farm-scale decision-making, existing tools rely on conceptual models for the provision and valuation of ecosystem services that flow from natural capital assets in agricultural systems. To usefully apply a natural capital approach to farm-scale agroforestry decision-making, we should look to develop adequate conceptual models for agroforestry systems. Underpinned by evidence-based reviews, these models could be useful for improving consistency in the valuation of agroforestry assets, guiding decision-making at the farm or paddock scale and supporting development of quantitative decision-making tools.

Author Contributions: Conceptualization, Z.E.M., M.A.H., T.P.B., A.P.O., D.T., and J.R.E.; investigation, Z.E.M.; writing—original draft preparation, Z.E.M.; writing—review and editing, T.P.B., A.P.O., D.T., J.R.E., and M.A.H.; supervision, M.A.H. and T.P.B.

Funding: This project was supported by an Australian Research Council Industrial Transformation Training Centre grant ICI150100004, the Department of Agriculture and Water Resources Rural Research for Profit Program, and a cooperative funding grant from the Forest Practices Authority.

Acknowledgments: We thank Daniel Mendham and William Jackson for constructive comments on an early draft of the manuscript and Private Forests Tasmania for introductions to landowners.

Conflicts of Interest: The authors declare no conflict of interest. The funders had no role in the design of the study; in the collection, analyses, or interpretation of data; in the writing of the manuscript; or in the decision to publish the results.

References

1. Smith, P.; Gregory, P.J.; van Vuuren, D.; Obersteiner, M.; Havlík, P.; Rounsevell, M.; Woods, J.; Stehfest, E.; Bellarby, J. Competition for land. *Philos. Trans. R. Soc. B* **2010**, *365*, 2941–2957. [CrossRef]
2. Sánchez-Bayo, F.; Wyckhuys, K.A. Worldwide decline of the entomofauna: A review of its drivers. *Biol. Conserv.* **2019**, *232*, 8–27. [CrossRef]
3. Parris, K. Impact of agriculture on water pollution in OECD countries: Recent trends and future prospects. *Int. J. Water Resour. Dev.* **2011**, *27*, 33–52. [CrossRef]
4. Mackay, A. Impacts of intensification of pastoral agriculture on soils: Current and emerging challenges and implications for future land uses. *N. Z. Vet. J.* **2008**, *56*, 281–288. [CrossRef]
5. World Resources Institute. *Creating a Sustainable Food Future*; World Resources Institute: Washington, DC, USA, 2018.
6. Reid, R.; Wilson, G. *Agroforestry in Australia and New Zealand*; Goddard and Dobson: Box Hill, Australia, 1985.
7. Smith, J.; Pearce, B.D.; Wolfe, M.S. Reconciling productivity with protection of the environment: Is temperate agroforestry the answer? *Renew. Agric. Food. Syst.* **2012**, *28*, 80–92. [CrossRef]
8. Bastin, J.-F.; Finegold, Y.; Garcia, C.; Mollicone, D.; Rezende, M.; Routh, D.; Zohner, C.M.; Crowther, T.W. The global tree restoration potential. *Science* **2019**, *365*, 76–79. [CrossRef]
9. Lewis, S.L.; Wheeler, C.E.; Mitchard, E.T.A.; Koch, A. Restoring natural forests is the best way to remove atmospheric carbon. *Nature* **2019**, *568*, 25–28. [CrossRef]
10. Black, A.W.; Frost, F.; Forge, K. *Extension and Advisory Strategies for Agroforestry*; Rural Industries Research and Development Corporation: Barton, Australia, 2000.
11. Stewart, H. *Victorian Farm Forestry Inventory Scoping Study*; Farm Forest Growers Victoria Incorporated: Mansfield, Australia, 2009.
12. Race, D.; Curtis, A. Adoption of farm forestry in Victoria: Linking policy with practice. *Australas. J. Environ. Manag.* **2007**, *14*, 166–178. [CrossRef]
13. Cary, J.W.; Wilkinson, R.L. Perceived profitability and farmers' conservation behaviour. *J. Agric. Econ.* **1997**, *48*, 13–21. [CrossRef]
14. Fleming, A.; O'Grady, A.P.; Mendham, D.; England, J.; Mitchell, P.; Moroni, M.; Lyons, A. Understanding the values behind farmer perceptions of trees on farms to increase adoption of agroforestry in Australia. *Agron. Sustain. Dev.* **2019**, *39*, 9. [CrossRef]
15. Pannell, D. Social and economic challenges in the development of complex farming systems. *Agrofor. Syst.* **1999**, *45*, 395–411. [CrossRef]
16. Atkinson, G.; Pearce, D. Measuring sustainable development. In *Handbook of Environmental Economics*; Bromley, D.W., Ed.; Blackwell: Oxford, UK, 1995.

17. Jansson, A.; Hammer, M.; Folke, C.; Costanza, R. Investing in natural capital: Why, what, and how? In *Investing in Natural Capital: The Ecological Economics Approach to Sustainability*; Jansson, A., Hammer, M., Folke, C., Costanza, R., Eds.; Island Press: Washington, DC, USA, 1994.

18. CBD. *Global Biodiversity Outlook 4: A Mid-Term Assessment of Progress towards the Implementation of the Strategic Plan for Biodiversity 2011–2020*; CBD: Montreal, QC, Canada, 2014.

19. Smith, P.; Clark, H.; Dong, H.; Elsiddig, E.; Haberl, H.; Harper, R.; House, J.; Jafari, M.; Masera, O.; Mbow, C. Agriculture, forestry and other land use (AFOLU). In *Climate Change 2014: Mitigation of Climate Change. IPCC Working Group III Contribution to AR5*; Cambridge University Press: Cambridge, UK; New York, NY, USA, 2014.

20. Jackson, W.; Argent, R.; Bax, N.; Bui, E.; Clark, G.; Coleman, S.; Cresswell, I.; Emmerson, K.; Evans, K.; Hibberd, M.; et al. Overview of State and Trends of Inland Water. In *Australia State of the Environment 2016*; Australian Government Department of the Environment and Energy: Canberra, Australia, 2016.

21. Natural Capital Coalition. Natural Capital Protocol. Available online: http://naturalcapitalcoalition.org/protocol (accessed on 31 October 2018).

22. United Nations; European Commission; Organisation for Economic Co-Operation and Development; World Bank. *System of Environmental-Economic Accounting Central Framework*; United Nations Statistics Division: New York, NY, USA, 2014.

23. Wentworth Group. *Accounting for Nature—A Scientific Method for Constructing Environmental Asset Condition Accounts*; Wentworth Group: Sydney, Australia, 2016.

24. Pearce, D. An intellectual history of environmental economics. *Annu. Rev. Energ. Environ.* **2002**, *27*, 57–81. [CrossRef]

25. TEEB. The Economics of Ecosystems and Biodiversity: Mainstreaming the Economics of Nature: A Synthesis of the Approach, Conclusion and Recommendations of TEEB. In Proceedings of the 10th meeting of the Conference of Parties to the CBD, Nagoya, Japan, 18–29 October 2010.

26. The World Bank. *WAVES Annual Report 2017*; The World Bank: Washington, DC, USA, 2017.

27. TEEB. *TEEB for Agriculture & Food: Scientific and Economic Foundations*; TEEB: Geneva, Switzerland, 2018.

28. FAO. *System of Environmental-Economic Accounting for Agriculture, Forestry and Fisheries: SEEA AFF White Cover Final*; FAO: New York, NY, USA, 2016.

29. Ascui, F.; Cojoianu, T. *Natural Capital Credit Risk Assessment in Agricultural Lending: An Approach Based on the Natural Capital Protocol*; Natural Capital Finance Alliance: Oxford, UK, 2019.

30. FAO. *Natural Capital Impacts in Agriculture—Supporting Better Decision Making*; FAO: Rome, Italy, 2015.

31. Czúcz, B.; Keith, H.; Jackson, B.; Maes, J.; Driver, A.; Nicholson, E.; Bland, L. *Discussion Paper 2.3: Proposed Typology of Condition Variables for Ecosystem Accounting and Criteria for Selection of Condition Variables. Paper Submitted to the SEEA EEA Technical Committee as input to the Revision of the Technical Recommendations in Support of the System on Environmental-Economic Accounting. Version of 13 March 2019*; United Nations: New York, NY, USA, 2019; p. 23.

32. Boyd, J.; Banzhaf, S. What are ecosystem services? The need for standardized environmental accounting units. *Ecol. Econ.* **2007**, *63*, 616–626. [CrossRef]

33. Daily, G. *Nature's Services: Societal Dependence on Natural Ecosystems*; Island Press: Washington, DC, USA, 1997.

34. De Groot, R.S.; Wilson, M.A.; Boumans, R.M.J. A typology for the classification, description and valuation of ecosystem functions, goods and services. *Ecol. Econ.* **2002**, *41*, 393–408. [CrossRef]

35. Fisher, B.; Turner, R.K.; Morling, P. Defining and classifying ecosystem services for decision making. *Ecol. Econ.* **2009**, *68*, 643–653. [CrossRef]

36. Haines-Young, R.; Potschin, M.B. Common International Classification of Ecosystem Services (CICES) V5.1 and Guidance on the Application of the Revised Structure. EEA Framework Contract No EEA/IEA/09/003. 2018. Available online: www.cices.eu (accessed on 10 September 2018).

37. Millennium Ecosystem Assessment. *Ecosystems and Human Wellbeing: Synthesis*; Island Press: Washington, DC, USA, 2005.

38. Nahlik, A.M.; Kentula, M.E.; Fennessy, M.S.; Landers, D.H. Where is the consensus? A proposed foundation for moving ecosystem service concepts into practice. *Ecol. Econ.* **2012**, *77*, 27–35. [CrossRef]

39. Wallace, K.J. Classification of ecosystem services: Problems and solutions. *Biol. Conserv.* **2007**, *139*, 235–246. [CrossRef]

40. United Nations. *Technical Recommendations in Support of the System of Environmental-Economic Accounting 2012—Experimental Ecosystem Accounting*; United Nations: New York, NY, USA, 2017.

41. Obst, C.G.; van de Ven, P.; Tebrake, J.; St Lawrence, J.; Edens, B. Valuation and Accounting Treatments: Issues and Options in Accounting for Ecosystem Degradation and Enhancement (Draft). In Proceedings of the 2019 Forum of Experts in SEEA Experimental Ecosystem Accounting, Glen Cove, NY, USA, 26–27 June 2019.

42. Haines-Young, R.; Potschin, M.B. Common International Classification of Ecosystem Services (CICES): Consultation on Version 4, August–December 2012. EEA Framework Contract No EEA/IEA/09/003. 2013. Available online: www.cices.eu (accessed on 15 September 2018).

43. Asbjornsen, H.; Hernandez-Santana, V.; Liebman, M.; Bayala, J.; Chen, J.; Helmers, M.; Ong, C.; Schulte, L.A. Targeting perennial vegetation in agricultural landscapes for enhancing ecosystem services. *Renew. Agric. Food. Syst.* **2014**, *29*, 101–125. [CrossRef]

44. Jose, S. Agroforestry for ecosystem services and environmental benefits: An overview. *Agrofor. Syst.* **2009**, *76*, 1–10. [CrossRef]

45. Ovando, P.; Campos, P.; Oviedo, J.L.; Caparrós, A. Ecosystem accounting for measuring total income in private and public agroforestry farms. *For. Policy Econ.* **2016**, *71*, 43–51. [CrossRef]

46. Kay, S.; Graves, A.; Palma, J.H.; Moreno, G.; Roces-Díaz, J.V.; Aviron, S.; Chouvardas, D.; Crous-Duran, J.; Ferreiro-Domínguez, N.; de Jalón, S.G. Agroforestry is paying off—Economic evaluation of ecosystem services in European landscapes with and without agroforestry systems. *Ecosyst. Serv.* **2019**, *36*, 100896. [CrossRef]

47. Campos, P.; Oviedo, J.L.; Álvarez, A.; Mesa, B.; Caparrós, A. The role of non-commercial intermediate services in the valuations of ecosystem services: Application to cork oak farms in Andalusia, Spain. *Ecosyst. Serv.* **2019**, *39*, 100996. [CrossRef]

48. Alkemade, R.; Burkhard, B.; Crossman, N.D.; Nedkov, S.; Petz, K. Quantifying ecosystem services and indicators for science, policy and practice. *Ecol. Indic.* **2014**, *37*, 161–162. [CrossRef]

49. Burkhard, B.; Crossman, N.; Nedkov, S.; Petz, K.; Alkemade, R. Mapping and modelling ecosystem services for science, policy and practice. *Ecosyst. Serv.* **2013**, *4*, 1–3. [CrossRef]

50. Crossman, N.D.; Burkhard, B.; Nedkov, S.; Willemen, L.; Petz, K.; Palomo, I.; Drakou, E.G.; Martín-Lopez, B.; McPhearson, T.; Boyanova, K.; et al. A blueprint for mapping and modelling ecosystem services. *Ecosyst. Serv.* **2013**, *4*, 4–14. [CrossRef]

51. Burkhard, B.; Kroll, F.; Müller, F.; Windhorst, W. Landscapes' capacities to provide ecosystem services—A concept for land-cover based assessments. *Landsc. Online* **2009**, *15*, 1–22. [CrossRef]

52. Eigenbrod, F.; Armsworth, P.R.; Anderson, B.J.; Heinemeyer, A.; Gillings, S.; Roy, D.B.; Thomas, C.D.; Gaston, K.J. The impact of proxy-based methods on mapping the distribution of ecosystem services. *J. Appl. Ecol.* **2010**, *47*, 377–385. [CrossRef]

53. Egoh, B.N.; Drakou, E.; Dunbar, M.B.; Maes, J.; Willemen, L. *Indicators for Mapping Ecosystem Services: A Review*; Publications Office of the European Union: Luxembourg, 2012.

54. Martínez-Harms, M.J.; Balvanera, P. Methods for mapping ecosystem service supply: A review. *Int. J. Biodivers. Sci. Ecosyst. Serv. Manag.* **2012**, *8*, 17–25. [CrossRef]

55. Metzger, M.J.; Rounsevell, M.D.A.; Acosta-Michlik, L.; Leemans, R.; Schröter, D. The vulnerability of ecosystem services to land use change. *Agric. Ecosyst. Environ.* **2006**, *114*, 69–85. [CrossRef]

56. Naidoo, R.; Balmford, A.; Costanza, R.; Fisher, B.; Green, R.E.; Lehner, B.; Malcolm, T.R.; Ricketts, T.H. Global mapping of ecosystem services and conservation priorities. *Proc. Natl. Acad. Sci. USA* **2008**, *105*, 9495. [CrossRef]

57. Volk, M. Modelling ecosystem services—Challenges and promising future directions. *Sustain. Water Qual. Ecol.* **2013**, *1–2*, 3–9. [CrossRef]

58. Kareiva, P. *Natural Capital: Theory and Practice of Mapping Ecosystem Services*; Oxford University Press: Oxford, UK, 2011.

59. Lonsdorf, E.; Kremen, C.; Ricketts, T.; Winfree, R.; Williams, N.; Greenleaf, S. Modelling pollination services across agricultural landscapes. *Ann. Bot.* **2009**, *103*, 1589–1600. [CrossRef] [PubMed]

60. Battaglia, M.; Sands, P.; White, D.; Mummery, D. CABALA: A linked carbon, water and nitrogen model of forest growth for silvicultural decision support. *For. Ecol. Manag.* **2004**, *193*, 251–282. [CrossRef]

61. Ensis. SPIF: The Scenario Planning and Investment Framework Tool. In *Commerical Environmental Forestry: Integrating Trees into Landscapes for Multiple Benefits*; Summary Technical Report June 2006; Ensis (the joint forces of CSIRO and Scion): Victoria, Australia, 2006.

62. Private Forests Tasmania. *The Farm Forestry Toolbox Version 5.0: An Aid to Successfully Growing Trees on Farms*; Private Forests Tasmania: Tasmania, Australia, 2008.

63. Young, A.; Menz, K.M.; Muraya, P.; Smith, C. *SCUAF-Version 4: A Model to Estimate Soil Changes under Agriculture, Agroforestry and Forestry*; ACIAR: Canberra, Australia, 1998; p. 49.

64. van der Werf, W.; Keesman, K.; Burgess, P.; Graves, A.; Pilbeam, D.; Incoll, L.; Metselaar, K.; Mayus, M.; Stappers, R.; van Keulen, H. Yield-SAFE: A parameter-sparse, process-based dynamic model for predicting resource capture, growth, and production in agroforestry systems. *Ecol. Eng.* **2007**, *29*, 419–433. [CrossRef]

65. Keating, B.A.; Carberry, P.S.; Hammer, G.L.; Probert, M.E.; Robertson, M.J.; Holzworth, D.; Huth, N.I.; Hargreaves, J.N.; Meinke, H.; Hochman, Z. An overview of APSIM, a model designed for farming systems simulation. *Eur. J. Agron.* **2003**, *18*, 267–288. [CrossRef]

66. Warner, A. *Farm Forestry Toolbox Version 5.0: Helping Australian Growers to Manage Their Trees: A Report for the RIRDC/L & WA/FWPRDC Joint Venture Agroforestry Program*; Rural Industries Research and Development Corporation: Canberra, Australia, 2007.

67. USDA Forest Service. i-Tree: Tools for Assessing and Managing Forests and Community Trees. Available online: https://www.itreetools.org/about.php (accessed on 3 February 2019).

68. Sandhu, H.S.; Wratten, S.D.; Cullen, R.; Case, B. The future of farming: The value of ecosystem services in conventional and organic arable land. An experimental approach. *Ecol. Econ.* **2008**, *64*, 835–848. [CrossRef]

69. Porter, J.; Costanza, R.; Sandhu, H.; Sigsgaard, L.; Wratten, S. The value of producing food, energy, and ecosystem services within an agro-ecosystem. *Ambio* **2009**, *38*, 186–193. [CrossRef]

70. Crossman, N.D.; Connor, J.D.; Bryan, B.A.; Summers, D.M.; Ginnivan, J. Reconfiguring an irrigation landscape to improve provision of ecosystem services. *Ecol. Econ.* **2010**, *69*, 1031–1042. [CrossRef]

71. Petz, K.; van Oudenhoven, A.P. Modelling land management effect on ecosystem functions and services: A study in the Netherlands. *Int. J. Biodivers. Sci. Ecosyst. Serv. Manag.* **2012**, *8*, 135–155. [CrossRef]

72. Fisher, I. *The Nature of Capital and Income*; The Macmillan Company: New York, NY, USA, 1906.

73. Krutilla, J.V. Conservation reconsidered. *Am. Econ. Rev.* **1967**, *57*, 777–786.

74. Fenichel, E.P.; Abbott, J.K.; Yun, S.D. Chapter 3—The nature of natural capital and ecosystem income. In *Handbook of Environmental Economics*; Dasgupta, P., Pattanayak, S.K., Smith, V.K., Eds.; Elsevier: Amsterdam, The Netherlands, 2018; Volume 4, pp. 85–142.

75. Pezzey, J.C.; Toman, M.A. Sustainability and its economic interpretations. In *Scarcity and Growth Revisited—Natural Resources and the Environment in the New Millenium*; Resources for the Future: Washington, DC, USA, 2005.

76. Pearce, D.; Moran, D. *The Economic Value of Biodiversity*; Routledge: London, UK, 2013.

77. Pascual, U.; Muradian, R.; Brander, L.; Gómez-Baggethun, E.; Martín-López, B.; Verma, M.; Armsworth, P.; Christie, M.; Cornelissen, H.; Eppink, F. The economics of valuing ecosystem services and biodiversity. In *The Economics of Ecosystems and Biodiversity: Ecological and Economic Foundations*; Taylor and Francis: London, UK, 2010; pp. 183–256. [CrossRef]

78. Thompson, D.; George, B. Financial and economic evaluation of agroforestry. In *Agroforestry for Natural Resource Management*; CSIRO Publishing: Canberra, Australia, 2009; pp. 283–308.

79. Stainback, G.A.; Alavalapati, J.R.R.; Shrestha, R.K.; Larkin, S.; Wong, G. Improving Environmental Quality in South Florida through Silvopasture: An Economic Approach. *J. Agric. Appl. Econ.* **2004**, *36*, 481–489. [CrossRef]

80. Kulshreshtha, S.; Kort, J. External economic benefits and social goods from prairie shelterbelts. *Agrofor. Syst.* **2009**, *75*, 39–47. [CrossRef]

81. Oviedo, J.L.; Huntsinger, L.; Campos, P. The Contribution of Amenities to Landowner Income: Cases in Spanish and Californian Hardwood Rangelands. *Rangel. Ecol. Manag.* **2017**, *70*, 518–528. [CrossRef]

82. Morse, R.; Calderone, N. The value of honey bee pollination in the United States. *Bee Cult.* **2000**, *128*, 1–15.

83. Wilson, S.J. *Ontario's Wealth, Canada's Future: Appreciating the Value of the Greenbelt's Eco-Services*; David Suzuki Foundation: Vancouver, BC, Canada, 2008.

84. Winfree, R.; Gross, B.J.; Kremen, C. Valuing pollination services to agriculture. *Ecol. Econ.* **2011**, *71*, 80–88. [CrossRef]

85. Alam, M.; Olivier, A.; Paquette, A.; Dupras, J.; Revéret, J.P.; Messier, C. A general framework for the quantification and valuation of ecosystem services of tree-based intercropping systems. *Agrofor. Syst.* **2014**, *88*, 679–691. [CrossRef]

86. Polyakov, M.; Pannell, D.J.; Pandit, R.; Tapsuwan, S.; Park, G. Capitalized amenity value of native vegetation in a multifunctional rural landscape. *Am. J. Agric. Econ.* **2015**, *97*, 299–314. [CrossRef]

87. Shrestha, R.K.; Alavalapati, J.R.R. Valuing environmental benefits of silvopasture practice: A case study of the Lake Okeechobee watershed in Florida. *Ecol. Econ.* **2004**, *49*, 349–359. [CrossRef]

88. de Jalón, S.G.; Graves, A.; Palma, J.H.; Williams, A.; Upson, M.; Burgess, P.J. Modelling and valuing the environmental impacts of arable, forestry and agroforestry systems: A case study. *Agrofor. Syst.* **2018**, *92*, 1059–1073. [CrossRef]

89. National Australia Bank. Natural Value. Available online: https://www.nab.com.au/about-us/corporate-responsibility/environment/natural-value (accessed on 20 June 2018).

90. Cojoianu, T.F.; Ascui, F. Developing an evidence base for assessing natural capital risks and dependencies in lending to Australian wheat farms. *J. Sustain. Financ. Invest.* **2018**, *8*, 95–113. [CrossRef]

91. Ogilvy, S. Developing the ecological balance sheet for agricultural sustainability. *Sustain. Acc. Manag. Policy. J.* **2015**, *6*, 110–137. [CrossRef]

92. Campos, P.; Rodríguez, Y.; Caparrós, A. Towards the dehesa total income accounting: Theory and operative Monfragüe study cases. *For. Syst.* **2001**, *10*. [CrossRef]

93. Caparrós, A.; Campos, P.; Montero, G. An Operative Framework for Total Hicksian Income Measurement: Application to a Multiple-Use Forest. *Environ. Resour. Econ.* **2003**, *26*, 173–198. [CrossRef]

94. Dominati, E.; Mackay, A.; Green, S.; Patterson, M. A soil change-based methodology for the quantification and valuation of ecosystem services from agro-ecosystems: A case study of pastoral agriculture in New Zealand. *Ecol. Econ.* **2014**, *100*, 119–129. [CrossRef]

95. Baker, T.P.; Moroni, M.T.; Mendham, D.S.; Smith, R.; Hunt, M.A. Impacts of windbreak shelter on crop and livestock production. *Crop. Pasture. Sci.* **2018**, *69*, 785–796. [CrossRef]

96. Cleugh, H. Effects of windbreaks on airflow, microclimates and crop yields. *Agrofor. Syst.* **1998**, *41*, 55–84. [CrossRef]

97. Bird, P.R.; Bicknell, D.; Bulman, P.A.; Burke, S.J.A.; Leys, J.F.; Parker, J.N.; Van Der Sommen, F.J.; Voller, P. The role of shelter in Australia for protecting soils, plants and livestock. *Agrofor. Syst.* **1992**, *20*, 59–86. [CrossRef]

98. Olander, L.; Mason, S.; Warnell, K.; Tallis, H. *Building Ecosystem Services Conceptual Models*; NESP Conceptual Model Series No. 1; Duke University: Durham, NC, USA, 2018.

99. England, J.R.; O'Grady, A.P.; Fleming, A.; Marais, Z.; Mendham, D. Trees on farms to support natural capital: An evidence-based review for grazed dairy systems. *Sci. Total Environ.* **2019**, unpublished work.

100. Müller, A.; Knoke, T.; Olschewski, R. Can existing estimates for ecosystem service values inform forest management? *Forests* **2019**, *10*, 132. [CrossRef]

101. Abadi, A.; Lefroy, T.; Cooper, D.; Hean, R.; Davies, C. *Profitability of Medium to Low Rainfall Agroforestry in the Cropping Zone*; Rural Industries Research and Development Corporation Publication: Barton, Australia, 2003.

102. Graves, A.R.; Burgess, P.J.; Liagre, F.; Terreaux, J.-P.; Borrel, T.; Dupraz, C.; Palma, J.; Herzog, F. Farm-SAFE: The process of developing a plot-and farm-scale model of arable, forestry, and silvoarable economics. *Agrofor. Syst.* **2011**, *81*, 93–108. [CrossRef]

103. Digital Agriculture Service. DAS Rural Intelligence Platform. Available online: https://digitalagricultureservices.com/platform (accessed on 20 August 2019).

104. FarmMap4D. FarmMap4D Spatial Hub Factsheet: Turning Big Data into Better Decisions. Available online: http://www.farmmap4d.com.au/wp-content/uploads/2018/02/Turning-big-data-into-better-decisionsV2.pdf (accessed on 20 August 2019).

Review

Temperate Agroforestry Systems and Insect Pollinators: A Review

Gary Bentrup [1,*], **Jennifer Hopwood** [2], **Nancy Lee Adamson** [3] **and Mace Vaughan** [4]

[1] Department of Agriculture, U.S. Forest Service, National Agroforestry Center, Lincoln, NE 68583, USA
[2] Xerces Society for Invertebrate Conservation, Omaha, NE 68134, USA; jennifer.hopwood@xerces.org
[3] Xerces Society for Invertebrate Conservation, Greensboro, NC 27401, USA; nancy.adamson@xerces.org
[4] Xerces Society for Invertebrate Conservation, Portland, OR 97232-1324, USA; mace.vaughan@xerces.org
*   Correspondence: gary.bentrup@usda.gov; Tel.: +1-(402)-437-5178

Received: 8 October 2019; Accepted: 1 November 2019; Published: 5 November 2019

Abstract: Agroforestry can provide ecosystem services and benefits such as soil erosion control, microclimate modification for yield enhancement, economic diversification, livestock production and well-being, and water quality protection. Through increased structural and functional diversity in agricultural landscapes, agroforestry practices can also affect ecosystem services provided by insect pollinators. A literature review was conducted to synthesize information on how temperate agroforestry systems influence insect pollinators and their pollination services with particular focus on the role of trees and shrubs. Our review indicates that agroforestry practices can provide three overarching benefits for pollinators: (1) providing habitat including foraging resources and nesting or egg-laying sites, (2) enhancing site and landscape connectivity, and (3) mitigating pesticide exposure. In some cases, agroforestry practices may contribute to unintended consequences such as becoming a sink for pollinators, where they may have increased exposure to pesticide residue that can accumulate in agroforestry practices. Although there is some scientific evidence suggesting that agroforestry practices can enhance crop pollination and yield, more research needs to be conducted on a variety of crops to verify this ecosystem service. Through a more comprehensive understanding of the effects of agroforestry practices on pollinators and their key services, we can better design agroforestry systems to provide these benefits in addition to other desired ecosystem services.

Keywords: alley cropping; bees; forest farming; hedgerows; pollinators; pollination; riparian buffers; shelterbelts; windbreaks

---

1. Introduction

Plant pollination by animals is one of the most important ecosystem services and is essential in both natural and agricultural landscapes. An estimated 85% of the world's flowering plants depend on animals—mostly insects—for pollination [1]. Insect pollination is critical to food security and roughly 35% of global crop production is dependent on pollination by animals [2,3]. Pollinators are also a keystone group in most terrestrial ecosystems, necessary for plant reproduction, and important for wildlife food webs [4,5]. Insect pollinators include bees, wasps, flies, beetles, butterflies, and moths, but some bird and bat species pollinate as well [6]. Although all pollinators play important roles, bees are considered particularly essential for pollination of agricultural crops [7,8] as well as for wild plants in temperate climates [6]. Globally, insect pollinators are in decline, with some estimates that 40% of invertebrate pollinator species may be at risk of extinction worldwide [9]. Threats such as the loss, degradation, and fragmentation of habitat (e.g., [5,10,11]); introduced species (e.g., [12,13]); the use of pesticides (e.g., [4,14–16]; and diseases and parasites (e.g., [17,18]) all contribute to pollinator decline. An annual value of global crop output (estimated at $235 to $577 billion $US in 2015) is at risk due to pollinator loss [9]. Threats to pollinators may have profound consequences for ecosystem health as well as our food systems [9,19,20].

Managing existing habitat and restoring additional habitat for pollinators has been demonstrated to increase their abundance and diversity (e.g., [13,21,22]). Agroforestry is the intentional integration of trees and/or shrubs with herbaceous crops and/or livestock in an agricultural production system [23]. By adding structural and functional diversity to agricultural landscapes, agroforestry can provide pollinator habitat and support pollinator services [24]. In temperate regions, agroforestry systems include many different permutations such as windbreaks, riparian buffers, alley cropping, hedgerows, shelterbelts, silvopasture, and forest farming [23]. Depending on the situation and application, these practices can provide protection for topsoil, livestock, and crops; increase crop and livestock productivity; reduce inputs of energy and chemicals; increase water use efficiency; improve air and water quality; sequester carbon; and enhance biodiversity [23,24]. Although agroforestry is rarely implemented for pollinator habitat or crop pollination services, there are opportunities to incorporate these services when designing multifunctional practices. A systematic review of existing scientific literature on the topic is a key first step. While several reviews have focused on biodiversity-based ecosystem services of agroforestry systems (e.g., [24,25]), none have examined in detail the effects of agroforestry on insect pollinators and pollination services. The objective of this paper is to review the role of temperate agroforestry systems in supporting insect pollinators and pollination services with a focus on the role of trees and shrubs in these systems.

2. Literature Review

Our systematic review followed recommended procedures outlined in the PRISMA statement [26]. An initial screening of research papers published in peer-reviewed English-language scientific journals was performed within Web of Science and Scopus databases. The literature was searched using a Boolean defined by logical strings containing one of the keywords "agroforestry", "alley cropping", "windbreak", "shelterbelt", "hedgerow", "forest farming", "multi-story cropping", "silvopasture", and "riparian buffer", in conjunction with at least one pollinator keyword. Pollinator keywords used in the queries included "bees", "flies", "moths", "wasps", "beetles", "pollinator", and "pollination". In order to have a comprehensive foundation for this review, the search was conducted for the earliest electronic records available to mid-2019. Review articles were included if they contained original research. We also examined reference lists within review articles to identify relevant papers.

To supplement this review, key articles on reducing pesticide drift and runoff using agroforestry practices were included even though some of these papers did not explicitly use pollinator terms (e.g., bees, butterflies, flies) in the text. It is recognized that reducing pesticide exposure pathways will have benefits for insect pollinators and we wanted to capture pesticide mitigation research regarding agroforestry practices. We also included papers that provided information on tree and shrub species used in temperate agroforestry practices where available. These papers were identified based on the authors' experience and knowledge regarding the topic area.

These initial records were screened for relevance to insect pollinators and pollination within temperate regions. These remaining articles were reviewed in whole and 134 were included in the final database. The PRISMA flowchart summarizing the search results and screening workflow is shown in Figure 1. Limiting our search to English articles in peer-reviewed sources in electronic databases exposes this review to the risk of language and publication bias [27]. To counter this potential bias, we included a broad variety of articles as possible to capture the dominant themes within the boundaries of the review objectives.

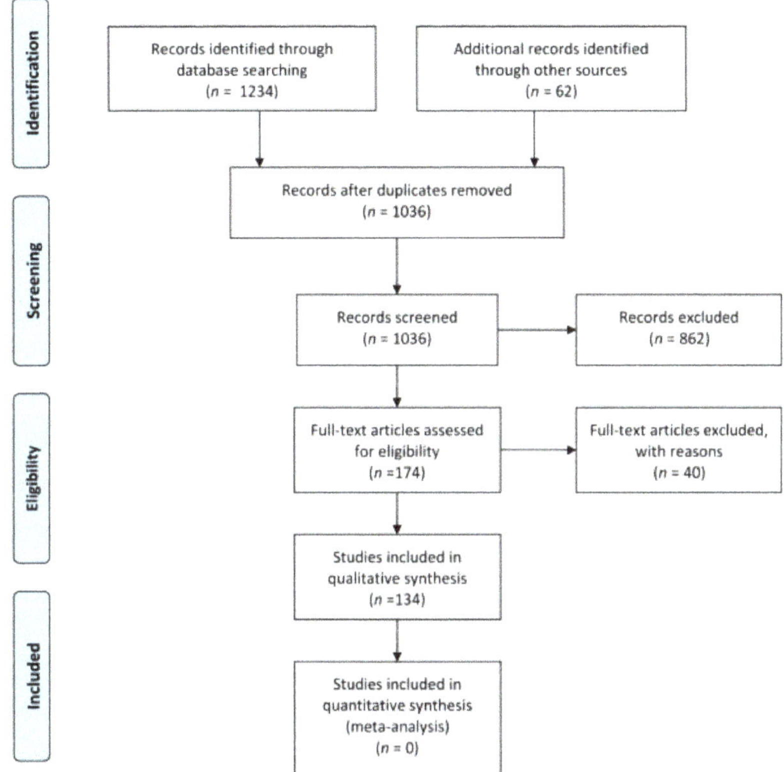

Figure 1. Flow diagram depicting the search and selection of the study process according to the Preferred Reporting Items for Systematic Reviews and Meta-Analyses (PRISMA) statement.

The majority of research articles in this review database was focused on windbreaks, shelterbelts, and hedgerows, with limited papers available on riparian buffers and alley cropping. There were few or no studies on silvopasture, forest farming, or multi-story cropping regarding insect pollinators in temperate regions. Our review indicates that agroforestry practices in general can provide three overarching functions for pollinators: (1) providing habitats, including foraging resources and nesting or egg-laying sites, (2) enhancing site and landscape connectivity, and (3) mitigating pesticide exposure. Available scientific evidence suggests that agroforestry practices can enhance crop pollination and yields; however, few studies have been conducted on this ecosystem service. Table 1 and Figure 2 provides a summary of this review.

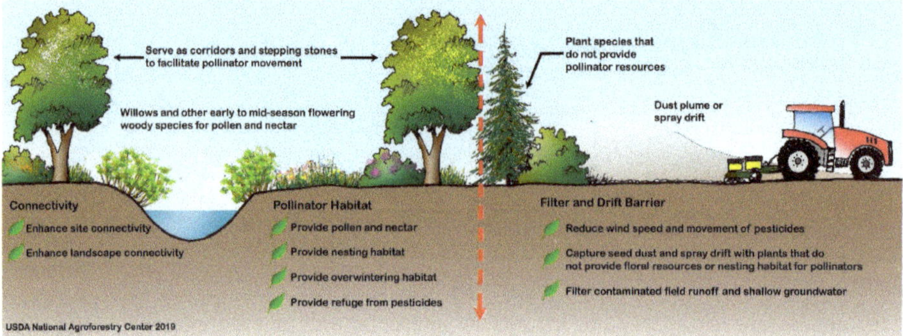

Figure 2. Conceptual diagram illustrating typical functions that a generic agroforestry practice can provide to insect pollinators.

3. Foraging Resources

3.1. Pollen and Nectar

Pollinators require a diversity of flowering plants over the foraging season to provide nectar and pollen resources to meet their nutritional needs [9]. Studies have documented diverse pollinator use of these floral resources in agroforestry practices, particularly if planted specifically to promote pollinators. Hedgerows, riparian buffers, windbreaks, and other linear agroforestry practices are used by bees [28–42], butterflies [36,37,42–49], moths [49–52], and flies [22,37,41,42,53,54]. Based on available evidence, woody species used in temperate agroforestry practices can play an important role in providing floral resources; however, if the agroforestry practice lacks pollinator-suitable floral resources, pollinator use is limited. For instance, Macdonald et al. [55] found limited pollinator use of shelterbelts in New Zealand that were predominantly comprised of Monterey pine (*Pinus radiata* D. Don) and Monterey cypress (*Hesperocyparis macrocarpa* (Hartw.) Bartel) (wind-pollinated exotic species).

Tree and shrub species offer abundant nectar with relatively high sugar contents such as maple (*Acer* spp.), horsechestnut (*Aesculus* spp.), basswood (*Tilia* spp.), willow (*Salix* spp.), brambles (*Rubus* spp.), cherry and plum (*Prunus* spp.), and serviceberry (*Amelanchier* spp.) [56–61]. For example, sugar content in horsechestnut (*Aesculus hippocastanum* L.) ranges from 0.58–3.57 mg/flower/24h, while black locust (*Robinia pseudoacacia* L.) ranges from 0.76–4.0 mg/flower/24h [62]. For comparison, white clover (*Trifolium repens* L.) ranges from 0.01–0.20 mg/flower/24h and alfalfa (*Medicago sativa* L.) ranges from 0.07–0.25 mg/flower/24h [62].

Pollen is a protein-rich resource that is used by native bees, honey bees, and some wasps to feed their brood or to provision their eggs. Pollen is also used by some adult and larval flies and beetles as a food source. Key woody species that can provide pollen with high concentrations of amino acids, sterols, trace minerals, and other nutritionally important compounds for bees and other pollinators include willow, maple, cherry and plum, brambles, chestnut (*Castanea* spp.), and ash (*Fraxinus* spp.) [60,61,63–66]. Some bees are pollen specialists (oligolectic), wholly dependent on specific shrubs and trees in certain families, such as willows, dogwoods (Cornaceae), heaths like blueberry and huckleberry (Ericaceae), buckthorns such as New Jersey tea (Rhamnaceae), and roses (Rosaceae) [67,68]. Other research has found apple (*Malus* spp.), hawthorn (*Crataegus* spp.), and elderberry (*Sambucus* spp.) to be attractive for pollinator foraging [31,33]. This list is not inclusive and additional woody species are likely to provide suitable forage resources even if they are not specifically mentioned in the literature reviewed.

Trees and shrubs in temperate regions often flower early in the spring and can deliver some of the first pollen and nectar resources of the season, boosting early-season pollinator populations [57,59–61,65,69,70]. In Michigan, US, Wood et al. [70] determined that willows, maples, and *Prunus* spp. provided over 90% of the pollen collected in April by social and solitary bees. Flower density and subsequent nectar availability can be higher in some tree and shrub species compared to herbaceous species [57,58]. For instance, during peak flowering season, gray willow (*Salix cinerea* L.) can produce 334,178 flowers/m^2 and oneseed hawthorn (*Crataegus monogyna* Jacq.) 19,003 flowers/m^2 compared to sea aster (*Aster tripolium* L.) 9,565 flowers/m^2 and buttercup (*Ranunculus acris* L.) 688 flowers/m^2 [58]. Respectively, nectar productivity for these species is 3612, 584, 169, and 50 kg/ha cover/year. Spatially, agroforestry practices that include a diversity of flowering woody and herbaceous species can deliver a high density of floral resources relative to the land area occupied due to vertical layering [22,31,59,71,72]. One study documented approximately two and four times greater nectar per unit area in hedgerows compared to woodlands and pasture, respectively [35]. In regards to native versus exotic plants, one study showed that wild bees, and managed bees in some cases, prefer to forage on native plants in hedgerows over co-occurring weedy, exotic plants [28]. Management that reduces floral resources such as hedgerow trimming has been shown to have a negative impact on pollinators [73,74].

3.2. Resins and Oils

Bees also collect resins and floral oils from trees and other plants to aid in nest construction and provisioning immatures [75–78]. Some tunnel-nesting native bees use tree resins to seal off their nests [78], while other bees use tree resins and embedded sand grains and plant materials or debris to create nest cells [79]. Honey bees use resins to make propolis to seal unwanted holes in their hives [80]. Propolis has antibacterial properties that help reduce disease transmission and parasite invasion [81]. Poplar trees (*Populus* spp.) are a common source for these resins [82–85]. Due to their fast growth rate, low disease and pest issues, lower shading effect, and marketable products (i.e., biofeedstock, lumber), poplar trees are often used in temperate agroforestry applications. Other tree species that provide resin sources include: pine (*Pinus* spp.), birch (*Betula* spp.), elm (*Ulmus* spp.), alder (*Alnus* spp.), beech (*Fagus* spp.), and horsechestnut [82,84–86]. Depending on application, these tree species have traits or products that can be useful in agroforestry, such as being evergreen (pine), nitrogen-fixing (alder), syrup production (birch), lumber (elm, pine), mast (beech), and dense shade (horsechestnut).

3.3. Microclimate Modification

Pollinator behavior, foraging, and resulting pollination services are strongly influenced by weather conditions (e.g., ambient temperature, wind speed, precipitation) [87,88]. Temperature and wind speed are two primary weather variables that agroforestry practices can influence. Agroforestry practices can reduce air movement and modify temperatures in a cropped area. Daytime air temperatures are several degrees warmer within a certain distance downwind of windbreaks (8 to 10 times the windbreak height) [89]. These elevated temperatures can increase pollinator activity and pollination, particularly in vegetable- and fruit-growing regions, where air temperatures at pollination time can be below optimum [90,91]. The vertical structure and shaded sites found within tree-based practices may offer a diversity of niches that allow pollinators to find suitable sites for thermal regulation, which is becoming increasingly important under climate change [9]. One study found that agricultural landscapes that have a higher proportion of hedgerows and other semi-natural habitats (i.e., 17% compared to 2%) reduced the detrimental effects of warmer temperatures on native bee species' richness and abundance [92].

Additional thermoregulation considerations for managed honey bees may be addressed by agroforestry plantings. Honey bees expend energy to cool themselves and their hives during hot weather. If the hives are shaded, that energy can be diverted to honey production and hive maintenance activities. Trees and shrubs can be used to shade beehives, especially if the hives are placed on the north or northeast sides of the woody plantings to receive maximum shading during the summer heat [93]. Windbreaks and other woody buffers can provide protection from winter temperatures and winds if the hives are located on the leeward side, helping reduce winter mortality [94]. A study in

Kansas, USA, documented this protective service, with overwintered bee populations being up to 52% higher when hives were protected by windbreaks [95].

Foraging in moderate-to-high winds increases energetic costs for pollinators and can reduce pollination efficiency [88,96]. Agroforestry practices can reduce wind speeds, which increases pollinator efficiency and allows pollinators to forage during wind events that would normally reduce or prohibit foraging [42,91]. The protective effect on insect flight extends up to a distance equal to about nine times the height of the windbreak [97,98] and the sheltered zone contains higher numbers of pollinating insects [99], increasing pollination and fruit set [91,100].

3.4. Nesting and Egg-Laying Sites

The availability of nesting and egg-laying sites is an important element of pollinator habitat, although less is known about this component compared to foraging needs [20,101–103]. In general, pollinator populations benefit most from flower-rich foraging areas if suitable nesting or egg-laying sites are nearby [9].

Approximately 30% of native bees in North America are above-ground cavity-nesting species that build their nests in hollow tunnels in the soft pithy centers of twigs of some plants, in abandoned wood-boring beetle tunnels, or in tunnels where some species excavate themselves into wood [77,101]. Hedgerows have been shown to increase the availability of nesting sites for above-ground nesting species [22,104] and older hedgerows had a higher incidence of above-ground nesting bees [104]. A modeling study calculated a higher nesting potential for cavity-nesting species in landscapes with agroforestry compared to landscapes without agroforestry [105]. Incorporating woody species with soft pithy centers such as sumac (*Rhus* spp.), elderberry, and brambles may be beneficial in providing nesting sites [77,101].

The majority of North American native bees are solitary ground nesters that excavate underground tunnels for nesting, which can be negatively impacted by tillage in agricultural fields [106,107]. The presence of trees and shrubs may provide protected ground-nesting areas that have limited soil disturbance. Hedgerows may provide suitable ground-nesting habitat and increase diversity of ground-nesting bees [22,71,104]; however, another study did not find enhanced nesting rates for ground-nesting bees in hedgerows [102].

Bumble bees construct nests in small cavities, often in old rodent burrows, either underground or beneath fallen plant matter. They may select nest sites at the interface between fields and linear woody habitat such as hedgerows and windbreaks [108,109]. One study documented bumble bee nest densities twice as great in these linear woody habitats when compared with grassland and other woodland habitats [110], while another study found hedgerows to be less preferred when compared to herbaceous field margins and grasslands for nest-searching bumble bee queens [111].

Hedgerows and other agroforestry practices can provide egg-laying sites, larval host plants, and overwintering sites for lepidopteran (butterfly and moth) species [44,52,112]. In the U.S. mid-Atlantic region, woody species used as larval host plants were found to support 10 times more lepidopteran species than herbaceous plants [12]. Some of the most highly used plant genera by lepidopteran species include willows, birch, poplar, cherry, plum, and oaks (*Quercus* spp.) [12]. Lepidopteran species and other pollinators, including beetles, overwinter under bark and leaf litter found in hedgerows [44,112,113].

Table 1. The role of temperate agroforestry practices and woody plants on insect pollinators and pollination.

Habitat Component or Ecosystem Service	Summary	Key References
Resins and oils	Honey bees harvest resins from tree buds, particularly poplar (*Populus* spp.), to make propolis, which provides antimicrobial and structural benefits for the colony. Other tree species, including pine (*Pinus* spp.), birch (*Betula* spp.), elm (*Ulmus* spp.), alder (*Alnus* spp.), beech (*Fagus* spp.), and horsechestnut (*Aesculus* spp.) are sources of resin. Some solitary bees also collect plant resins to include in brood cell linings and others use oils in brood cell provisions.	[75,76,78,82–84,86]
Early-season pollen and nectar	Woody species in temperate regions can provide important early-season sources of pollen and nectar.	[57,59–61,65,69,70]
Pollen protein quality	Willows (*Salix* spp.), cherry, and plum (*Prunus* spp.), and other woody species, offer pollen with high nutritive value.	[60,61,63–65]
Nectar sugar density	Tree and shrub flowers can provide nectar with relatively high sugar content and high flower densities. Hedgerows can provide greater nectar per unit area compared to woodlands and pastures.	[35,56–62]
Butterfly and moth larval hosts	Woody plants are important host plants for the larvae of many lepidopteran species (moths and butterflies). Some of the most highly used plant genera by lepidopteran species include oaks (*Quercus* spp.), cherry and plum (*Prunus* spp.), willows (*Salix* spp.), birch (*Betula* spp.), and poplar (*Populus* spp.).	[12,44,52,112]
Ground-nesting	Agroforestry practices can offer stable sites for ground-nesting bees and wasps in frequently disturbed agricultural landscapes.	[22,32,71,104,108–110]
Cavity-nesting	Shrub species with pithy centers such as elderberry (*Sambucus* spp.), sumac (*Rhus* spp), and brambles (*Rubus* spp.) can provide hollow tunnels for above-ground cavity-nesting bees. Dead trees and branches left in an agroforestry practice can also provide nesting sites.	[22,104,105]
Overwintering	Lepidopteran (butterflies and moths), Coleopteran (beetles), and other pollinators overwinter under bark and leaf litter found in hedgerows and other woody plantings.	[44,112,113]
Microclimate modification: wind	Windbreaks and other agroforestry practices can reduce winds and desiccation of pollen and floral parts, thereby enhancing pollinator foraging. Windbreaks can protect insect flight up to a distance equal to about 9 times its height. Agroforestry practices can help reduce winter mortality in honey bee hives by providing protection from winter winds.	[94,95,98,99]
Microclimate modification: temperature	Trees and shrubs can shade honey bee hives and reduce summer temperatures. Daytime air temperatures are several degrees warmer within a certain distance downwind of windbreaks and these elevated temperatures can increase pollinator activity and pollination if air temperatures at pollination time are below optimum.	[89,91,93]
Connectivity	Hedgerows and other linear agroforestry practices can facilitate pollinator movement across agricultural and urban landscapes at multiple spatial scales. These practices can provide spatially-distributed habitat that is within the foraging range of many pollinators, including short-distance foragers.	[22,53,104,114–120]
Barrier	Hedgerows may act as a barrier to pollinator dispersal and pollen transfer, depending on the landscape context and pollinator species. The orientation of plant rows may influence whether a hedgerow functions as a barrier or corridor.	[116,121–124]

Table 1. *Cont.*

Habitat Component or Ecosystem Service	Summary	Key References
Pesticide spray drift mitigation	Agroforestry practices can reduce pesticide exposure to pollinators by reducing spray drift from coming onto or leaving a farm by capturing particles and reducing wind speed. Windbreaks can reduce drift by up to 80% to 90%. Agroforestry buffers that are 2.5–3 m tall, with 40-50% porosity and fine, evergreen foliage are generally the most effective for drift prevention.	[125–133]
Pesticide runoff mitigation	Agroforestry practices can reduce pesticide exposure to pollinators by helping to capture pesticide runoff, prevent or slow pesticide movement through soil, and help break down some pesticides.	[134–138]
Refuge from pesticides	Agroforestry practices may serve as a safe haven for pollinators from pesticides, if adequately protected from spray drift. No-spray buffer zones may be necessary to protect the agroforestry planting.	[126,130,139–142]
Pesticide accumulation	Plants used in agroforestry practices can become contaminated with pesticides through aerial deposition and uptake through root systems. Plants contaminated by neonicotinoids through non-target drift of treated seed-coating dust during crop planting can negatively impact pollinators.	[143–146]
Adaptation to climate change	Agroforestry practices may offer ecological niches that allow pollinators to find suitable sites for thermal regulation under increasing temperatures and may serve as corridors and stepping stones to facilitate pollinator range shifts due to climate change. Landscapes that have a higher proportion of semi-natural habitats, including hedgerows and other woody plantings, may decrease the detrimental effects of warmer temperatures on pollinators.	[92,147,148]
Crop pollination	Agroforestry practices can provide increased pollination services and crop yields, including higher crop quality, although few studies have been conducted to document this direct agronomic benefit.	[30,46,149–151]

4. Habitat Connectivity

Habitat fragmentation due to agricultural intensification, urban development, and other human activities is negatively impacting pollinators and pollination services at multiple spatial scales [9]. For example, Garibaldi et al. [152] estimated that fruit set of pollinator-dependent crops decreased by 16% at 1 km distance from the nearest pollinator habitat.

4.1. Site Connectivity

Based on field-level studies and modeling efforts, windbreaks, hedgerows, and alley cropping systems can provide pollinator habitat close to crops and at a scale that can benefit foraging [22,29,30,53,103,104,117–120,149]. For instance, Morandin and Kremen [22] observed higher native bee and honey bee numbers in fields adjacent to hedgerows than in fields adjacent to control edges. They also found a decrease in bee numbers as the distance from a hedgerow increased (i.e., 0 to 200 m), suggesting that habitat may need to be closely spaced across a field to promote exportation of bees into a field to support pollination [22]. Windbreaks are typically planted at intervals of 10 to 15 times windbreak height (H) to reduce wind erosion and enhance crop yields through microclimate modification [90,91,153,154]. For example, windbreaks 18 m tall may be spaced 180 m to 270 m across a field (center of field 90 m to 135 m from a windbreak), placing potential habitat within the foraging range of many pollinators, including short-distance foragers [155,156]. None of the studies reviewed found evidence that hedgerows or windbreaks reduced pollinators in crops or crop pollination by concentrating pollinators in the agroforestry practice. Habitat connectivity benefits can be higher when this semi-natural habitat is added to more homogenous and intensely managed fields and landscapes [21,157,158].

4.2. Landscape Connectivity

At the landscape scale, habitat connectivity is important for sustaining pollinator abundance, diversity, reproduction, and dispersal [9]. Agroforestry practices can support connectivity by serving as habitat corridors or stepping stones that facilitate pollinator movement across fragmented landscapes. Evidence documenting this function includes hedgerow-promoted movement of butterflies [123], moths [114], flies [53] and bees [115,116]; and butterfly travel along windbreaks [121] and riparian buffers [48]. A network of hedgerows was found to support wild bee species' richness and functional diversity [71] and the establishment and maintenance of populations at the landscape scale [157]. Syrphid flies were more abundant in forest-connected hedgerows than in forest edges with isolated hedges being intermediate [53]. Connectivity at the landscape scale may also benefit crop production as forest-connected hedgerows were documented to produce more high quality strawberries (*Fragaria x ananassa* Duch.) than isolated hedgerows [151]. Another study found a strong positive relationship between the percentage of hedgerows at the landscape scale (1 km radius) and pollinator visits, resulting in an increase in crop pollination of almost 70% when hedgerow cover increased from 1% to 6% [46].

4.3. Barrier

Agroforestry plantings may act as barriers to some pollinators, inhibiting movement between habitats [121,122]. Hedgerows may channel pollinator movement, which could enhance connectivity but restrict movement across hedgerows, isolating some plant populations [116]. The orientation of plant rows may influence hedgerows' abilities to promote movement or act as barriers [122,123] and keeping hedgerow height lower than 2 m may minimize the barrier effect on butterflies [45]. Pollen flow can also potentially be reduced across hedgerows [116] and possibly other tree-row plantings. Krewenka et al. [159] found that bee foraging was not impacted by hedgerows; however, another study found that bombyliid flies had reduced pollen transfer [124].

5. Pesticide Exposure Mitigation

Pesticides, particularly insecticides, can have acute toxicity leading to pollinator mortality and sublethal effects on growth, health, and behavior [160]. Across agricultural to urban landscapes, pollinators may come into contact with pesticides through numerous exposure pathways, including direct and residue contact, pollen and nectar contaminated by systemic insecticides, contaminated nesting materials, and contaminated water [9]. Agroforestry practices can affect the various exposure pathways and thereby influence the exposure and risk of pesticides to pollinators. Based on existing research, spray drift reduction and runoff mitigation are the primary ways that agroforestry can impact pesticide exposure.

5.1. Spray Drift Mitigation

Pesticide spray drift can be divided into thermal drift (lighter droplets transported to high altitude), vapor drift (volatilization from target), and droplet drift (droplets moved off-target by ambient wind) [129]. By reducing wind speeds and trapping particles, windbreaks and other linear agroforestry plantings can decrease pesticide droplet drift by up to 80%–90% and thereby reduce direct exposure to pollinators [126–129,137,142,161,162]. These practices can be used to minimize drift from either leaving or coming onto a site. Although agroforestry practices may influence thermal or vapor drift, no studies were found on this topic. Hedgerows and windbreaks that are 2.5 m–3 m tall, with 40%–50% porosity and fine, evergreen foliage have been shown to be the most effective for droplet drift reduction [125,127,131–133]. Hedgerows with porosity of nearly 75% have also been found to be effective in reducing drift by more than 80% [129]. Fine, evergreen, coniferous foliage can capture two to four times that of broadleaf species, with the additional benefit of trapping pesticides in early spring before deciduous plants have leafed out [126,131,132,142]. Leaves with hairy, resinous, and coarse surfaces can capture more particles than plants with smooth leaves [125,132].

5.2. Runoff Mitigation

In addition to spray drift, pesticides in solution, seed coating dust, or attached to soil particles can be transported off target by runoff [137]. Agroforestry practices can help reduce pollinator contact with pesticides by reducing runoff and contaminated surface water and by breaking down pesticides into less toxic forms. The perennial vegetation in agroforestry practices can help intercept pesticide-laden runoff, increase infiltration, and aid in phytoremediation of pesticides [24,134–138]. Based on a review of the available studies, Pavlidis and Tsihrintzis [134] documented a 40% to 100% reduction of pesticides (including herbicides) in runoff using agroforestry systems. A meta-analysis by Zhang et al. [136] highlights how sediment captured by vegetative buffers helps improve pesticide removal, particularly those pesticides that are strongly hydrophobic such as pyrethroids and many organophosphates. Plants and their rhizosphere microorganisms vary in their ability to degrade or immobilize pesticides. North American native trees with documented effectiveness in capturing pesticide run-off or immobilizing pesticides within their woody tissue include poplar, willow, birch, alder, black locust, and sycamore (*Platanus* spp.) [134,135].

5.3. Pesticide Accumulation

The same factors that make agroforestry practices effective buffers may lead to pesticide accumulation and pose danger for pollinators, particularly from systemic insecticides and those with long residual activity such as neonicotinoids [143,146]. Nectar and pollen of early-flowering woody species may become contaminated by systemic action of neonicotinoids or through non-target drift of treated seed-coating dust during crop planting [144]. Pesticide droplets, particles, or pesticides adhering to dust can also accumulate in the foliage or at the base of agroforestry practices [145], which pollinators may ingest or carry back to the nest [146]. This evidence suggests that if an agroforestry practice is to function as a buffer from pesticides with long residual activity, it will be important to choose plants that are not attractive to pollinators. If long residual activity is not a concern, avoid using species that flower when pesticides are typically applied.

5.4. Refuge from Pesticides

Agroforestry practices may serve as refugia for pollinators and other beneficial insects if they are well protected from pesticides. No-spray buffer zones adjacent to agroforestry practice have been shown to be an effective strategy to protect these plantings from pesticide deposition [126,130,139–142]. In one study, spray drift deposition in hedgerows was reduced by 72% when a 12 m no-spray buffer zone was used next to the hedgerows [140]. At the landscape level, increasing the proportion of non-cropped habitat (i.e., riparian forest buffers, woodlands) in an agricultural area has been shown to buffer the effects of pesticides on native bees [163].

6. Crop Pollination Services

As described in the previous sections, scientific evidence demonstrates the conservation effects that linear agroforestry practices can provide to insect pollinators, including greater pollinator abundance and richness. Although these benefits should translate into enhanced pollination services leading to increased crop yields and quality, few studies have been conducted to document this direct agronomic benefit. Two studies demonstrated positive effects on canola yields due to hedgerows at local and landscapes scales [46,149], although these studies were conducted with potted plants. The observed yield effects might have resulted from concentration of pollinators and may not scale up to whole fields. For instance, one study found positive effects of hedgerows on pollination that did not result in increased yields in winter oilseed rape (*Brassica napus* L.) [30], while the other study showed no local effects on crop pollination in sunflower (*Helianthus annuus* L.) [34]. In apple orchards, researchers found increased pollinator abundance adjacent to an artificial windbreak, which led to a 20%–30% increase in fruit set with no reduction in fruit size [150]. While the artificial windbreak was created out of coir netting, this study may suggest potential yield increases due to pollinator activity in apple orchards with planted windbreaks.

Many factors are likely to influence the ability of agroforestry practices to promote crop pollination services, including specific pollinator attributes, field size, crop type, vegetative composition of the agroforestry practice, and landscape context [9]. The diversity of interacting variables makes it challenging to conduct studies and develop robust guidance for producers. For instance, the ratio of agroforestry practice to crop area in order to supply sufficient pollination service is largely unexplored [164]. One study demonstrated that native bees can provide full pollination services for watermelon (*Citrullus lanatus* Thunb.) when around 30% of the land within 1.2 km of a field is in natural habitat [165], which could be an approximate analog to an agroforestry practice. Regarding landscape context, one study found an increase in quality and quantity of strawberries grown adjacent to forest-connected hedgerows, as compared to isolated hedgerows or grass margins [151]. Plants placed at forest-connected hedgerows produced more high-quality strawberries with 90% classified as "marketable". In comparison, only 75% of strawberries from plants at isolated hedgerows, 48% of strawberries from plants on grassy margins, and 41% of strawberries from self-pollinated control plants were classified as marketable. Based on market prices of 2016, the increase in economic value between strawberries produced at grassy margins and forest-connected hedgerows amounted to 61% [151]. Cost-benefit studies that assess the benefits of an agroforestry practice for pollination services compared to the costs of installation and maintenance, opportunity costs, and costs of potential unintended negative effects are very limited. One study on hedgerows determined that seven years would be required for farmers to recover implementation costs based on the estimated yield benefits from both pollination and pest control to the crop [149]. With numerous interacting ecosystem services, cost-benefit analyses should consider the suite of agronomic services in order to provide comprehensive economic assessment.

7. Conclusions

Agroforestry has been identified as a multifunctional land use approach that can balance production with environmental stewardship [166]. Combining woody vegetation with cropping and livestock production via agroforestry systems increases total production, enhances food and nutrition security, and protects natural resources, while helping to mitigate and adapt to the effects of climate change [23,166]. Capitalizing on insect-based ecosystem services, agroforestry offers emerging opportunities to support pollinators and other beneficial insects and their services including crop pollination and biological pest management. Based on the available scientific literature, linear agroforestry practices (i.e., windbreaks, hedgerows, riparian buffers, alley cropping) in temperate regions can aid pollinator conservation by providing habitat, including foraging resources and nesting or egg-laying sites, enhancing site and landscape connectivity, and mitigating pesticide exposure. This evidence provides considerations for enhancing agroforestry in supporting pollinators such as the importance of early- to mid-season flowering woody species to provide pollen and nectar. From this review, efforts are currently underway by the Xerces Society for Invertebrate Conservation and the USDA National Agroforestry Center to develop regionally-based recommendations for agricultural producers.

There are still important knowledge gaps on how to design and manage these agroforestry practices to deliver pollinator conservation services. Key gaps to investigate include optimized design and placement of agroforestry practices to benefit pollinators (e.g., best plant combinations to provide foraging resources in space and time, orientation, and distance to target crop), designs to support habitat connectivity (e.g., spacing within the farm landscape, minimizing potential barrier effects), and ongoing maintenance for the long-term health of the practice, balanced with the impacts of management on pollinators (e.g., efforts to increase nesting habitat, effects of brush removal). Regarding silvopasture and forest farming practices, the knowledge base is very limited in temperate regions and could benefit from initial research that assesses opportunities and constraints of providing pollinator conservation benefits with these forest-based practices. For silvopasture, the interactive effects of stocking rate/timing, canopy density, and plant selection will be important variables to assess; while in forest farming, a better understanding of pollination needs of many forest-farmed crops is particularly needed. Although limited, some evidence suggests that agroforestry practices could accumulate pesticides and may inadvertently expose pollinators to higher concentrations. Future investigations should address this concern by better documentation of exposure risk and translocation of different pesticides in plant tissues, nectar, and pollen of agroforestry species to determine if some woody plants pose a greater risk. In addition, research on agroforestry species that best support phytoremediation processes that break down pesticides and analyses of agroecology approaches that combine best management practices reducing pesticide use and exposure along with other farm goals is needed.

Based on this review, agroforestry practices with appropriate species and management can support better pollinator conservation than homogenous fields and farms without agroforestry. However, for producers in temperate regions interested in relying on agroforestry to provide full crop pollination services from insects, agronomic evidence is currently limited for decision making. Crop type, landscape context, plant material used in the agroforestry practice, ratio of agroforestry area to crop area, and spatial distribution of the agroforestry practice are critical variables needing further research in order to better understand the financial impacts of agroforestry on pollination services. Although investigating pollination services is challenging, this information will be key for advancing the use of agroforestry to support this ecosystem service. In addition, cost-benefit studies that assess the economic benefits of an agroforestry practice for crop pollination compared to implementation, management, and opportunity costs need to be conducted. A primary advantage for using agroforestry to support pollinators is that these practices are often being implemented for other production and ecosystem services. These practices often inherently provide some pollinator benefits and with additional considerations during design and management, the effectiveness of agroforestry practices for pollinator conservation and pollination services should be enhanced.

Author Contributions: G.B. conceived and designed the review. G.B., J.H. and N.L.A. collected the literature and conducted the systematic review of available data on the subject matter. G.B., J.H. and N.L.A. wrote the first draft of the manuscript. M.V. reviewed and helped revised various versions of the manuscript.

Funding: This work was supported by funding provided by a contribution agreement with the United Stated Department of Agriculture (USDA)-Natural Resources Conservation Service (NRCS) to the Xerces Society.

Acknowledgments: We would like to acknowledge David Inouye and two anonymous reviewers whose comments and feedback helped to improve the paper.

Conflicts of Interest: The authors declare no conflict of interest.

References

1. Ollerton, J.; Winfree, R.; Tarrant, S. How many flowering plants are pollinated by animals? *Oikos* **2011**, *120*, 321–326. [CrossRef]
2. Klein, A.M.; Vaissière, B.E.; Cane, J.H.; Steffan-Dewenter, I.; Cunningham, S.A.; Kremen, C.; Tscharntke, T. Importance of pollinators in changing landscapes for world crops. *Proc. R. Soc. B Biol. Sci.* **2007**, *274*, 303–313. [CrossRef] [PubMed]
3. Eilers, E.J.; Kremen, C.; Greenleaf, S.S.; Garber, A.K.; Klein, A.-M. Contribution of pollinator-mediated crops to nutrients in the human food supply. *PLoS ONE* **2011**, *6*, e21363. [CrossRef] [PubMed]
4. Kearns, C.A.; Inouye, D.W. Pollinators, flowering plants, and conservation biology. *BioScience* **1997**, *47*, 297–307. [CrossRef]
5. Potts, S.G.; Biesmeijer, J.C.; Kremen, C.; Neumann, P.; Schweiger, O.; Kunin, W.E. Global pollinator declines: Trends, impacts and drivers. *Trends Ecol. Evol.* **2010**, *25*, 345–353. [CrossRef] [PubMed]
6. Potts, S.G.; Imperatriz-Fonseca, V.; Ngo, H.T.; Aizen, M.A.; Biesmeijer, J.C.; Breeze, T.D.; Dicks, L.V.; Garibaldi, L.A.; Hill, R.; Settele, J.; et al. Safeguarding pollinators and their values to human well-being. *Nature* **2016**, *540*, 220–229. [CrossRef]
7. McGregor, S.E. *Insect Pollination of Cultivated Crop Plants*, 1st ed.; United States Department of Agriculture: Washington, DC, USA, 1976.
8. Garibaldi, L.A.; Steffan-Dewenter, I.; Winfree, R.; Aizen, M.A.; Bommarco, R.; Cunningham, S.A.; Kremen, C.; Carvalheiro, L.G.; Harder, L.D.; Afik, O.; et al. Wild pollinators enhance fruit set of crops regardless of honey bee abundance. *Science* **2013**, *339*, 1608–1611. [CrossRef]
9. IPBES. *Assessment Report on Pollinators, Pollination and Food Production*; Intergovernmental Science-Policy Platform on Biodiversity and Ecosystem Services: Bonn, Germany, 2016.
10. Kremen, C.; Williams, N.M.; Thorp, R.W. Crop pollination from native bees at risk from agricultural intensification. *Proc. Natl. Acad. Sci. USA* **2002**, *99*, 16812–16816. [CrossRef]
11. Williams, N.M.; Kremen, C. Resource distributions among habitats determine solitary bee offspring production in a mosaic landscape. *Ecol. Appl.* **2007**, *17*, 910–921. [CrossRef]
12. Tallamy, D.W.; Shropshire, K.J. Ranking lepidopteran use of native versus introduced plants. *Conserv. Biol.* **2009**, *23*, 941–947. [CrossRef]
13. Fiedler, A.K.; Landis, D.A.; Arduser, M. Rapid shift in pollinator communities following invasive species removal. *Restor. Ecol.* **2012**, *20*, 593–602. [CrossRef]
14. Kevan, P.G. Pollinators as bioindicators of the state of the environment: Species, activity and diversity. *Agric. Ecosyst. Environ.* **1999**, *74*, 373–393. [CrossRef]
15. Dover, J.; Sotherton, N.; Gobbett, K. Reduced pesticide inputs on cereal field margins: The effects on butterfly abundance. *Ecol. Entomol.* **1990**, *15*, 17–24. [CrossRef]
16. Whitehorn, P.R.; O'Connor, S.; Wackers, F.L.; Goulson, D. Neonicotinoid pesticide reduces bumble bee colony growth and queen production. *Science* **2012**, *336*, 351–352. [CrossRef]
17. Colla, S.R.; Otterstatter, M.C.; Gegear, R.J.; Thomson, J.D. Plight of the bumble bee: Pathogen spillover from commercial to wild populations. *Biol. Conserv.* **2006**, *129*, 461–467. [CrossRef]

18. Cameron, S.A.; Lozier, J.D.; Strange, J.P.; Koch, J.B.; Cordes, N.; Solter, L.F.; Griswold, T.L. Patterns of widespread decline in North American bumble bees. *Proc. Natl. Acad. Sci. USA* **2011**, *108*, 662–667. [CrossRef]

19. Kearns, C.A.; Inouye, D.W.; Waser, N.M. Endangered mutualisms: The conservation of plant-pollinator interactions. *Annu. Rev. Ecol. Syst.* **1998**, *29*, 83–112. [CrossRef]

20. Steffan-Dewenter, I.; Westphal, C. The interplay of pollinator diversity, pollination services and landscape change. *J. Appl. Ecol.* **2008**, *45*, 737–741. [CrossRef]

21. Klein, A.-M.; Brittain, C.; Hendrix, S.D.; Thorp, R.; Williams, N.; Kremen, C. Wild pollination services to California almond rely on semi-natural habitat. *J. Appl. Ecol.* **2012**, *49*, 723–732. [CrossRef]

22. Morandin, L.A.; Kremen, C. Hedgerow restoration promotes pollinator populations and exports native bees to adjacent fields. *Ecol. Appl.* **2013**, *23*, 829–839. [CrossRef]

23. Schoeneberger, M.M.; Bentrup, G.; Patel-Weynand, T. *Agroforestry: Enhancing Resiliency in U.S. Agricultural landscapes under Changing Conditions*; U.S. Department of Agriculture, U.S. Forest Service: Washington, DC, USA, 2017.

24. Jose, S. Agroforestry for ecosystem services and environmental benefits: An overview. *Agrofor. Syst.* **2009**, *76*, 1–10. [CrossRef]

25. Udawatta, R.P.; Rankoth, L.M.; Jose, S. Agroforestry and biodiversity. *Sustainability* **2019**, *11*, 2879. [CrossRef]

26. Moher, D. Preferred reporting items for systematic reviews and meta-analyses: The PRISMA statement. *Ann. Intern. Med.* **2009**, *151*, 264. [CrossRef] [PubMed]

27. O'Brien, A.M.; Mc Guckin, C. *The Systematic Literature Review Method: Trials and Tribulations of Electronic Database Searching at Doctoral Level*; SAGE Publications: London, UK, 2016.

28. Morandin, L.A.; Kremen, C. Bee preference for native versus exotic plants in restored agricultural hedgerows. *Restor. Ecol.* **2013**, *21*, 26–32. [CrossRef]

29. Garratt, M.P.D.; Senapathi, D.; Coston, D.J.; Mortimer, S.R.; Potts, S.G. The benefits of hedgerows for pollinators and natural enemies depends on hedge quality and landscape context. *Agric. Ecosyst. Environ.* **2017**, *247*, 363–370. [CrossRef]

30. Sutter, L.; Albrecht, M.; Jeanneret, P. Landscape greening and local creation of wildflower strips and hedgerows promote multiple ecosystem services. *J. Appl. Ecol.* **2018**, *55*, 612–620. [CrossRef]

31. Minarro, M.; Prida, E. Hedgerows surrounding organic apple orchards in north-west Spain: Potential to conserve beneficial insects. *Agric. For. Entomol.* **2013**, *15*, 382–390. [CrossRef]

32. Ponisio, L.C.; Gaiarsa, M.P.; Kremen, C. Opportunistic attachment assembles plant-pollinator networks. *Ecol. Lett.* **2017**, *20*, 1261–1272. [CrossRef]

33. Rollin, O.; Bretagnolle, V.; Decourtye, A.; Aptel, J.; Michel, N.; Vaissiere, B.E.; Henry, M. Differences of floral resource use between honey bees and wild bees in an intensive farming system. *Agric. Ecosyst. Environ.* **2013**, *179*, 78–86. [CrossRef]

34. Sardiñas, H.S.; Kremen, C. Pollination services from field-scale agricultural diversification may be context-dependent. *Agric. Ecosyst. Environ.* **2015**, *207*, 17–25. [CrossRef]

35. Timberlake, T.P.; Vaughan, I.P.; Memmott, J. Phenology of farmland floral resources reveals seasonal gaps in nectar availability for bumblebees. *J. Appl. Ecol.* **2019**, *56*, 1585–1596. [CrossRef]

36. Varah, A.; Jones, H.; Smith, J.; Potts, S.G. Enhanced biodiversity and pollination in UK agroforestry systems. *J. Sci. Food Agric.* **2013**, *93*, 2073–2075. [CrossRef] [PubMed]

37. Cole, L.J.; Brocklehurst, S.; Robertson, D.; Harrison, W.; McCracken, D.I. Exploring the interactions between resource availability and the utilisation of semi-natural habitats by insect pollinators in an intensive agricultural landscape. *Agric. Ecosyst. Environ.* **2017**, *246*, 157–167. [CrossRef]

38. Hannon, L.E.; Sisk, T.D. Hedgerows in an agri-natural landscape: Potential habitat value for native bees. *Biol. Conserv.* **2009**, *142*, 2140–2154. [CrossRef]

39. Hanley, M.E.; Franco, M.; Dean, C.E.; Franklin, E.L.; Harris, H.R.; Haynes, A.G.; Rapson, S.R.; Rowse, G.; Thomas, K.C.; Waterhouse, B.R.; et al. Increased bumblebee abundance along the margins of a mass flowering crop: Evidence for pollinator spill-over. *Oikos* **2011**, *120*, 1618–1624. [CrossRef]

40. Kremen, C.; M'Gonigle, L.K.; Ponisio, L.C. Pollinator community assembly tracks changes in floral resources as restored hedgerows mature in agricultural landscapes. *Front. Ecol. Evol.* **2018**, *6*, 170. [CrossRef]

41. M'Gonigle, L.K.; Ponisio, L.C.; Cutler, K.; Kremen, C. Habitat restoration promotes pollinator persistence and colonization in intensively managed agriculture. *Ecol. Appl.* **2015**, *25*, 1557–1565. [CrossRef]

42. Whitaker, D.; Carroll, A.; Montevecchi, W. Elevated numbers of flying insects and insectivorous birds in riparian buffer strips. *Can. J. Zool. Rev. Can. Zool.* **2000**, *78*, 740–747. [CrossRef]

43. Sobczyk, D. Butterflies (Lepidoptera) of young midfield shelterbelts. *Pol. J. Ecol.* **2004**, *52*, 449–453.

44. Dover, J.; Sparks, T. A review of the ecology of butterflies in British hedgerows. *J. Environ. Manag.* **2000**, *60*, 51–63. [CrossRef]

45. Luppi, M.; Dondina, O.; Orioli, V.; Bani, L. Local and landscape drivers of butterfly richness and abundance in a human-dominated area. *Agric. Ecosyst. Environ.* **2018**, *254*, 138–148. [CrossRef]

46. Dainese, M.; Montecchiari, S.; Sitzia, T.; Sigura, M.; Marini, L. High cover of hedgerows in the landscape supports multiple ecosystem services in Mediterranean cereal fields. *J. Appl. Ecol.* **2017**, *54*, 380–388. [CrossRef]

47. Dover, J.W.; Sparks, T.H.; Greatorex-Davies, J.N. The importance of shelter for butterflies in open landscapes. *J. Insect Conserv.* **1997**, *1*, 89–97. [CrossRef]

48. Meier, K.; Kuusemets, V.; Luig, J.; Mander, U. Riparian buffer zones as elements of ecological networks: Case study on *Parnassius mnemosyne* distribution in Estonia. *Ecol. Eng.* **2005**, *24*, 531–537. [CrossRef]

49. Rosas-Ramos, N.; Banos-Picon, L.; Tormos, J.; Asis, J.D. The complementarity between ecological infrastructure types benefits natural enemies and pollinators in a Mediterranean vineyard agroecosystem. *Ann. Appl. Biol.* **2019**, *175*, 193–201. [CrossRef]

50. Froidevaux, J.S.P.; Broyles, M.; Jones, G. Moth responses to sympathetic hedgerow management in temperate farmland. *Agric. Ecosyst. Environ.* **2019**, *270*, 55–64. [CrossRef]

51. Merckx, T.; Feber, R.E.; Mclaughlan, C.; Bourn, N.A.D.; Parsons, M.S.; Townsend, M.C.; Riordan, P.; Macdonald, D.W. Shelter benefits less mobile moth species: The field-scale effect of hedgerow trees. *Agric. Ecosyst. Environ.* **2010**, *138*, 147–151. [CrossRef]

52. Merckx, T.; Marini, L.; Feber, R.E.; Macdonald, D.W. Hedgerow trees and extended-width field margins enhance macro-moth diversity: Implications for management. *J. Appl. Ecol.* **2012**, *49*, 1396–1404. [CrossRef]

53. Haenke, S.; Kovács-Hostyánszki, A.; Fründ, J.; Batáry, P.; Jauker, B.; Tscharntke, T.; Holzschuh, A. Landscape configuration of crops and hedgerows drives local syrphid fly abundance. *J. Appl. Ecol.* **2014**, *51*, 505–513. [CrossRef]

54. Schirmel, J.; Albrecht, M.; Bauer, P.-M.; Sutter, L.; Pfister, S.C.; Entling, M.H. Landscape complexity promotes hoverflies across different types of semi-natural habitats in farmland. *J. Appl. Ecol.* **2018**, *55*, 1747–1758. [CrossRef]

55. Macdonald, K.J.; Kelly, D.; Tylianakis, J.M. Do local landscape features affect wild pollinator abundance, diversity and community composition on Canterbury farms? *N. Z. J. Ecol.* **2018**, *42*, 262–268. [CrossRef]

56. Stubbs, C.S.; Jacobson, H.A.; Osgood, E.A.; Drummond, F.A. *TB148: Alternative Forage Plants for Native (Wild) Bees Associated with Lowbush Blueberry, Vaccinium spp., in Maine*; Maine Agricultural & Forest Experiment Station: Orono, ME, USA, 1992.

57. Loose, J.L.; Drummond, F.A.; Stubbs, C.; Woods, S. *Conservation and Management of Native Bees in Cranberry*; Maine Agricultural & Forest Experiment Station: Orono, ME, USA, 2005.

58. Baude, M.; Kunin, W.E.; Boatman, N.D.; Conyers, S.; Davies, N.; Gillespie, M.A.K.; Morton, R.D.; Smart, S.M.; Memmott, J. Historical nectar assessment reveals the fall and rise of floral resources in Britain. *Nature* **2016**, *530*, 85–88. [CrossRef] [PubMed]

59. Somme, L.; Moquet, L.; Quinet, M.; Vanderplanck, M.; Michez, D.; Lognay, G.; Jacquemart, A.-L. Food in a row: Urban trees offer valuable floral resources to pollinating insects. *Urban Ecosyst.* **2016**, *19*, 1149–1161. [CrossRef]

60. Batra, S.W.T. Red maple (*Acer rubrum* l.), an important early spring food resource for honey bees and other insects. *J. Kans. Entomol. Soc.* **1985**, *58*, 169–172.

61. Ostaff, D.P.; Mosseler, A.; Johns, R.C.; Javorek, S.; Klymko, J.; Ascher, J.S. Willows (*Salix* spp.) as pollen and nectar sources for sustaining fruit and berry pollinating insects. *Can. J. Plant Sci.* **2015**, *95*, 505–516. [CrossRef]

62. Crane, E.; Walker, P. Some nectar characteristics of certain important honey sources. *Pszczel. Zesz. Naukowe* **1985**, *29*, 29–45.

63. Tasei, J.-N.; Aupinel, P. Nutritive value of 15 single pollens and pollen mixes tested on larvae produced by bumblebee workers (*Bombus terrestris*, Hymenoptera: Apidae). *Apidologie* **2008**, *39*, 397–409. [CrossRef]

64. Di Pasquale, G.; Salignon, M.; Le Conte, Y.; Belzunces, L.P.; Decourtye, A.; Kretzschmar, A.; Suchail, S.; Brunet, J.-L.; Alaux, C. Influence of pollen nutrition on honey bee health: Do pollen quality and diversity matter? *PLoS ONE* **2013**, *8*, e72016. [CrossRef]

65. Russo, L.; Danforth, B. Pollen preferences among the bee species visiting apple (*Malus pumila*) in New York. *Apidologie* **2017**, *48*, 806–820. [CrossRef]

66. Filipiak, M. Key pollen host plants provide balanced diets for wild bee larvae: A lesson for planting flower strips and hedgerows. *J. Appl. Ecol.* **2019**, *56*, 1410–1418. [CrossRef]

67. Fowler, J. Specialist bees of the Northeast: Host plants and habitat conservation. *Northeast. Nat.* **2016**, *23*, 305–320. [CrossRef]

68. Dötterl, S.; Vereecken, N.J. The chemical ecology and evolution of bee–flower interactions: A review and perspectives. *Can. J. Zool.* **2010**, *88*, 668–697. [CrossRef]

69. Carvell, C.; Bourke, A.F.G.; Dreier, S.; Freeman, S.N.; Hulmes, S.; Jordan, W.C.; Redhead, J.W.; Sumner, S.; Wang, J.; Heard, M.S. Bumblebee family lineage survival is enhanced in high-quality landscapes. *Nature* **2017**, *543*, 547–549. [CrossRef] [PubMed]

70. Wood, T.J.; Kaplan, I.; Szendrei, Z. Wild bee pollen diets reveal patterns of seasonal foraging resources for honey bees. *Front. Ecol. Evol.* **2018**, *6*, 210. [CrossRef]

71. Ponisio, L.C.; M'Gonigle, L.K.; Kremen, C. On-farm habitat restoration counters biotic homogenization in intensively managed agriculture. *Glob. Chang. Biol.* **2016**, *22*, 704–715. [CrossRef]

72. Donkersley, P. Trees for bees. *Agric. Ecosyst. Environ.* **2019**, *270–271*, 79–83. [CrossRef]

73. Staley, J.T.; Botham, M.S.; Amy, S.R.; Hulmes, S.; Pywell, R.F. Experimental evidence for optimal hedgerow cutting regimes for Brown hairstreak butterflies. *Insect Conserv. Divers.* **2018**, *11*, 213–218. [CrossRef]

74. Staley, J.T.; Botham, M.S.; Chapman, R.E.; Amy, S.R.; Heard, M.S.; Hulmes, L.; Savage, J.; Pywell, R.F. Little and late: How reduced hedgerow cutting can benefit Lepidoptera. *Agric. Ecosyst. Environ.* **2016**, *224*, 22–28. [CrossRef]

75. Policarová, J.; Cardinal, S.; Martins, A.C.; Straka, J. The role of floral oils in the evolution of apid bees (Hymenoptera: Apidae). *Biol. J. Linn. Soc.* **2019**, *128*, 486–497. [CrossRef]

76. Portman, Z.M.; Orr, M.C.; Griswold, T. A review and updated classification of pollen gathering behavior in bees (Hymenoptera, Apoidea). *J. Hymenopt. Res.* **2019**, *71*, 171–208. [CrossRef]

77. Cane, J.H.; Griswold, T.; Parker, F.D. Substrates and materials used for nesting by North American *Osmia* bees (Hymenoptera: Apiformes: Megachilidae). *Ann. Entomol. Soc. Am.* **2007**, *100*, 350–358. [CrossRef]

78. Wcislo, W.T.; Cane, J.H. Floral resource utilization by solitary bees (Hymenoptera: Apoidea) and exploitation of their stored foods by natural enemies. *Annu. Rev. Entomol.* **1996**, *41*, 257–286. [CrossRef] [PubMed]

79. O'Brien, M. Notes on *Dianthidium Simile* (Cresson) (Hymenoptera: Megachilidae) in Michigan. *Gt. Lakes Entomol.* **2018**, *40*, 1.

80. Simone-Finstrom, M.; Spivak, M. Propolis and bee health: The natural history and significance of resin use by honey bees. *Apidologie* **2010**, *41*, 295–311. [CrossRef]

81. Simone-Finstrom, M.; Borba, R.S.; Wilson, M.; Spivak, M. Propolis counteracts some threats to honey bee health. *Insects* **2017**, *8*, 46. [CrossRef] [PubMed]

82. Greenaway, W.; Scaysbrook, T.; Whatley, F.R. The composition and plant origins of propolis: A report of work at Oxford. *Bee World* **1990**, *71*, 107–118. [CrossRef]

83. Bankova, V.S.; de Castro, S.L.; Marcucci, M.C. Propolis: Recent advances in chemistry and plant origin. *Apidologie* **2000**, *31*, 3–15. [CrossRef]

84. König, B. Plant sources of propolis. *Bee World* **1985**, *66*, 136–139. [CrossRef]

85. Drescher, N.; Klein, A.-M.; Schmitt, T.; Leonhardt, S.D. A clue on bee glue: New insight into the sources and factors driving resin intake in honeybees (*Apis mellifera*). *PLoS ONE* **2019**, *14*, e0210594. [CrossRef]

86. Ghisalberti, E.L. Propolis: A review. *Bee World* **1979**, *60*, 59–84. [CrossRef]

87. Corbet, S.A. Pollination and the weather. *Israel J. Plant Sci.* **1990**, *39*, 13–30.

88. Vicens, N.; Bosch, J. Weather-dependent pollinator activity in an apple orchard, with special reference to *Osmia cornuta* and *Apis mellifera* (Hymenoptera: Megachilidae and Apidae). *Environ. Entomol.* **2000**, *29*, 413–420. [CrossRef]

89. McNaughton, K.G. Effects of windbreaks on turbulent transport and microclimate. *Agric. Ecosyst. Environ.* **1988**, *22*, 17–39. [CrossRef]

90. Baldwin, C.S. The influence of field windbreaks on vegetable and specialty crops. *Agric. Ecosyst. Environ.* **1988**, *22*, 191–203. [CrossRef]

91. Norton, R.L. Windbreaks: Benefits to orchard and vineyard crops. *Agric. Ecosyst. Environ.* **1988**, *22*, 205–213. [CrossRef]

92. Papanikolaou, A.D.; Kühn, I.; Frenzel, M.; Schweiger, O. Semi-natural habitats mitigate the effects of temperature rise on wild bees. *J. Appl. Ecol.* **2017**, *54*, 527–536. [CrossRef]

93. Hill, D.B.; Webster, T.C. Apiculture and forestry (bees and trees). *Agrofor. Syst.* **1995**, *29*, 313–320. [CrossRef]

94. Haydak, M.H. Wintering of bees in Minnesota. *J. Econ. Entomol.* **1958**, *51*, 332–334. [CrossRef]

95. Merrill, J.H. Value of winter protection for bees. *J. Econ. Entomol.* **1923**, *16*, 125–130. [CrossRef]

96. Brittain, C.; Kremen, C.; Klein, A.-M. Biodiversity buffers pollination from changes in environmental conditions. *Glob. Chang. Biol.* **2013**, *19*, 540–547. [CrossRef]

97. Lewis, T. Patterns of distribution of insects near a windbreak of tall trees. *Ann. Appl. Biol.* **1970**, *65*, 213–220. [CrossRef]

98. Pinzauti, M. The influence of the wind on nectar secretion from the melon and on the flight of bees: The use of an artificial wind-break. *Apidologie* **1986**, *17*, 63–72. [CrossRef]

99. Pasek, J.E. Influence of wind and windbreaks on local dispersal of insects. *Agric. Ecosyst. Environ.* **1988**, *22–23*, 539–554. [CrossRef]

100. Peri, P.L.; Bloomberg, M. Windbreaks in southern Patagonia, Argentina: A review of research on growth models, windspeed reduction, and effects on crops. *Agrofor. Syst.* **2002**, *56*, 129–144. [CrossRef]

101. Potts, S.G.; Vulliamy, B.; Roberts, S.; O'Toole, C.; Dafni, A.; Ne'eman, G.; Willmer, P. Role of nesting resources in organising diverse bee communities in a Mediterranean landscape. *Ecol. Entomol.* **2005**, *30*, 78–85. [CrossRef]

102. Sardiñas, H.S.; Ponisio, L.C.; Kremen, C. Hedgerow presence does not enhance indicators of nest-site habitat quality or nesting rates of ground-nesting bees. *Restor. Ecol.* **2016**, *24*, 499–505. [CrossRef]

103. Sardiñas, H.S.; Tom, K.; Ponisio, L.C.; Rominger, A.; Kremen, C. Sunflower (*Helianthus annuus*) pollination in California's Central Valley is limited by native bee nest site location. *Ecol. Appl.* **2016**, *26*, 438–447. [CrossRef]

104. Kremen, C.; M'Gonigle, L.K. Small-scale restoration in intensive agricultural landscapes supports more specialized and less mobile pollinator species. *J. Appl. Ecol.* **2015**, *52*, 602–610. [CrossRef]

105. Kay, S.; Kühn, E.; Albrecht, M.; Sutter, L.; Szerencsits, E.; Herzog, F. Agroforestry can enhance foraging and nesting resources for pollinators with focus on solitary bees at the landscape scale. *Agrofor. Syst.* **2019**, 1–9. [CrossRef]

106. Shuler, R.E.; Roulston, T.H.; Farris, G.E. Farming practices influence wild pollinator populations on squash and pumpkin. *J. Econ. Entomol.* **2005**, *98*, 790–795. [CrossRef]

107. Kim, J.; Williams, N.; Kremen, C. Effects of Cultivation and Proximity to Natural Habitat on Ground-nesting Native Bees in California Sunflower Fields. *J. Kans. Entomol. Soc.* **2006**, *79*, 309–320. [CrossRef]

108. Svensson, B.; Lagerlöf, J.; Svensson, B.G. Habitat preferences of nest-seeking bumble bees (Hymenoptera: Apidae) in an agricultural landscape. *Agric. Ecosyst. Environ.* **2000**, *77*, 247–255. [CrossRef]

109. Kells, A.R.; Goulson, D. Preferred nesting sites of bumblebee queens (Hymenoptera: Apidae) in agroecosystems in the UK. *Biol. Conserv.* **2003**, *109*, 165–174. [CrossRef]

110. Osborne, J.L.; Martin, A.P.; Shortall, C.R.; Todd, A.D.; Goulson, D.; Knight, M.E.; Hale, R.J.; Sanderson, R.A. Quantifying and comparing bumblebee nest densities in gardens and countryside habitats. *J. Appl. Ecol.* **2008**, *45*, 784–792. [CrossRef]

111. Lye, G.; Park, K.; Osborne, J.; Holland, J.; Goulson, D. Assessing the value of Rural Stewardship schemes for providing foraging resources and nesting habitat for bumblebee queens (Hymenoptera: Apidae). *Biol. Conserv.* **2009**, *142*, 2023–2032. [CrossRef]

112. Maudsley, M. A review of the ecology and conservation of hedgerow invertebrates in Britain. *J. Environ. Manag.* **2000**, *60*, 65–76. [CrossRef]

113. Pywell, R.F.; James, K.L.; Herbert, I.; Meek, W.R.; Carvell, C.; Bell, D.; Sparks, T.H. Determinants of overwintering habitat quality for beetles and spiders on arable farmland. *Biol. Conserv.* **2005**, *123*, 79–90. [CrossRef]

114. Coulthard, E.; McCollin, D.; Littlemore, J. The use of hedgerows as flight paths by moths in intensive farmland landscapes. *J. Insect Conserv.* **2016**, *20*, 345–350. [CrossRef]

115. Cranmer, L.; McCollin, D.; Ollerton, J. Landscape structure influences pollinator movements and directly affects plant reproductive success. *Oikos* **2012**, *121*, 562–568. [CrossRef]

116. Klaus, F.; Bass, J.; Marholt, L.; Müller, B.; Klatt, B.; Kormann, U. Hedgerows have a barrier effect and channel pollinator movement in the agricultural landscape. *J. Landsc. Ecol.* **2015**, *8*, 22–31. [CrossRef]

117. Graham, J.B.; Nassauer, J.I. Wild bee abundance in temperate agroforestry landscapes: Assessing effects of alley crop composition, landscape configuration, and agroforestry area. *Agrofor. Syst.* **2019**, *93*, 837–850. [CrossRef]

118. Foeldesi, R.; Kovacs-Hostyanszki, A. Hoverfly (Diptera: Syrphidae) community of a cultivated arable field and the adjacent hedgerow near Debrecen, Hungary. *Biologia* **2014**, *69*, 381–388. [CrossRef]

119. Rands, S.A.; Whitney, H.M. Field margins, foraging distances and their impacts on nesting pollinator success. *PLoS ONE* **2011**, *6*, e25971. [CrossRef] [PubMed]

120. Walther-Hellwig, K.; Frankl, R. Foraging habitats and foraging distances of bumblebees, *Bombus* spp. (Hym., Apidae), in an agricultural landscape. *J. Appl. Entomol.* **2000**, *124*, 299–306. [CrossRef]

121. Dover, J.; Fry, G.L.A. Experimental simulation of some visual and physical components of a hedge and the effects on butterfly behaviour in an agricultural landscape. *Entomol. Exp. Appl.* **2001**, *100*, 221–233. [CrossRef]

122. Wratten, S.; Bowie, M.; Hickman, J.; Evans, A.; Sedcole, J.; Tylianakis, J. Field boundaries as barriers to movement of hover flies (Diptera: Syrphidae) in cultivated land. *Oecologia* **2003**, *134*, 605–611. [CrossRef]

123. Ouin, A.; Burel, F. Influence of herbaceous elements on butterfly diversity in hedgerow agricultural landscapes. *Agric. Ecosyst. Environ.* **2002**, *93*, 45–53. [CrossRef]

124. Campagne, P.; Affre, L.; Baumel, A.; Roche, P.; Tatoni, T. Fine-scale response to landscape structure in *Primula vulgaris* Huds.: Does hedgerow network connectedness ensure connectivity through gene flow? *Popul. Ecol.* **2009**, *51*, 209–219. [CrossRef]

125. Ucar, T.; Hall, F.R.; Tew, J.E.; Hacker, J.K. Wind tunnel studies on spray deposition on leaves of tree species used for windbreaks and exposure of honey bees. *Pest Manag. Sci.* **2003**, *59*, 358–364. [CrossRef]

126. Ucar, T.; Hall, F.R. Windbreaks as a pesticide drift mitigation strategy: A review. *Pest Manag. Sci.* **2001**, *57*, 663–675. [CrossRef]

127. Otto, S.; Loddo, D.; Baldoin, C.; Zanin, G. Spray drift reduction techniques for vineyards in fragmented landscapes. *J. Environ. Manag.* **2015**, *162*, 290–298. [CrossRef] [PubMed]

128. Otto, S.; Lazzaro, L.; Finizio, A.; Zanin, G. Estimating ecotoxicological effects of pesticide drift on nontarget arthropods in field hedgerows. *Environ. Toxicol. Chem.* **2009**, *28*, 853–863. [CrossRef] [PubMed]

129. Lazzaro, L.; Otto, S.; Zanin, G. Role of hedgerows in intercepting spray drift: Evaluation and modelling of the effects. *Agric. Ecosyst. Environ.* **2008**, *123*, 317–327. [CrossRef]

130. Longley, M.; Čilgi, T.; Jepson, P.C.; Sotherton, N.W. Measurements of pesticide spray drift deposition into field boundaries and hedgerows: 1. Summer applications. *Environ. Toxicol. Chem.* **1997**, *16*, 165–172. [CrossRef]

131. Wenneker, M.; Heijne, B.; van de Zande, J.C. Effect of natural windbreaks on drift reduction in orchard spraying. *Commun. Agric. Appl. Biol. Sci.* **2005**, *70*, 961–969.

132. Chen, L.; Liu, C.; Zhang, L.; Zou, R.; Zhang, Z. Variation in tree species ability to capture and retain airborne fine particulate matter (pm 2.5). *Sci. Rep.* **2017**, *7*, 1–11.

133. Mercer, G.N. Modelling to determine the optimal porosity of shelterbelts for the capture of agricultural spray drift. *Environ. Model. Softw.* **2009**, *24*, 1349–1352. [CrossRef]

134. Pavlidis, G. Pollution control by agroforestry systems: A short review. *Eur. Water* **2017**, *59*, 297–301.

135. Pavlidis, G.; Tsihrintzis, V.A. Environmental benefits and control of pollution to surface water and groundwater by agroforestry systems: A review. *Water Resour. Manag.* **2018**, *32*, 1–29. [CrossRef]

136. Zhang, X.; Liu, X.; Zhang, M.; Dahlgren, R.A.; Eitzel, M. A review of vegetated buffers and a meta-analysis of their mitigation efficacy in reducing nonpoint source pollution. *J. Environ. Qual.* **2010**, *39*, 76–84. [CrossRef]

137. Reichenberger, S.; Bach, M.; Skitschak, A.; Frede, H.-G. Mitigation strategies to reduce pesticide inputs into ground- and surface water and their effectiveness; A review. *Sci. Total Environ.* **2007**, *384*, 1–35. [CrossRef] [PubMed]

138. Chaudhry, Q.; Blom-Zandstra, M.; Gupta, S.; Joner, E.J. Utilising the synergy between plants and rhizosphere microorganisms to enhance breakdown of organic pollutants in the environment. *Environ. Sci. Pollut. Res. Int.* **2005**, *12*, 34–48. [CrossRef] [PubMed]

139. Longley, M.; Sotherton, N.W. Measurements of pesticide spray drift deposition into field boundaries and hedgerows: 2. Autumn applications. *Environ. Toxicol. Chem.* **1997**, *16*, 173–178. [CrossRef]

140. Kjær, C.; Bruus, M.; Bossi, R.; Løfstrøm, P.; Andersen, H.V.; Nuyttens, D.; Larsen, S.E. Pesticide drift deposition in hedgerows from multiple spray swaths. *J. Pestic. Sci.* **2014**, *39*, 14–21.

141. Davis, B.N.K.; Williams, C.T. Buffer zone widths for honeybees from ground and aerial spraying of insecticides. *Environ. Pollut.* **1990**, *63*, 247–259. [CrossRef]

142. Felsot, A.S.; Unsworth, J.B.; Linders, J.B.H.J.; Roberts, G.; Rautman, D.; Harris, C.; Carazo, E. Agrochemical spray drift; assessment and mitigation-a review. *J. Environ. Sci. Health Part. B Pestic. Food Contam. Agric. Wastes* **2011**, *46*, 1–23. [CrossRef] [PubMed]

143. Botías, C.; David, A.; Hill, E.M.; Goulson, D. Contamination of wild plants near neonicotinoid seed-treated crops, and implications for non-target insects. *Sci. Total Environ.* **2016**, *566*, 269–278. [CrossRef]

144. Long, E.Y.; Krupke, C.H. Non-cultivated plants present a season-long route of pesticide exposure for honey bees. *Nat. Commun.* **2016**, *7*, 1–12. [CrossRef]

145. Zaady, E.; Katra, I.; Shuker, S.; Knoll, Y.; Shlomo, S. Tree belts for decreasing aeolian dust-carried pesticides from cultivated areas. *Geosciences* **2018**, *8*, 286. [CrossRef]

146. Krupke, C.H.; Hunt, G.J.; Eitzer, B.D.; Andino, G.; Given, K. Multiple routes of pesticide exposure for honey bees living near agricultural fields. *PLoS ONE* **2012**, *7*, e29268. [CrossRef]

147. Gilchrist, A.; Barker, A.; Handley, J.F. Pathways through the landscape in a changing climate: The role of landscape structure in facilitating species range expansion through an urbanised region. *Landsc. Res.* **2016**, *41*, 26–44. [CrossRef]

148. Krosby, M.; Tewksbury, J.; Haddad, N.M.; Hoekstra, J. Ecological connectivity for a changing climate. *Conserv. Biol.* **2010**, *24*, 1686–1689. [CrossRef] [PubMed]

149. Morandin, L.A.; Long, R.F.; Kremen, C. Pest control and pollination cost-benefit analysis of hedgerow restoration in a simplified agricultural landscape. *J. Econ. Entomol.* **2016**, *109*, 1020–1027. [CrossRef] [PubMed]

150. Smith, B.D.; Lewis, T. The effects of windbreaks on the blossom-visiting fauna of apple orchards and on yield. *Ann. Appl. Biol.* **1972**, *72*, 229–238. [CrossRef]

151. Castle, D.; Grass, I.; Westphal, C. Fruit quantity and quality of strawberries benefit from enhanced pollinator abundance at hedgerows in agricultural landscapes. *Agric. Ecosyst. Environ.* **2019**, *275*, 14–22. [CrossRef]

152. Garibaldi, L.A.; Steffan-Dewenter, I.; Kremen, C.; Morales, J.M.; Bommarco, R.; Cunningham, S.A.; Carvalheiro, L.G.; Chacoff, N.P.; Dudenhöffer, J.H.; Greenleaf, S.S.; et al. Stability of pollination services decreases with isolation from natural areas despite honey bee visits. *Ecol. Lett.* **2011**, *14*, 1062–1072. [CrossRef]

153. Kort, J. Benefits of windbreaks to field and forage crops. *Agric. Ecosyst. Environ.* **1988**, *22*, 165–190. [CrossRef]

154. Ticknor, K.A. Design and use of field windbreaks in wind erosion control systems. *Agric. Ecosyst. Environ.* **1988**, *22*, 123–132. [CrossRef]

155. Gathmann, A.; Tscharntke, T. Foraging ranges of solitary bees. *J. Anim. Ecol.* **2002**, *71*, 757–764. [CrossRef]

156. Moisan-DeSerres, J.; Chagnon, M.; Fournier, V. Influence of windbreaks and forest borders on abundance and species richness of native pollinators in lowbush blueberry fields in Québec, Canada. *Can. Entomol.* **2015**, *147*, 432–442. [CrossRef]

157. Ponisio, L.C.; de Valpine, P.; M'Gonigle, L.K.; Kremen, C. Proximity of restored hedgerows interacts with local floral diversity and species' traits to shape long-term pollinator metacommunity dynamics. *Ecol. Lett.* **2019**, *22*, 1048–1060. [CrossRef] [PubMed]

158. Diekoetter, T.; Billeter, R.; Crist, T.O. Effects of landscape connectivity on the spatial distribution of insect diversity in agricultural mosaic landscapes. *Basic Appl. Ecol.* **2008**, *9*, 298–307. [CrossRef]

159. Krewenka, K.M.; Holzschuh, A.; Tscharntke, T.; Dormann, C.F. Landscape elements as potential barriers and corridors for bees, wasps and parasitoids. *Biol. Conserv.* **2011**, *144*, 1816–1825. [CrossRef]

160. Stanley, J.; Preetha, G. *Pesticide Toxicity to Non-Target Organisms: Exposure, Toxicity and Risk Assessment Methodologies*; Springer: Dordrecht, The Netherlands, 2016.

161. Bischof, M.; Coffey, T.; Drennan, M. The effectiveness of riparian hedgerows at intercepting drift from aerial pesticide application. *J. Environ. Qual.* **2019**, *48*, 1481–1488.

162. Davis, B.N.K.; Brown, M.J.; Frost, A.J.; Yates, T.J.; Plant, R.A. The effects of hedges on spray deposition and on the biological impact of pesticide spray drift. *Ecotoxicol. Environ. Saf.* **1994**, *27*, 281–293. [CrossRef]

163. Park, M.G.; Blitzer, E.J.; Gibbs, J.; Losey, J.E.; Danforth, B.N. Negative effects of pesticides on wild bee communities can be buffered by landscape context. *Proc. R. Soc. B Biol. Sci.* **2015**, *282*, 20150299. [CrossRef]
164. Venturini, E.M.; Drummond, F.A.; Hoshide, A.K.; Dibble, A.C.; Stack, L.B. Pollination reservoirs for wild bee habitat enhancement in cropping systems: A review. *Agroecol. Sustain. Food Syst.* **2017**, *41*, 101–142. [CrossRef]
165. Kremen, C.; Williams, N.M.; Bugg, R.L.; Fay, J.P.; Thorp, R.W. The area requirements of an ecosystem service: Crop pollination by native bee communities in California. *Ecol. Lett.* **2004**, *7*, 1109–1119. [CrossRef]
166. Nair, P.R. The coming of age of agroforestry. *J. Sci. Food Agric.* **2007**, *87*, 1613–1619. [CrossRef]

 *forests*

Article

Phosphorus Availabilities Differ between Cropland and Forestland in Shelterbelt Systems

Mihiri C.W. Manimel Wadu [1], Fengxiang Ma [1,2] and Scott X. Chang [1,*

[1] 442 Earth Sciences Building, Department of Renewable Resources, University of Alberta, Edmonton, AB T6G 2E3, Canada; mihiri.wadu@gmail.com (M.C.W.M.W.); mafengxiang@bjfu.edu.cn (F.M.)

[2] School of Science, Beijing Forestry University, Beijing 100083, China

*   Correspondence: Scott.chang@ualberta.ca; Tel.: +1-780-492-6375

Received: 6 October 2019; Accepted: 5 November 2019; Published: 8 November 2019

Abstract: Shelterbelt systems play pivotal roles in providing goods and services to the rural community and the society at large, but phosphorus (P) cycling in shelterbelt systems is poorly studied, while P cycling and availability would be linked to the ecological function and services of shelterbelt systems. This study was conducted to understand how long-term (>30 years) land-use between cropland and forestland in shelterbelt systems affect soil P status. We investigated modified Kelowna ($P_{Kelowna}$) and Mehlich-3 ($P_{Mehlich}$) extractable P, P fractions (by sequential chemical fractionation), P sorption properties in the 0–10 and 10–30 cm soils and their relationship in six pairs of the cropland areas and adjacent forestland (each pair constitutes a shelterbelt system) in central Alberta. Both $P_{Kelowna}$ and $P_{Mehlich}$ in the 0–10 cm soil were greater in the cropland than in the forestland. The $P_{Kelowna}$ ranged from 10 to 170 and 2 to 57 mg kg^{-1} within the cropland areas and forestland, respectively. The inorganic P fraction in the 0–30 cm depth was significantly related to $P_{Kelowna}$ ($R^2 = 0.55$) and $P_{Mehlich}$ ($R^2 = 0.80$) in cropland, but organic P fraction was not significantly related with neither $P_{Kelowna}$ nor $P_{Mehlich}$. The iron (Fe) and aluminum (Al) associated P (Fe/Al-P) explained ~50% and ~45% of the variation of $P_{Kelowna}$ in the 0–30 cm soil in the cropland and forestland, respectively. The Fe/Al-P and organic P fractions in the 0–10 cm soil were greater in the cropland than in the forestland. The differences in availability and P forms depending on the land use type in shelterbelts suggest that P management needs to be land-use type-specific for shelterbelt systems.

Keywords: phosphorus; fractionation; sorption; cropland; forestland

1. Introduction

Agroforestry is an alternative land-use system to monocultural cropping systems to achieve sustainable agricultural production. Shelterbelt systems are one of the most important agroforestry systems. There was an estimated 0.2 million km of shelterbelt within the three Canadian Prairie Provinces of Alberta, Saskatchewan, and Manitoba [1]. Ecological benefits of shelterbelts, including modification of microclimate [2], influence on the quality of litter input [3], and effects on carbon (C) sequestration have been extensively studied [4,5]. In addition, competition with trees for soil available nutrients and water causing yield reduction in adjacent croplands in shelterbelt systems [6] and nitrate accumulation in deeper soil layers in croplands as compared to forestlands have also been reported [7]. However, how soil test phosphorus (P) and P availability differ between cropland and forestland in shelterbelt systems has not been studied, even though shelterbelts have been established on the Canadian prairies since the early nineteenth century and P is the second most limiting plant nutrient after nitrogen (N) in agricultural production systems.

Most soils in Alberta are low in soil-test P as quantified by the modified Kelowna ($P_{Kelowna}$) and Mehlich-3 ($P_{Mehlich}$) methods [8,9]. The Mehlich-3 method is more aggressive in extracting P fractions [10], even though the modified Kelowna method is most widely used in Alberta to determine soil available P [11]. A study by Manunta et al., [8] with more than 56,000 soil analysis, has shown that the majority of Alberta soils had a mean soil test P ($P_{Kelowna}$) of 25–30 mg kg^{-1} in the top 15 cm that is markedly below the agronomic threshold of 60 mg kg^{-1} for most crops [12]. Therefore significant efforts have been made to investigate soil test P and P fractionation in agricultural soils [13,14]. Accumulation of different soil P forms is related to the land-use type, and several abiotic and biotic processes control the accumulation of different soil P species within land-use types. For instance, organic matter inputs associated with tree growth following grassland afforestation affected soil P dynamics over time [15]. Moreover, long-term soil management results in the transformation of Fe oxides in the soil [16] and that inevitably alters the related P retention and bioavailability in the soil. Decreasing soil-water content in cropland with increasing proximity to the forestland in shelterbelt systems [17] may also decrease crop P uptake.

Phosphorus is expected to accumulate in agricultural soils where P-fertilizer is repeatedly applied [18], which would increase the potential loss of P into surface waters and cause eutrophication [19]. Phosphorus in the soil exists in several geochemical forms: exchangeable, Ca and Mg-bound (Ca/Mg-P), Fe and Al-bound (Fe/Al-P) and organically bound P (Org-P) [20]. The distribution of P forms is strongly dependent on soil type [21], and management practices affect P bioavailability in the soil [22], some of those effects are related to changes in soil pH as pH has a strong control on P availability. For instance, most of the P applied to pasture, and cultivated soils were transformed into Fe/Al-P and the Ca/Mg-P fractions, respectively [23], and P is known to be a nutrient that should be studied in with regard to the changes in land-use [24]. In this context, it is helpful to study various P forms that govern P availability in adjacent cropland and forestland in the same land unit to determine if any difference exists between the two land-use types.

Integrating trees into the agricultural landscape enhances organic matter input to the soil from litterfall [25], which can result in a high proportion of Org-P in the soil [26]. Although nearly 70% of P in forest soil was Org-P, crop production and fertilizer P application resulted in decreased Org-P concentrations and increased inorganic P in soils [27]. Inorganic and Org-P fractions in soils vary in terms of their bioavailability and mobility and Org-P forms in general increase with decreasing soil particle size in forest and cropland soils [28]. Although the Org-P and inorganic P fractions are likely to change due to prolonged P fertilizer application in cropland, the relationship between inorganic and organic P fractions and the plant-available P in soil is not always understood, and the land-use effect on soil P fractions has not been studied in shelterbelt systems.

This study was conducted to evaluate the extent to which agricultural practices have altered the inorganic and organic fractions of P in the land cultivated for crop production as compared to the adjacent forestland in a shelterbelt system in Central Alberta, no such study has been done in the past. The objectives of this research were to determine the effect of land-use type on inorganic and organic P fractions and to identify the P fraction mostly related to the available P based on the study of selected shelterbelt systems in central Alberta. Improved knowledge of the effect of land-use type on the distribution of P forms will inform better management practices of soil P in shelterbelt systems that could benefit agricultural production as well as environmental quality.

2. Materials and Methods

2.1. Site Description and Soil Sampling

This study was conducted at six sites (or six replications) located in three counties in central Alberta: Sturgeon, Thorhild, and Lacombe (Figure S1). Two sites were selected from each county and each site comprised of a cropland and forestland pair, allowing a pairwise comparison to be made. Within each pair, the site conditions (e.g., soil properties and climatic conditions) for the two land-use types (cropland vs forestland) were similar before the land diverged into cropland and forestland. In the sites located in the north end (Thorhild) and south end (Lacombe) sampling area, the mean annual air temperature based on the 1981–2010 tri-decadal climate period was 1.9 °C and 2.4 °C, respectively, and mean annual precipitation was 463 mm and 448 mm, respectively [29]. Having the six pairs distributed widely allows our study results to be applicable over a broad geographic region, rather than just for a single site. Luvisols (Boralfs in the USDA soil classification system [30], same below), Dark Gray Chernozems (Boralfic Boroll), and Black Chernozems (Udic Boroll) [30] were the dominant soils in the north, central and south areas, respectively, and the soils have a loam texture [31].

In each site, trees were planted in 1–2 rows (3–5 m wide) to establish the forestland around the edge of agricultural lands. Forestland was comprised of 20–50-year-old trees, and white spruce (*Picea glauca* Moench) was the dominant species [32]. Adjacent to the forestland, the cropland was typically rotated among barley (*Hordeum vulgare*), wheat (*Triticum aestivum*), canola (*Brassica napus*), or pea (*Pisum sativum*) crops. Agricultural management practices included annual fertilizer application up to 120 kg N ha^{-1} year^{-1} and 25 kg P ha^{-1} year^{-1} and minimum tillage [31]. At the minimum, those forestlands had been out of the influence of direct agricultural management (such as P fertilization) for 30 years.

Soil samples were collected after crop harvest in September–October 2012. First, a transect of 30–50 m long was established in both the cropland and forestland. The transect in the cropland was established at least one tree height (~30 m) away from the nearest tree to reduce the immediate impact of trees at the proximity to cropland, while the transect in the forestland was located in the center of the treed area. The paired cropland and forestland transects were located on the same ecosite with similar slope, elevation, landform, and drainage class as much as possible. Mineral soil samples were then collected along each transect from 0–10 and 10–30 cm depths using a 3.2 cm diameter corer, after removing the surficial organic layer in soil and fresh litter. A composite sample for each depth and each transect was obtained by mixing ten soil cores, and fresh soil samples were air-dried in the laboratory at room temperature (20–25 °C) and then passed through a 2 mm sieve.

2.2. Analysis of Basic Soil Chemical and Physical Properties

Soil pH was measured in a 1:2 (w:v) mixture of soil to 0.01 mol L^{-1} CaCl$_2$ solution with a PHH-200 pH meter (Omega Eng. Inc., Stamford CT, USA. Total C was measured with the use of an ECS 4010 Elemental Analyzer (Costech International Strumatzione, Florence, Italy). Soil samples were taken per depth in triplicate using metal rings (106 cm^3 volume) and oven-dried at 105 °C to determine soil bulk density. The hydrometer method [33] was used to determine soil texture by dispersing 40 g of soil (100 g for sandy or loamy sand soils) in 400 mL of Calgon$^®$ [(NaPO$_3$)$_6$] solution (50 g L^{-1}). Cation exchange capacity was determined by extracting 3 g of air-dried soil with 30 mL of 0.1 mol L^{-1} BaCl$_2$ solution. The sample was slowly shaken for 2 h (15 rpm), centrifuged for 23 min, filtered using Whatman No. 41 filter papers (20 μm pore size) and atomic absorption spectrophotometry was used to measure cations in the soil extract. Mehlich-3 extraction of soil samples was conducted to determine extractable P, Ca^{2+}, Mg^{2+}, Fe^{3+} and Al^{3+} [34]. A 2.5-g sample of soil was extracted using 25 mL of the Mehlich-3 solution that was composed of 0.2 mol L^{-1} CH$_3$COOH, 0.25 mol L^{-1} NH$_4$NO$_3$, 0.015 mol L^{-1} NH$_4$F, 0.013 mol L^{-1} HNO$_3$, and 0.001 mol L^{-1} EDTA. The soil and solution mixture was shaken for 5 min at 120 strokes per minute, and then soil extract was filtered with a Whatman No. 40 filter paper. A PerkinElmer Optima 3000 DV inductively coupled plasma mass spectrometer (ICP-MS) (PerkinElmer Inc., Shelton, CT) was used to measure cation concentrations in the soil extract.

2.3. Soil P extraction and Fractionation

The P$_{Kelowna}$ extraction was done by extracting soil using the modified Kelowna reagent that was composed of 0.015 mol L^{-1} NH$_4$F, 0.25 mol L^{-1} CH$_3$COONH$_4$, and 0.25 mol L^{-1} CH$_3$COOH [35]. A 2.5 g of soil was extracted with 25 mL of the reagent by shaking for 15 min, and then the molybdate-ascorbic acid method was used to measure P in the filtered solution colorimetrically at a wavelength of 882 nm [36] using a spectrophotometer (GENESYSTM 10 Series). The P$_{Mehlich}$ was determined together with the analysis of Mehlich extractable cations by extracting the soil with Mehlich-3 solution. The P adsorption maximum was estimated using a single point P adsorption (P$_{150}$) determined at

150 mg P L^{-1} concentration (made in 0.01 mol L^{-1} KCl) by extracting 2 g of air-dried soil with 20 mL of the 150 mg P L^{-1} solution [37]. The soil suspension was shaken for 24 h at room temperature after adding two drops of toluene to inhibit microbial activity and then centrifuged at 3000× g for 10 min. The soil suspension was filtered using a 0.45 µm Fisherbrand™ syringe filter and P concentration was determined using the molybdate-ascorbic acid method as described above. The P saturation index (Psi) was determined as the molar ratio of Mehlich-3 extractable P to Mehlich-3 extractable (Al + Fe) in mmol kg^{-1} [38].

Chemical fractionation of P was conducted by sequentially extracting 2 g of soil with 20 mL of extractants in the following order: deionized water, 0.1 mol L^{-1} NaOH, and 0.5 mol L^{-1} HCl to determine water-extractable P, Fe/Al-P, and Ca/Mg-P, respectively [39]. The amount of Org-P was calculated by the difference between the total P concentration in the digested and undigested NaOH extracts. Inorganic P determination in soil extracts and digests was carried out colorimetrically according to the molybdate-ascorbic acid method. The Kjeldahl digestion (HClO$_4$ and HNO$_3$ acid digestion) of soil samples [40] and the colorimetric method were used to measure pseudototal P in the soil. The sum of Water-P, Fe/Al-P, and Ca/Mg-P will be named as the inorganic P fraction hereafter. The difference between total P and the total extractable P was identified as Residual P.

2.4. Statistical Analysis

All statistical analyses were performed using the SAS software (SAS 9.2, SAS Institute Inc., Cary, NC, USA. Analysis of variance was conducted based on a split-plot design to determine the effects of land-use type (main plot) and soil depth (subplot) and their interaction on soil physical and chemical properties. Normality of distribution was tested using Shapiro–Wilk's and log transformation was applied to P$_{Kelowna}$, P$_{Mehlich}$, Water-P, Fe/Al-P, total P and Residual P to make the distribution normal prior to statistical analysis. The transformation was conducted to meet the assumptions (e.g., normality of distribution) for statistical analysis. However, untransformed data are presented in this paper. A Pearson correlation analysis was performed between P$_{150}$, soil properties, and concentrations of soil P fractions in the 0–10 cm layer of both cropland and forestland. Linear regression analysis was conducted between P$_{Kelowna}$, P$_{Mehlich}$, Psi and concentrations of P fractions in both soil layers in each land-use type. Tukey's multiple comparison test was conducted for multiple comparisons. Given the study was conducted across a wide geographic range with large variation in soil properties as well as vegetation composition, a p-value of 0.10 was used to evaluate the significance so that the risk for committing a type II error is reduced [31,41].

A principal component analysis (PCA) was conducted to reveal the differences between cropland and forestland within each site. The PCA was applied to P$_{Kelowna}$, P$_{Mehlich}$, concentrations of P fractions and Mehlich extractable cations of 0–10 cm layer in all sites using the PROC FACTOR procedure. A correlation matrix was used to derive principal components (PC), and the final component structure was unrotated. Only PCs with eigenvalues greater than one were considered.

3. Results

3.1. Subsection Soil Properties and P Status in Cropland and Forestland

There were no differences in soil properties in the 0–10 cm layer except that the bulk density was greater in cropland than in forestland (Table 1). The mean pH in the 0–10 cm soil was 6 in both the cropland and forestland. A land-use by depth interaction was found for P$_{Kelowna}$, P$_{Mehlich}$, P$_{150}$, and Psi. The P$_{Kelowna}$ ranged from 10 to 170 and 2 to 57 mg kg^{-1} in cropland and forestland, respectively. As expected P$_{Kelowna}$, P$_{Mehlich}$ and P$_{150}$ in the 0–10 cm layer were greater in cropland than in forestland, but this was not the case in the 10–30 cm layer (Figure 1a). Moreover, P$_{Kelowna}$ and P$_{Mehlich}$ were greater in the 0–10 cm than in the 10–30 cm layer only in cropland (Figure 1a). The P$_{Mehlich}$ was greater than P$_{Kelowna}$ in the soil samples measured when the total sampling depth (0–30 cm) was considered in each land-use type given the slope for the regression line was less than one (Figure S2). The extractability of

Mehlich-3 solution was approximately two-fold greater than that of the modified Kelowna solution (Figure S2). The inorganic P fraction was greater in the 0–10 cm depth of cropland compared to that of forestland (Figure 1c).

Table 1. Soil properties in the 0–10 cm layer of selected cropland and forestland in shelterbelt systems in central Alberta. Standard deviations are given in parenthesis ($n = 6$).

Land-Use Type	Soil Properties					
	Sand	Silt	Clay	Total C	pH	Bulk Density (Mg m^{-3})
	(g kg^{-1})					
Cropland	27.74(14.76)	46.73(5.88)	25.51(11.0)	47.88(9.75)	6.07(0.85)	1.47(0.10)a [2]
Forestland	29.50(10.07)	44.07(12.8)	26.44(6.96)	65.75(23.95)	6.04(0.67)	1.24(0.20)b

Land-use type	Soil Properties					
	Cation exchange capacity (cmolc kg^{-1})	Ca	Mg	Al	Fe	Mn
		(mg kg^{-1}) [1]				
Cropland	41.58(11.93)	4.83(0.84)	0.83(0.66)	0.57(0.18)	0.30(0.84)	0.06(0.84)
Forestland	42.91(12.80)	4.98(0.77)	0.69(0.18)	0.52(0.06)	0.24(0.05)	0.06(0.03)

[1] Mehlich-3 extractable concentration. [2] Lower case letters indicate significant difference between the land-use types at $p < 0.1$. No significant difference was found where there is no mean separation indicated.

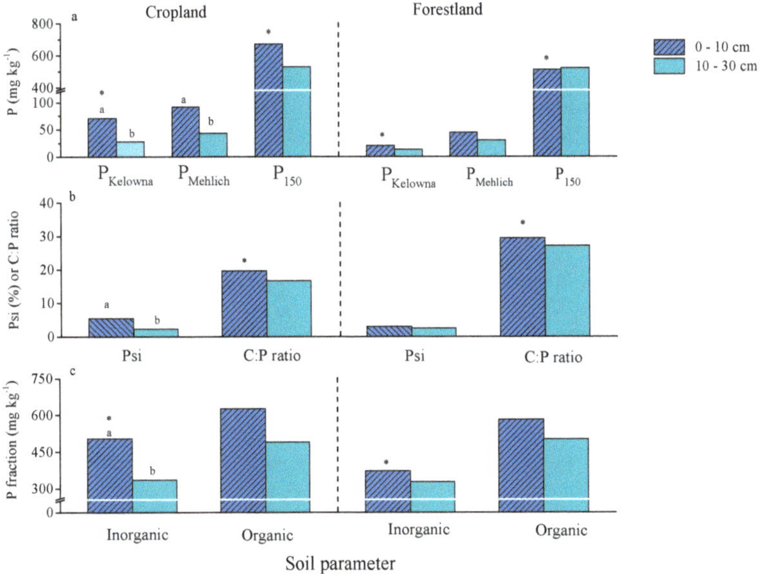

Figure 1. (**a**) The average concentrations of modified Kelowna extractable P (phosphorus), Mehlich-3 extractable P and P sorption index, (**b**) P saturation index and total C:P ratio, (**c**) inorganic P and organic P concentrations in 0–10 and 10–30 cm depths in studied cropland and forestland in shelterbelt systems in central Alberta. $P_{Kelowna}$ and $P_{Mehlich}$ are P extracted with modified Kelowna and Mehlich-3 extractions, P_{150}, P sorption index, Psi, P saturation index. Different lower-case letters indicate a significant difference between two depths for a particular soil parameter in cropland at $p < 0.01$ and asterisks indicate significant differences of particular soil parameters between two land-use types in the 0–10 cm depth at $p < 0.01$ ($n = 6$).

The P_{150} was different between the cropland and forestland in the 0–10 cm layer but did not differ between the two layers in each land-use type (Figure 1a). The P_{150} in the 0–10 cm layer was positively correlated with total C in cropland and Fe, Al and Fe/Al-P in the forestland (Table 2). The Psi was greater in the 0–10 cm than in the 10–30 cm layer only in the cropland. The C:P (based on total P) ratio was higher in the forestland than in the cropland while the C:P did not differ with depth in each land-use type (Figure 1b).

Table 2. Pearson correlation coefficients (r) and p values (in brackets) for correlation analysis between P sorption index (P_{150}) and some soil properties in the 0–10 cm layer of cropland and forestland in shelterbelt systems ($n = 6$).

					Soil Property				
Land-Use Type	Sand	Clay	Total C	pH	Al [1]	Fe [1]	Mn [1]	Fe/Al-P [1]	Org-P [1]
		(g kg^{-1})				(mg kg^{-1})			
Cropland	−0.77 (0.07)	0.77 (0.07)	0.94 (<0.01)	−0.82 (0.04)	NS	NS	NS	NS	−0.77 (0.07)
Forestland	−0.82 (0.04)	NS [2]	NS	−0.82 (0.04)	0.88 (0.01)	0.77 (0.07)	−0.82 (0.04)	0.77 (0.07)	NS

[1] Mehlich-3 extractable Al, Fe and Mn concentrations and P fractions associated with Fe and Al and organic compounds, respectively. [2] NS non-significant.

3.2. Soil P Fractionation in Cropland and Forestland

The Water-P was greater in the 0–10 cm than in the 10–30 cm layer in both land-use types (Table 3). Among all P fractions, a soil depth by land-use type interaction was found only for Fe/Al-P and Org-P. Both P fractions were greater in the 0–10 cm than in the 10–30 cm layer of the cropland but were not different in the forestland (Table 3). The inorganic P fraction in the 0–30 cm depth was significantly related to $P_{Kelowna}$ in both land-use types (Table 4). The Fe/Al-P explained about 50 and 45% of the variations of $P_{Kelowna}$ in the 0–30 cm layer in the cropland and forestland, respectively (Table 4). Moreover, the difference between $P_{Kelowna}$ and $P_{Mehlich}$ in each site was positively related to Ca/Mg-P, Fe/Al-P and Org-P in the 0–30 cm layer in the cropland but not in the forestland (Figure 3). The Psi was also positively related to inorganic P fraction in the cropland (0–30 cm) but not in the forestland (Figure 2b).

The PCA analysis of soil properties revealed differences between cropland and forestland in three sites based on the coefficients of the first two PCs (Figure 3). Two significant PC were extracted from the PCA, which explained 46% (PC1) and 20% (PC2) of the variance (Figure 3). Cropland and forestland sites were segregated along the two PCs reflecting differences between two land-use types with respect to soil P solubility, fractionation, Fe, Al, Ca and Mg concentration. The PC1 showed high loadings of $P_{Kelowna}$, $P_{Mehlich}$, Ca/Mg-P, Fe/Al-P, Org-P and Fe concentrations with a positive effect (Table 5). The PC2 was associated with Al and Ca with a negative and a positive effect, respectively.

Table 3. Average concentration of P fractions and total P in cropland and forestland in shelterbelt systems ($n = 6$).

Land-Use Type	Depth (cm)	Water-P	Ca/Mg-P [a]	Fe/Al-P [1]	Org-P [1]	Residual P	Total P
				(mg kg^{-1})			
Cropland	0–10	7.68 a	170.0 a	328.0 aA [2]	626.3 aA	1339.1 a	2467.0 a
	10–30	1.97 b	130.0 a	203.8 b	489.7 b	1272.2 a	2097.7 a
Forestland	0–10	6.82 a	149.6 a	215.6 aB	581.5 aB	1326.3 a	2279.9 a
	10–30	1.74 b	100.6 b	225.6 a	501.9 a	1049.9 a	1879.8 a

[1] Ca/Mg-P, Fe/Al-P, and Org-P denote P fractions associated with Ca and Mg, Fe and Al and organic compounds, respectively. [2] Lower and upper case letters in the same column indicate significant difference at $p < 0.1$ between two depth levels within a particular land-use type and between the two land-use types in the 0–10 cm soil, respectively.

Table 4. Linear regression models that describe the modified Kelowna and Mehlich-3 extractable P (Y) as a function of P fractions (X) over the 0–30 cm depth in cropland and forestland in shelterbelt systems ($n = 6$).

Land-Use Type	Soil Test P Method	P Fraction [1]	Linear Regression Model	R^2	*p*-Value
Cropland	Kelowna	Fe/Al-P	$Y = 0.22x - 9.77$	0.50	0.005
		Ca/Mg-P	$Y = 0.41x - 20.14$	0.24	0.051
		Inorganic P	$Y = 0.19x - 30.68$	0.55	0.003
	Mehlich-3	Fe/Al-P	$Y = 0.37x - 31.27$	0.67	<0.001
		Ca/Mg-P	$Y = 0.88x - 65.52$	0.48	0.007
		Inorganic P	$Y = 0.32x - 70.98$	0.80	<0.001
Forestland	Kelowna	Fe/Al-P	$Y = 0.09x - 4.96$	0.45	0.009
		Inorganic P	$Y = 0.08x - 12.75$	0.47	<0.001
		Org-P	$Y = 0.09x - 32.52$	0.24	0.051
	Mehlich-3	Ca/Mg-P	$Y = 0.39x - 12.23$	0.22	0.061
		Inorganic P	$Y = 0.15x - 16.61$	0.29	0.031

[1] Ca/Mg-P, Fe/Al-P, and Org-P stand for P fractions associated with Ca and Mg, Fe and Al and organic compounds, respectively.

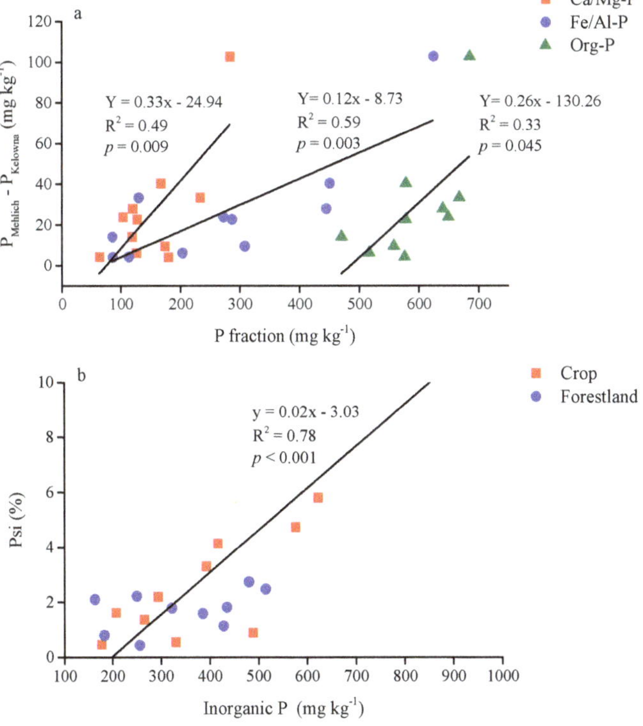

Figure 2. The relationship between (**a**) the differences between P extracted by Mehlich-3 and modified Kelowna methods and the concentration of different P fractions in the 0–30 cm depth in cropland, and (**b**) the inorganic P concentration and the P saturation index in the 0–30 cm soil in selected cropland and forestland in shelterbelt systems in central Alberta. Ca/Mg-P, Fe/Al-P, and Org-P indicate P fractions associated with Ca and Mg, Fe and Al and organic compounds, respectively. $P_{Kelowna}$ and $P_{Mehlich}$ are P extracted with modified Kelowna and Mehlich-3 methods, respectively. Psi stands for P saturation index.

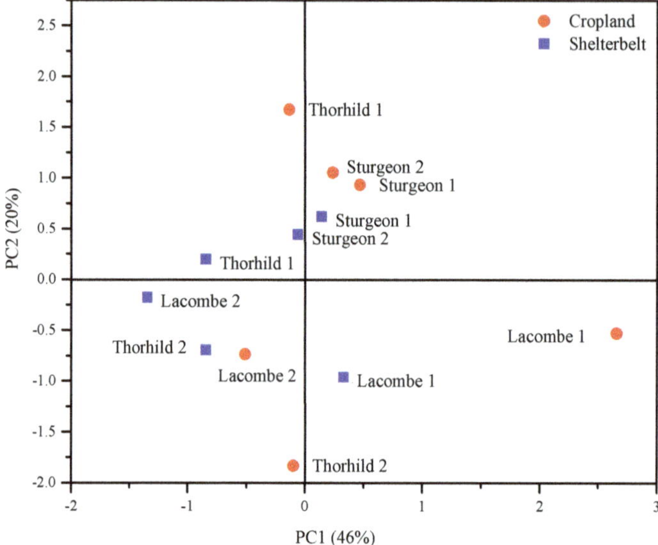

Figure 3. Principal component analysis of soil parameters related to P status in soils of six cropland and forestland in shelterbelt systems.

Table 5. Principal components extracted from a correlation matrix based on soil properties in six shelterbelt systems located in central Alberta.

Soil Property [a]	PC1 [b]	PC2
$P_{Kelowna}$	0.76	0.20
$P_{Mehlich}$	0.89	0.29
Fe/Al-P	0.92	−0.24
Ca/Mg-P	0.72	0.35
Org-P	0.78	0.37
Fe	0.70	−0.22
Al	0.42	−0.84
Ca	−0.30	0.78
Eigenvalue	4.11	1.82
% Variance	46	20

[a] $P_{Kelowna}$ and $P_{Mehlich}$ refer to P extracted using water, modified Kelowna and Mehlich-3 extractants, respectively, Fe/Al-P, Ca/Mg-P, Org-P, indicate P associated with Fe and Al, Ca and Mg, organic compounds, respectively, and Fe, Al, Ca and Mg are Mehlich-3 extractable cations. [b] Values in the table are the component loading of each variable with each principal component. Bold values indicate the strongest component loadings.

4. Discussion

4.1. Greater P Build-up in the Surface Soil in the Cropland than in the Forestland

The $P_{Kelowna}$ in the 0–10 cm layer of cropland was twice as high as that in the forestland, indicating that soil P was built up by agricultural practices such as P-fertilizing and minimum tillage (Figure 1a), suggesting greater P solubility in the cropland than in the forestland. The limited mobility of P in soils leads to the accumulation of P in surface soils [42] and increased plant-available P in the soil was related to the long-term P fertilizer application [43]. In addition to P inputs, the lower C:P ratio in the 0–10 cm layer of cropland may result in organic P mineralization [44], which would further enhance available P in the cropland than in the forestland. Fine roots of *Picea glauca* might be present up to the 45 cm depth in the soil [45], and thus greater root exploitation and lack of P input caused lower soil test P in forestland than in cropland. The similar soil test P in both soil layers indicated the utilization of P over the 0–30 cm depth by *Picea* species in the studied forestland. We acknowledge that these results are in line with what we expected from soils that have diverged in land use 30–40 years ago and that croplands have more P because of fertilization practices and the tendency of soils to adsorb PO_4^{3-}.

The upper-level agronomic threshold of $P_{Kelowna}$ in the top 15 cm of soil is 60 mg kg^{-1} for most crops grown in Alberta [12]. Three cropland sites (one site in Sturgeon county and both sites in Lacombe county) contained high $P_{Kelowna}$ (60–170 mg kg^{-1}) in the 0–10 cm layer while the other three sites had $P_{Kelowna}$ (10–46 mg kg^{-1}) lower than the agronomic threshold. The low $P_{Kelowna}$ in some cropland sites is consistent with findings reported by [8] where the mean extractable P concentrations varied between 25 and 30 mg kg^{-1} in the top 15 cm soil for the majority of eco-districts in Alberta.

Phosphorus sorption capacity in soils is related to the clay content [46], and the strong relationship between clay content and P_{150} (Table 2) in cropland implied that clay particles were more reactive and more effective in retaining P in the cropland than in the forestland. Interestingly, P_{150} was positively related to Al in the forestland (Table 2), and this calls for further investigations on the compositional differences of Al forms given the similar pH and Al concentrations between the two land-use types. Unlike in the forestland, greater $P_{Kelowna}$ with high P_{150} suggests saturation of P sorption sites in some croplands or reflects the degree of desorption of retained P in cropland soils (Figure S3). Although the inherent P sorption capacity seemed to be similar in the two land-use types (Figure S3), P input caused the saturation of P sorption sites in the cropland in contrast to that in the forestland. The high Psi and soil test P are related to the ecological risk associated with inorganic P loss (surface run-off and/or leaching) from the cropland [9,47]. Excessive P loss from cropland soils can decrease the quality of surface water. Although the increasing inorganic P resulted in increasing Psi in cropland (Figure 2b), soil test P in forestland was not influenced by the adjacent cropland with high Psi and soil test P that was greater than the agronomic threshold. The two land-use types had the same ecosite type and were on a flat topography. In the cropland, both soil test P, P_{150} and Psi were related to the inorganic P fraction and indicated P build-up in the surface soil.

4.2. Soil Test P ($P_{Kelowna}$) is Related to Fe/Al-P

The greater inorganic P concentration in the 0–10 cm soil in the cropland than in the forestland was a remarkable difference between two land-use types (Figure 1c) caused by agricultural practices. Phosphorus fertilizer treatments showed a positive effect on the inorganic P fractions in-long-term crop rotation plots on a Luvisolic soil and a Chernozemic soil in Alberta, indicating that most inorganic P fractions were related to P fertility in the long-run [13]. Moreover, inorganic P was increased in the 0–20 cm soil in managed grasslands [48] and croplands as compared to natural forestlands [49]. The increased availability of inorganic P in the cropland was evident in the strong relationship between inorganic P fraction and $P_{Kelowna}$ and the greater amount of P that was extracted by the modified Kelowna method in the cropland than in the forestland (i.e., 58 vs. 46% of the total P). Although the Org-P comprised a larger fraction of total P than the inorganic P in the cropland (Table 3), $P_{Kelowna}$ was not related to Org-P in the cropland (Table 4). A high level of the recalcitrance of the organic P pool that acts as a sink for inorganic P has been reported [50]. The weak correlation between available P and Org-P in Alberta soils [14] and the strongest correlation between P_{150} and soil C are consistent with the lack of relationships between $P_{Kelowna}$ and Org-P in the cropland in our study. The stable soil organic matter pool may be related to the low availability of Org-P in *Picea* based forestlands in shelterbelt systems [14].

The readily extractable P associated with Fe and Al strongly contributed to $P_{Kelowna}$ in the cropland and forestland. However, it was interesting to note that P_{150} was positively correlated with Fe and Al in the forestland but not in the cropland (Table 2), indicating that Fe and Al played a greater role in affecting P availability in the soil under forestland than that in the cropland soil. However, the geochemical investigation of Fe/Al-P in the forestland may provide better insights into the P behavior in the forestland. The weak relationship between $P_{Kelowna}$ or $P_{Mehlich}$ and Ca/Mg-P (Table 4) cannot be related to the low solubility of calcium phosphates since soil pH was around 6 in both land-use types where calcium phosphates are soluble [51]. In addition, Mehlich 3 and Kelowna are acid extractants, which can dissolve more Ca-P and the above relationships weak. However, the extractability of P by the Mehlich-3 solution was greater than the modified Kelowna solution with increasing Fe/Al-P and Ca/Mg-P concentrations in the 0–30 cm soil layer in the cropland (Figure 2a), indicating the greater solubility of the inorganic P fraction in the cropland than in the forestland, likely related to the external P input and the lower organic C content in the cropland soil that decreases P fixation by organic matter. Therefore, the fraction of inorganic P that was not extracted by the modified Kelowna method (i.e., non-available P) may consist of extractable P that could be released into the soil solution through desorption as the amount of available P decreases in the soil [50]. The importance of plant non-available P forms as P reserves may contribute to the P nutrition of crops [52] since desorption, mineralization, and weathering may change non-available P forms into available forms in the cropland. Unlike inorganic P, the lack of relationship between Org-P and soil test P in the cropland and forestland may be due to the recalcitrant nature of the Org-P fraction in the soil (Figure 2a and Table 4). Given the low soil test P, replenishment of inorganic P by mineralization of soil organic P fraction seems to be important to provide available P for trees in the forestland as the Org-P is the largest fraction of the extractable P in the soil.

The PC1 in the PCA plots indicated that croplands were distinct from the forestland, but such difference between land-use types was site-dependent. The least difference between the cropland and the forestland was found in sites located in Sturgeon County (Figure 3) while, the cropland and forestland in Lacombe and Thorhild counties were mainly differentiated by PC1 and PC2, respectively (Figure 3). According to the PC1, the differentiation between cropland and forestland in different counties was increased in the following order: Sturgeon > Thorhild > Lacombe. The absence of clear segregation between the cropland and forestland in all sites supported that P availability and forms depend on the inherent characteristic of each site [53]. Moreover, accumulation of P species in soils with different land-uses was attributed to the collective influence of biotic and abiotic factors such as variation between soil, site, and management factors within a land use [54]. The PC1 tended to be positively contributed by extractable P and Fe/Al-P having the largest PC loading (Table 5). Therefore, in addition to the strong positive relationship between Fe/Al-P fraction and $P_{Kelowna}$ (i.e., available P) (Table 4), the Fe/Al-P fraction is also related to the difference between the studied shelterbelt systems regarding P status in soil (Table 5).

5. Conclusions

Our study provides insights into the linkages among soil test P, P fractions and P sorption properties in cropland and forestland in shelterbelt systems. The difference in the P status between the cropland and the forestland in shelterbelt systems was dependent on the site location but soil test P was mainly governed by soluble inorganic P that was associated with Fe and Al in both the cropland and the forestland. A direct impact of the P build-up in the cropland on the P status in forestland was not evident in this study as seen by low soil test P in the forestland. However, given the low soil test P in the forestland as compared to the cropland, competition for available P may exist between trees and crops at the boundary between the cropland and the forestland. Such relationships are worth to be studied in the future. Moreover, considering a large amount of P associated with the organic fraction in cropland and forestland, further exploitation of organically bound P as a potential source of available P is worthwhile in future P management plans. Improved knowledge of the sources and types of organic P and factors controlling organic P mineralization is needed in developing management strategies to promote organic P availability for plant production.

Supplementary Materials: The following are available online at http://www.mdpi.com/1999-4907/10/11/1001/s1, Figure S1: Map of Canada and the Province of Alberta showing the location of six study sites of shelterbelt systems. Figure S2: The relationship between Mehlich-3 extractable P and modified Kelowna extractable P in the 0–30 cm depth of selected cropland and forestland in shelterbelt systems in central Alberta. Figure S3: The relationship between P sorption index and P extracted by the modified Kelowna method.

Author Contributions: Conceptualization, S.X.C. and M.C.W.M.W.; methodology, M.C.W.M.W.; formal analysis, M.C.W.M.W.; resources, S.X.C.; writing—original draft preparation, M.C.W.M.W.; writing—review and editing, S.X.C., M.C.W.M.W., and F.M.; visualization, M.C.W.M.W.; funding acquisition, S.X.C.

Funding: This research was funded by the Natural Science and Engineering Research Council of Canada (NSERC), grant number 249664-2013.

Acknowledgments: We thank the China Scholarship Council (grant #201406515044) for supporting FXM's visit to Canada and Mark Baah-Acheamfour for assistance in sample collection.

Conflicts of Interest: The authors declare no conflict of interest.

References

1. Thevathasan, N.V.; Gordon, A.M.; Bradley, R. Agroforestry research and development in Canada: The way forward. In *The Future of Global Land Use*; Nair, P.K.R., Garrity, D.P., Eds.; Springer: Dordrecht, The Netherlands, 2012; pp. 247–283.

2. McNaughton, K.G. Effects of windbreaks on turbulent transport and microclimate. *Agric. Ecosyst. Environ.* **1988**, *22–23*, 17–39. [CrossRef]

3. Lorenz, K.; Lal, R. Soil organic carbon sequestration in agroforestry systems. *Rev. Agron. Sustain. Dev.* **2014**, *34*, 443–454. [CrossRef]

4. Kort, J.; Turnock, R. Carbon reservoir and biomass in Canadian prairie shelterbelts. *Agrofor. Syst.* **1998**, *44*, 175–186. [CrossRef]

5. Dhillon, G.S.; Van Rees, K.C.J. Soil organic carbon sequestration by shelterbelt agroforestry systems in Saskatchewan. *Can. J. Soil Sci.* **2017**, *97*, 394–409. [CrossRef]

6. Kowalchuk, T.E.; de Jong, E. Shelterbelts and their effect on crop yield. *Can. J. Soil Sci.* **1995**, *75*, 543–550. [CrossRef]

7. Qiao, Y.; Fan, J.; Wang, Q. Effects of farmland shelterbelts on accumulation of soil nitrate in agro-ecosystems of an oasis in the Heihe River Basin, China. *Agric. Ecosyst. Environ.* **2016**, *235*, 182–192. [CrossRef]

8. Manunta, P.; Kryzanowski, L.; Keyes, D. *Preliminary Assessment of Available Soil P in Alberta: Status and Trends*; Soil Quality Program, Conservation and Development Branch; Alberta Agriculture, Food and Rural Development: Edmonton, AB, Canada, 2000; p. 64.

9. Paterson, B.A.; Olson, B.M.; Bennett, D.R. *Alberta Soil Phosphorus Limits Project*; Volume 1: Summary and recommendations; Alberta Agriculture, Food and Rural Development: Lethbridge, AB, Canada, 2006; p. 82.

10. Zhang, M.; Wright, R.; Heaney, D.; Vanderwe, D. Comparison of different phosphorus extraction and determination methods using manured soils. *Can. J. Soil Sci.* **2004**, *84*, 469–475. [CrossRef]

11. McKenzie, R.H.; Kryzanowski, L.; Cannon, K.; Solberg, E.; Penney, D.; Coy, G.; Heaney, D.; Harapiak, J.; Flore, N. *Field Evaluation of Laboratory Tests for Soil Phosphorus*; Alberta Agricultural Research Institute Report No. 90 M230; Alberta Innovation and Science: Edmonton, AB, Canada, 1995.

12. Howard, A.E. Agronomic thresholds for soil phosphorus in Alberta: A review. In *Alberta Soil Phosphorus Limits Project*; Alberta Agriculture, Food and Rural Development: Lethbridge, AB, Canada, 2016; Volume 5, p. 42.

13. McKenzie, R.H.; Stewart, J.W.B.; Dormaar, J.F.; Schaalje, G.B. Long-term crop rotation and fertilizer effects on phosphorus transformations: I. In a Chernozemic soil. *Can. J. Soil Sci.* **1992**, *72*, 569–579. [CrossRef]

14. McKenzie, R.H.; Bremer, E. Relationship of soil phosphorus fractions to phosphorus soil tests and fertilizer response. *Can. J. Soil Sci.* **2003**, *83*, 443–449. [CrossRef]

15. Chirino-Valle, M.R.D.; Condron, L.M. Impact of different tree species on soil phosphorus immediately following grassland afforestation. *J. Soil Sci. Plant Nutr.* **2016**, *16*, 477–489. [CrossRef]

16. Li, J.; Richter, D.; Mendoza, A.; Heine, P. Four-decade responses of soil trace elements to an aggrading old-field forest: B, Mn, Zn, Cu, and Fe. *Ecology* **2008**, *89*, 2911–2923. [CrossRef]

17. Liversley, S.J.; Gregory, P.J.; Buresh, R.J. Competition in tree row Agroforestry systems. 3. Soil water distribution and dynamics. *Plant Soil* **2004**, *264*, 129–139. [CrossRef]

18. Bünemann, E.K.; Heenan, D.P.; Marschner, P.; McNeill, A.M. Long-term effects of crop rotation, stubble management and tillage on soil phosphorus dynamics. *Soil Res.* **2006**, *44*, 611–618. [CrossRef]

19. Sharpley, A.N.; Kleinman, P.J.A.; McDowell, R.W.; Gitau, M.; Bryant, R.B. Modeling phosphorus transport in agricultural watersheds: Processes and possibilities. *J. Soil Water Conserv.* **2002**, *57*, 425–439.

20. Hedley, M.J.; Stewart, J.W.B.; Chauhan, B.S. Changes in inorganic and organic soil phosphorus fractions by cultivation practices and by laboratory incubations. *Soil Sci. Soc. Am. J.* **1982**, *46*, 970–976. [CrossRef]

21. Tiessen, H.; Stewart, J.W.B.; Cole, C.V. Pathways of phosphorus transformations in soils of differing pedogenesis. *Soil Sci. Soc. Am. J.* **1984**, *48*, 853–858. [CrossRef]

22. Motavalli, P.P.; Miles, R.J. Soil phosphorus fractions after 111 years of animal manure and fertilizers applications. *Biol. Fertil. Soils* **2002**, *36*, 35–42. [CrossRef]

23. Castillo, M.S.; Wright, A.L. Microbial activity and phosphorus availability in a subtropical soil under different land uses. *World J. Agric. Sci.* **2008**, *4*, 314–320.

24. Garcia-Montiel, D.C.; Nelly, C.H.; Melillo, J.; Thomas, S.; Steudler, P.A.; Cerri, C.C. Soil phosphorus transformation following forest clearing for pasture in the Brazilian Amazon. *Soil Sci. Soc. Am. J.* **2000**, *64*, 1792–1804. [CrossRef]

25. Paul, K.I.; Polglase, P.J.; Nyakuengama, J.G.; Khanna, P.K. Change in soil carbon following afforestation. *For. Ecol. Manag.* **2002**, *168*, 241–257. [CrossRef]

26. Frossard, E.; Stewart, J.W.B.; St. Arnaud, R.J. Distribution and mobility of phosphorus in grassland and forest soils of Saskatchewan. *Can. J. Soil Sci.* **1989**, *69*, 401–416. [CrossRef]

27. Vaithiyanathan, P.; Correll, D.L. The Rhode River Watershed: Phosphorus Distribution and Export in Forest and Agricultural Soils. *J. Environ. Qual.* **1992**, *21*, 280–288. [CrossRef]

28. Frossard, E.; Condron, L.M.; Oberson, A.; Sinaj, S.; Fardeau, J.C. Processes governing phosphorus availability in temperate soils. *J. Environ. Qual.* **2000**, *29*, 15–23. [CrossRef]

29. Environment Canada. Alberta Weather Condition. 2012. Available online: http://weather.gc.ca/forecast/canada/indexe.html?id=AB (accessed on 2 May 2017).

30. Soil Classification Working Group. *The Canadian System of Soil Classification*; NRC Research Press: Ottawa, ON, Canada, 1998; p. 187.

31. Baah-Acheamfour, M.; Chang, S.X.; Cameron, N.C.; Bork, E.W. Carbon pool size and stability are affected by trees and grassland cover types within agroforestry systems of western Canada. *Agric. Ecosyst. Environ.* **2015**, *213*, 105–113. [CrossRef]

32. Lim, S.S.; Baah-Acheamfour, M.; Choi, W.J.; Arshad, M.A.; Fatemi, F.; Banerjee, S.; Carlyle, C.N.; Bork, E.W.; Park, H.J.; Chang, S.X. Soil organic carbon stocks in three Canadian agroforestry systems from surface organic to deeper mineral soils. *For. Ecol. Manag.* **2018**, *417*, 103–109. [CrossRef]

33. Carter, M.R.; Gregorich, E.G. *Soil Sampling and Methods of Analysis*, 2nd ed.; Francis and Taylor Group LLC/CRC Press: Boca Raton, FL, USA, 2006; pp. 334887–342742.

34. Mehlich, A. Mehlich 3 soil test extractant: A modification of Mehlich 2 extractant. *Commun. Soil Sci. Plant Anal.* **1984**, *15*, 1409–1416. [CrossRef]

35. Qian, P.; Schoenau, J.J.; Karamanos, R.E. Simultaneous extraction of available phosphorus and potassium with a new soil test: A modification of the Kelowna extraction. *Commun. Soil Sci. Plant Anal.* **1994**, *25*, 627–635. [CrossRef]

36. Murphy, J.; Riley, J.P. A modified single solution methods for the determination of phosphate in natural water. *Anal. Chim. Acta* **1962**, *27*, 31–36. [CrossRef]

37. Ige, D.V.; Akinremi, O.O.; Flaten, D.N.; Ajiboye, B.; Kashem, M.A. Phosphorus sorption capacity of alkaline Manitoba soils and its relationship to soil properties. *Can. J. Soil Sci.* **2005**, *85*, 417–426. [CrossRef]

38. Kleinman, P.J.A.; Sharpley, A.N. Estimation soil phosphorus sorption saturation from Mehlich-3 data. *Commun. Soil Sci. Plant Anal.* **2002**, *33*, 1825–1839. [CrossRef]

39. Nair, V.D.; Graetz, D.A.; Portier, K.M. Forms of phosphorus in soil profiles from dairies of south Florida. *Soil Sci. Soc. Am. J.* **1995**, *59*, 1244–1249. [CrossRef]

40. Sparks, D.L.; Page, A.L.; Helmke, P.A.; Loeppert, R.H. *Methods of Soil Analysis Part 3—Chemical Methods*; SSSA Book Ser. 5.3; SSSA, ASA: Madison, WI, USA, 1996.

41. Baah-Acheamfour, M.; Cameron, N.C.; Bork, E.W.; Chang, S.X. Forest and perennial herbland cover reduce microbial respiration but increase root respiration in agroforestry systems. *Agric. For. Meteorol.* **2020**, *280*, 107790. [CrossRef]

42. Sharpley, A.N. Depth of surface soil-runoff interaction as affected by rainfall, soil slope and management. *Soil Sci. Soc. Am. J.* **1985**, *49*, 1010–1015. [CrossRef]

43. McCollum, R.E. Buildup and decline in soil-phosphorus—30-year trends on a Typic Umprabuult. *Agron. J.* **1991**, *83*, 77–85. [CrossRef]

44. Spohn, M.; Kuzyakov, Y. Phosphorus mineralization can be driven by microbial need for carbon. *Soil Biol. Biochem.* **2013**, *61*, 69–75. [CrossRef]

45. Safford, L.O.; Bell, S. Biomass of fine roots in a white spruce plantation. *Can. J. For. Res.* **1972**, *2*, 169–172. [CrossRef]

46. Soon, Y.K. Solubility and retention of phosphate in soils of the northwestern Canadian prairie. *Can. J. Soil Sci.* **1991**, *71*, 453–463. [CrossRef]

47. Sims, J.T.; Maguire, R.O.; Leyten, A.B.; Gartley, K.L.; Pautler, M.C. Evaluation of Mehlich 3 as an agri-environmental soil phosphorus test for the mid-Atlantic United States of America. *Soil Sci. Soc. Am. J.* **2002**, *66*, 2016–2032. [CrossRef]

48. Von Sperber, C.; Stallforth, R.; Du Preez, C.; Amelung, W. Changes in soil phosphorus pools during prolonged arable cropping in semiarid grasslands. *Eur. J. Soil Sci.* **2017**, *68*, 462–471. [CrossRef]

49. Yang, W.; Cheng, H.; Hao, F.; Ouyang, W.; Liu, S.; Lin, C. The influence of land-use change on the forms of phosphorus in soil profiles from the Sanjiang Plain of China. *Geoderma* **2012**, *189–190*, 207–214. [CrossRef]

50. Costa, M.; Gama-Rodrigues, A.; Gonçalves, J.; Gama-Rodrigues, E.; Sales, M.; Aleixo, S. Labile and non-labile fractions of phosphorus and its transformations in soil under eucalyptus plantations, Brazil. *Forests* **2016**, *7*, 15. [CrossRef]

51. Lindsay, W.L. *Chemical Equilibria in Soils*; John Wiley and Sons: New York, NY, USA, 1979.

52. Reddy, D.D.; Rao, A.S.; Takkar, P.N. Effects of repeated manure and fertilizer phosphorus additions on soil phosphorus dynamics under a soybean-wheat rotation. *Biol. Fertil. Soils* **1999**, *28*, 150–155. [CrossRef]

53. Xavier, F.A.D.S.; Almeida, E.F.; Cardoso, I.M. Soil phosphorus distribution in sequentially extracted fractions in tropical coffee-agroecosystems in the Atlantic Forest biome, Southeastern Brazil. *Nutr. Cycl. Agroecosyst.* **2011**, *89*, 31–44. [CrossRef]

54. Stutter, M.I.; Shand, C.A.; George, T.S.; Blackwell, M.S. Land use and soil factors affecting accumulation of phosphorus species in temperate soils. *Geoderma* **2015**, *257*, 29–39. [CrossRef]

 *forests*

Article

A Slash-And-Mulch Improved-Fallow Agroforestry System: Growth and Nutrient Budgets over Two Rotations

Aaron H. Joslin [1,*]**, Steel S. Vasconcelos** [2]**, Francisco de Assis Oliviera** [3]**, Osvaldo R. Kato** [2]**, Lawrence Morris** [1] **and Daniel Markewitz** [1]

[1]   Warnell School of Forestry and Natural Resources, The University of Georgia, Athens, GA 30602, USA; lmorris@uga.edu (L.M.); dmarke@uga.edu (D.M.)
[2]   Embrapa Amazônia Oriental, PA 66095-903 Belém, Brazil; steel.vasconcelos@embrapa.br (S.S.V.); osvaldo.kato@embrapa.br (O.R.K.)
[3]   University of Agricultural Sciences of Amazonia (UFRA), Institute of Agricultural Sciences (ICA), PA 66095-903 Belém, Brazil; fdeassis@gmail.com
*   Correspondence: aaron.hoyt.joslin@gmail.com; Tel.: +51-920-449-277

Received: 9 October 2019; Accepted: 11 November 2019; Published: 10 December 2019

Abstract: Agroforestry systems are important, globally affecting 1.2 billion people and covering 0.6 billion hectares. They are often cited for providing ecosystem services, such as augmenting soil fertility via N accumulation and increasing soil C stocks. Improved-fallow slash-and-mulch systems have the potential to do both, while reducing nutrient losses associated with burning. In the absence of burning, these systems also have the potential to grow trees through multiple rotations. This project collected soil, mulch, and biomass data over the course of one 9-year crop-fallow rotation and the first two years of the second rotation. A split-plot design was used to assess the effects of P + K fertilization and inclusion of an N-fixing tree species, *Inga edulis*, on crop and tree biomass production. Fertilization increased growth and nutrient accumulation during Rotation 1 by an average of 36%, ranging from 11% in *Parkia multijuga* to 52% in *Ceiba pentandra*. Residual P + K fertilization improved tree and crop growth 20 months into Rotation 2 by an average of 50%, ranging from 15% in *Cedrela odorata* to 73% in *Schizolobium amazonicum*. The improved-fallow slash-and-mulch system increased the rates of secondary succession biomass accumulation ($11–15$ Mg ha^{-1} yr^{-1}) by 41–64% compared to natural succession ($7–8$ Mg ha^{-1} yr^{-1}). Furthermore, P + K fertilization increased secondary-succession biomass accumulation by 9–24%. Nutrient accumulation through biomass production was adequate to replace nutrients exported via crop root and timber stem harvests.

Keywords: slash-and-mulch; improved-fallow; native trees; N-fixing trees; soil N; soil C; nutrient content; agroforestry system; Amazonia

---

1. Introduction

Agroforestry systems (AFS), agricultural land with >20% tree cover, account for 0.6 billion hectares and support approximately 1.2 billion people often with limited landholdings that depend on some form of these practices for their livelihoods [1,2]. AFS are commonly employed in Brazilian Amazonia by smallholding farmers (<100 ha) and could comprise up to 35% of land under cultivation in that region [3]. Slash-and-burn land preparation is a common way used by smallholding farmers to prepare land for cultivation, in which the standing vegetation is cut, left to dry, and then burned. Burning vegetation creates an initial pulse of available nutrients for crop production via ash, but leaching, erosion, and crop export lead to declining fertility resulting in crop field abandonment [4].

Slash-and-burn agricultural production is dependent on the length of time fallow vegetation can accumulate nutrients before being cleared and burned again for cropping, since nutrient accumulation in the fallow vegetation is the primary source of crop nutrients [5]. For example, N (~500 kg ha^{-1}) and P (~30 kg ha^{-1}) accumulation were five- and three-fold greater in 8-year-old compared to 3-year-old regenerating secondary forest [6]. Improved fallow (i.e., the planting of desired species) is one method used to enhance vegetative growth in secondary succession. Although much of the research into use of improved fallows in AFS has looked at cover crops and herbaceous plants [7], the use of tree species is also possible and in various tropical locations has increased nutrient accumulation [8–10].

Pioneer tree species (e.g., *Cecropia* spp.) that first establish on sites in natural forest succession are typically of low commercial timber value [11]. Species planted for improved fallow could generate income via timber sales, potentially resulting in increased fallow lengths or even the purchase of fertilizers. Improved fallow plantings with commercial timber (e.g., *Schizolobium amazonicum*) or biomass species (e.g., *Inga edulis*) may be harvested as early as six years after slashing [12]. In the absence of fire, slower growing species (e.g., *Parkia multijuga*) of higher value that are also planted might be sustained through successive crop–fallow cycles prior to harvest.

In addition to fast-growing nutrient accumulators or timber species, improved fallow can improve soil fertility by using nitrogen (N)-fixing species (i.e., *Inga* sp.). Use of N-fixing cover crops (i.e., *Centrosema macrocarpum*) and hedgerows as green manures are common in continuous cultivation systems, where incorporation of cover crop or hedgerow biomass adds N to the agroecosystem [7]. Alternatively, as it is done in improved fallow, N fixers are added to the fallow mix to stimulate N acquisition for subsequent release to the cropping phase after the fallow is cleared [7]. Enrichment plantings with various N-fixing tree species, including *I. edulis*, were shown to increase fallow biomass by 6–117%, compared to unimproved fallow [13]. In Panama, young forests (<12 years) were estimated to have the highest rates of N fixation along a 300-year forest age gradient, and N fixers were estimated to supply >50% of the N necessary for biomass growth [14]. Other research in mixed-species plantings of non-fixing *Eucalyptus* sp. with N-fixing *Acacia mearnsii* found 3x the foliar N concentration in eucalypt when grown in a 50–50 mix [15], and *Theobroma cacao* roots contained as much as 60% higher N concentrations when grown in the presence of *I. edulis* [16]. The tropical N-fixing tree species *Gliricidia sepium* was reported to supply up to 63% of the N content of neighboring grass species via root exudation [17]. These results indicate that N-fixing trees can increase neighbor N content.

In contrast, some recent studies in moist tropical forests did not support the notion of N-fixing trees facilitating the growth or nutrient acquisition of neighboring plants during secondary succession [18] or any effect on the total biomass accumulation of the secondary forest [19]. Rather than depleting agroecosystem N capital, as happens in slash-and-burn systems due to gaseous losses of N during burning [5], the inclusion of N-fixing tree species to enhance N stocks in AFS requires continued validation.

The slash-and-mulch technology was developed to reduce losses of soil fertility due to biomass burning and can fell, chop, and evenly distribute secondary forest vegetation using a mulching tractor [20]. Mulching the forest adds all biomass nutrients to the soil surface, which become available for future plant uptake, and, unlike burning, the mulching tractor can maneuver around planted trees.

In 2005, a field trial was initiated to develop an improved-fallow AFS that could take advantage of the slash-and-mulch technology and the potential for multiple rotations of tree growth that have economic value [21]. After mulching of a 7-year-old unimproved secondary forest, tree rows were planted on a 4 m spacing that would allow future passage of the mulching tractor. Trees included four timber species, i.e., *S. amazonicum*, a fast-growing pioneer that could be harvested after one rotation, and *Ceiba pentandra*, *Cedrela odorata*, and *P. multijuga*, all likely to require >2 rotations. Additionally, *I. edulis*, an N fixer, was co-planted in some improved-fallow treatments. The expectation was that *I. edulis* growth and mulching as a green manure would improve the N stocks of the soil. A P+K fertilization (Fert) treatment was also included, since these are also primary limiting elements along with N. The regional staple food crop *Manihot esculenta* was planted simultaneously with the trees.

During the first two years of Rotation 1, manioc growth increased significantly with P+K fertilization, but there was no response to the N fixer [21]. P+K fertilization also enhanced tree growth through Year 6, and estimates of planted tree biomass and nutrient accumulation indicated that inclusion of *I. edulis* increased the N stock [22]. The mulch layer at site initiation had a mean mass of 54 Mg ha^{-1} with ~300 kg N ha^{-1} [21] and decreased by Year 6 to 10–15 Mg ha^{-1} with N contents of ~130 kg ha^{-1} [22]. In year 6, mulch mass and N content showed no treatment effects. Mineral soil N at that time was not clearly elevated by the presence of *I. edulis*, but soil P concentration had clearly increased by 37% and 45% to ~6 and 4 μg/dm^{-3} in the 0–10 and 10–20 cm soil layers.

In this paper, we report on the end of the first rotation (Year 9) of this experiment and the initiation of the second rotation through years 10 and 11 that included driving the mulching tractor between planted tree rows, converting *I. edulis* trees to green manure, harvesting all *S. amazonicum* trees, and planting a second crop of manioc and trees as initially designed. Many AFS studies, including several from this region, have investigated 3–5 years after forest clearing [4,5,20,23], but only one decade-long study was identified, on shade-grown cacao [24].

Given the multiple-rotation nature of this experiment, we posed hypotheses relative to the end of the first rotation and others related to the second rotation. In May 2014 at the end of Rotation 1, we hypothesized that the improved-fallow plantings would generate higher biomass and nutrient stocks than the unimproved fallow prior to site conversion. Furthermore, tree growth during Rotation 1 would increase with P+K fertilization and the presence of *I. edulis* (Nfix). Since we previously observed increased aboveground stocks of N [25] under improved fallow, we hypothesized that agroecosystem N stocks would be higher with P+K fertilization and with the presence of *I. edulis* at the end of the first Rotation. In July 2014 after mulching to initiate the second rotation, we hypothesized that the growth of newly planted trees would be greater due to P+K fertilization at initiation, as residual effects of fertilization on tree growth have been previously observed up to 20 years after application [26]. We also expected greater tree growth with *I. edulis* due to green manure effects, and that manioc growth would be greater with Fert and N fix treatments in Rotation 2.

2. Materials and Methods

2.1. Site Description

The research was conducted at the Fazenda Experimental de Igarapé Açu (FEIGA) of the Universidade Federal Rural da Amazônia (UFRA) in the Municipality of Igarapé Açu (1°07′41″ S 47°47′15″ W), approximately 110 km East of Belém, Pará, Brazil (Figure 1a). This region, the Bragantina, is one of the oldest continually inhabited agricultural areas in Amazonia [4], and the landscape is now dominated by human activities, such as urban, row-crop farms, plantation forests, cattle ranches, and secondary forests.

Soils in the municipality of Igarapé Açu are predominantly Kandiudults [27]. Particle size analyses for 2 m soil profiles were recently performed in 2016 on soils from one plot from within each block and one of each treatment (n = 4). Each plot location was sampled at increments of 0–10, 10–20, 20–50, 50–100, 100–150, and 150–200 cm. Particle size distribution was determined using the hydrometer method [28] (Table S1). Prior bulk density (BD) measures found 1.2 g cm^{-3} from 0 to 5 cm and 1.4 g cm^{-3} from 5 to 10 cm [21]. Igarapé Açu has an average annual temperature of 26 °C and annual rainfall of 2500 mm [29]; the driest months are August–November, and the wettest months are January–May.

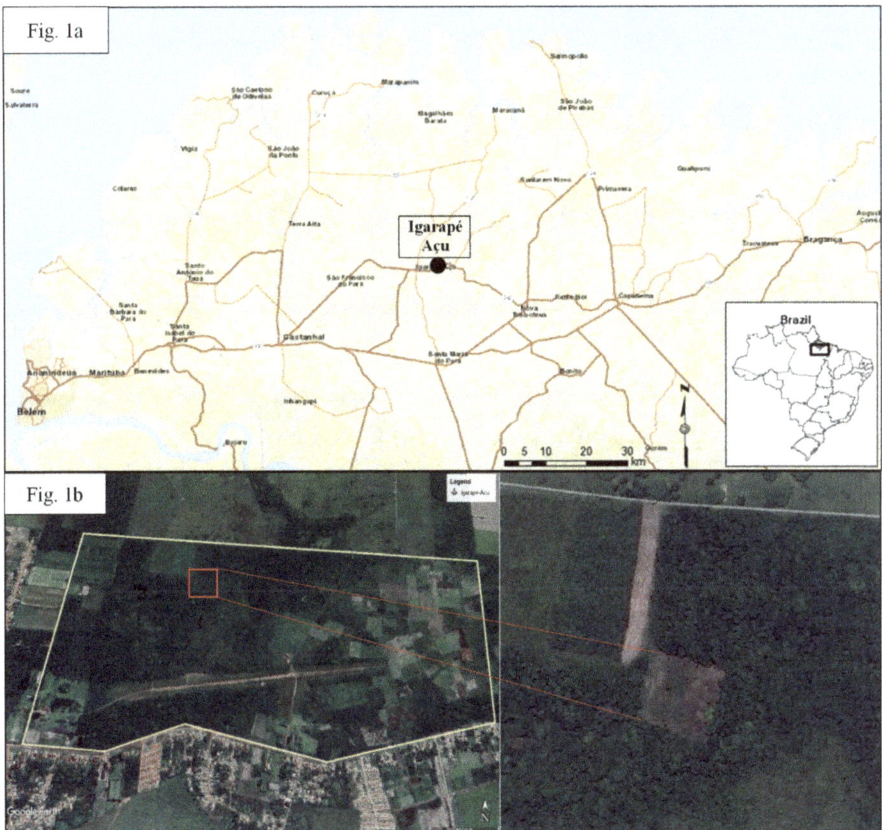

Figure 1. (**a**) Map of the Bragantina region, in the northeast of Pará State, Brazil. The study was located in the town in the center of the map, Igarapé Açu; (**b**) map of the Fazenda Experimental de Igarapé Açu (FEIGA) research station, with the one-hectare plot within FEIGA.

2.2. Species Descriptions

The five planted tree species are native to forests of the Bragantina region. *I. edulis* (Fabaceae) is the only known planted N fixer, although other N-fixing species may be present due to natural recruitment in secondary succession. *S. amazonicum* (Fabaceae) and *C. pentandra* (Malvaceae) are rapid-growing pioneers with soft wood. *P. multijuga* (Fabaceae) and *C. odorata* (Meliaceae) are slower growing tropical hardwoods (see [21] for a more complete species description).

All five tree species were planted with the food crop manioc (*M. esculenta*), which was sampled after 12 months and then harvested after 20 months during the first crop–fallow rotation. Manioc growth response has been reported [21]. The same tree species were planted with manioc within 10 days of site conversion via mulching tractor in July 2014 to begin the second rotation of the crop–fallow cycle. Manioc sampling of the second rotation was performed 12 months after site preparation, but the majority of the manioc plants were stolen prior to the 20-month sampling event, so that data are not present in this study.

2.3. Plot Establishment

Plot establishment of this site was previously described [21]. Briefly, in March of 2005, a 1 ha study site within FEIGA (Figure 1b) was selected and mulched. Experimental treatments were applied in June

2005, and the aforementioned species were planted in four experimental blocks (N = 4). Each block was divided into four plots (n = 16) that measured 24 × 24 m. Trees and manioc were planted simultaneously, with trees planted at 4 × 1.8 m spacing, for a total of 78 trees per plot, or 1354 trees ha^{-1}. The crop species *M. esculenta* was planted at 1 × 1 m spacing, or 10,000 stems ha^{-1} (Figure 2).

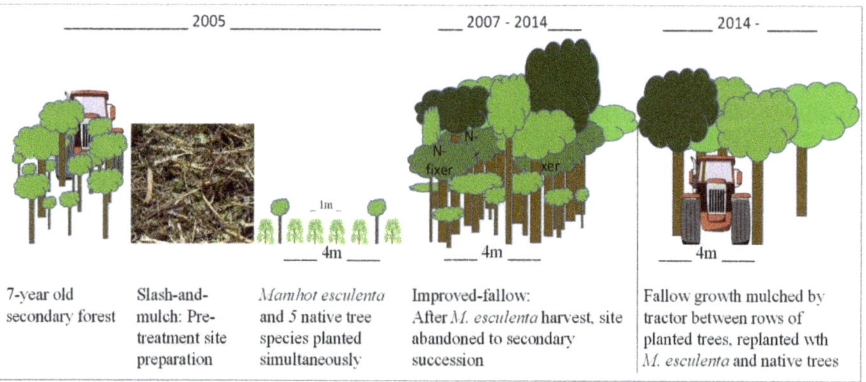

Figure 2. Pictorial representation of site selection, preparation, treatment application, and growth for Rotation 1 and subsequent site preparation for Rotation 2 of a slash-and-mulch improved-fallow agroforestry system.

A factorial combination of fertilization treatment and N-fixing species was assigned in a split-plot randomized complete block design. The two main plot fertilizer treatments consisted of no fertilization (PK−) and fertilization (PK+) by an application around the base of planted trees of 46 kg P ha^{-1} as P_2O_5 (100 kg Simple-Super Phosphate with 46% P) and of 30 kg K ha^{-1} as KCl (50 kg KCl with 60% K). Sub-plot treatments consisted of planting 26 trees each of the native species *C. odorata*, *C. pentandra*, and *S. amazonicum* together (I−) or in combination with the N-fixing species *I. edulis* and *P. multijuga* (I+). In the I+ treatment, half of the trees planted were *I. edulis*, while the remainder was divided evenly among the remaining species.

In July 2014, the site was prepared for the second rotation of the crop—fallow system via mulching tractor, as was done during the first rotation. However, the tractor passed between the 4 m-wide spacing between rows of trees, mulching the secondary vegetation but leaving the planted trees to continue growing. Unfortunately, many small trees did not survive the mulching event, either because they were mulched or because other vegetation fell on them. The same five species of trees were planted in both rotations.

2.4. Soil/Litter Sampling and Analysis

During both the end of Rotation 1 and the beginning of Rotation 2, mulch layer samples were collected for plot-level analysis from four random sample locations between rows 2 and 5 and between the 4th and the 10th trees of those rows to avoid edge effects of neighboring plots. Samples were collected within 1 m × 1 m polyvinyl chloride (PVC) quadrats, dried in a forced-air oven at 65 °C, weighed, and averaged per plot. Soil samples were taken from a 0–10 cm depth from the center of each framed location beneath where litter samples had been collected. These samples were air-dried and composited by plot.

All samples were ground to a powder using a vial roller or ball mill grinder (SPEX, Metuchen, NJ). Soil and litter samples were analyzed for carbon (C) and nitrogen (N) concentrations using a CN analyzer (CE Elantech, Lakewood, NJ, USA). Litter samples were analyzed for additional elements using the EPA-3050B (U.S. EPA, 1996) method [30]. Soil samples were analyzed for cations and P via Mehlich−1 double-acid extraction [31]. Extracts of 1 M KCl were analyzed via flow-injection spectrometry for NH_4 and NO_3 concentrations (OI Analytical, College Station, TX, USA).

2.5. Plant Biomass

Trees' diameters at breast height (DBH) and ground line (GLD) were measured in 2013, 2014, 2015, and 2016. Additionally, height was measured in 2014, 2015, and 2016 using a hypsometer (Hägloff, Långsele, Sweden) or a range pole for shorter trees. The diameter was measured using digital calipers for smaller girth trees and seedlings or slide calipers for larger diameters.

For seedlings and saplings that had not achieved DBH, the tree volume was estimated using the frustum formula:

$$V = \left(\pi \cdot \frac{h}{3} \cdot \left(R^2 + R \cdot r + r^2 \right) \right) \tag{1}$$

where h = length of the section measured, R = radius of the lower end of the section, r = radius of the upper end of the section

Planted-tree aboveground biomass was estimated by multiplying published wood density values [32] by the frustum volume estimate:

$$B = V \cdot \rho \tag{2}$$

where B is biomass, V is volume, and ρ is wood density.

Exceptions to this estimate are noted below.

S. amazonicum trees were harvested after mulching at the beginning of the second rotation. Nine trees, out of 42 survivors (21.5%), were selected for measurement of DBH and for diameter at each 5 m increment along the stem. Total length was measured, and branches were measured for diameter at the junction with the main stem. All leaves were collected, dried, and weighed. Discs of the stem were collected, dried, and weighed. The lengths of individual branches were not measured but estimated with a species-specific allometric equation for *S. amazonicum* (R^2 = 0.8916) derived from saplings sampled in this research:

$$L = 80.104 \cdot D - 49.66 \tag{3}$$

where L = branch length (cm), D = branch diameter (cm).

Volume estimates of *S. amazonicum* at the end of the first rotation (Year 9) used [33]:

$$V = 0.077476 + 0.517987 \cdot \left(DBH^2 \cdot H \right) \tag{4}$$

where V = volume (m^3), DBH = diameter at breast height (cm), H = height (m).

For trees under 2.0 m of height, the frustum volume Equation (1) was used and multiplied by the wood density [32] to generate the biomass of the stem. Stem biomass was then multiplied by the ratio of the stem biomass to total biomass measured from the nine sample trees mentioned above to generate a total biomass estimate. For trees with height above 2.0 m, biomass (B) was estimated using diameter at breast height (DBH) by the allometric equation [34]:

$$B = 0.076 \cdot DBH^{2.346} \tag{5}$$

I. edulis trees were cut at ground level in May 2014 prior to site conversion via mulching tractor, so that the biomass could be mulched and distributed as green manure with the rest of the secondary vegetation. Twelve trees, 7% of trees in the 2014 sample event, with at least one tree per I+ plot was selected for allometric analysis. These 12 trees were sampled for GLD and DBH, and the mass of each section was weighed wet. The wet weight was converted to wood biomass using data from Reference [35] in which 39% of wet weight was wood. Additionally, all leaves were collected per tree and dried in a forced-air oven at 65 °C until a constant weight. Height, GLD (cm), and DBH (cm) were used in regression equations to predict total biomass, with the best fit ($R^2 = 0.9127$) for trees of GLD < 13.5 cm:

$$B = 44.461 \cdot GLD^{2.2579} \tag{6}$$

where B is biomass (kg), and GLD is ground line diameter (cm); for trees with $GLD > 13.5$ cm ($R^2 = 0.9086$):

$$B = 138.39 \cdot DBH^{2.4706} \tag{7}$$

where B is in kg, and DBH = diameter at breast height (cm).

Since *I. edulis* trees were not measured for height in 2013 due to understory vegetation blocking sight of the tops of most fertilized trees, height, GLD, and DBH measurements from 2011 were used to find the best allometric equation to predict height for the measurements in 2013 and 2014. The best fit was ($R^2 = 0.8047$):

$$DBH = 233.6 \cdot H^{0.5172} \tag{8}$$

where H = height (cm).

I. edulis leaf biomass estimates were generated from the leaf-to-stem biomass ratio from the sampled trees, and the best fit was a linear regression ($R^2 = 0.8168$):

$$LM = (42.973 \cdot SB) - 2079.3 \tag{9}$$

where LM = leaf mass (g), SB = stem biomass (g).

For *C. odorata* saplings without DBH, biomass estimates were generated using the frustum volume (Equation 1) multiplied by the specific gravity of *C. odorata* (0.39 g cm^{-3}; [32]). For all trees above 2.0 m height, the allometric equation [36] was used to estimate the biomass:

$$B = 200.2 \cdot DBH^2 \cdot H^{0.5615} \tag{10}$$

To separate stem biomass from leaf biomass for trees under 200 cm of height, the leaf biomass was assumed to be 28% of the total biomass [37], and for trees above 200 cm, the allometric equation [36] was used to predict leaf biomass:

$$LB = 126.5 \cdot DBH^2 \cdot H^{0.2787} \tag{11}$$

Foliar and wood N content were estimated using published data: 3.07% N for leaf [38] and 0.109% N for wood [39].

P. multijuga biomass estimates for trees <2.0 m of height were generated using the frustum volume (Equation 1) multiplied by the specific gravity of *P. multijuga* (0.38 g cm^{-3}; [32]). Foliar N content was determined from the mean of published data, i.e., 2.362% [40–42], multiplied by leaf biomass estimates. Wood N content was estimated from published data, i.e., 0.109% [40], multiplied by wood biomass estimates. Wood mass was separated from leaf mass using a 7.3:1 wood/leaf ratio for *P. multijuga* from Central Amazonia [43]. The content of other macronutrients was determined using reported concentrations for wood [43] and leaf [39]. For trees with height >2.0 m, the following biomass estimate [43] was used:

$$B = 0.076 \cdot DBH^{2.346} \tag{12}$$

The following power function ($R^2 = 0.9218$) was used to estimate the height (H) of *P. multijuga* that could not be measured due to secondary vegetation interference:

$$H = 124.65 \cdot DBH^{0.8337} \tag{13}$$

C. pentandra volume was estimated using the frustum method (Equation 1). The volume was then multiplied by the basic density, 0.28 g cm^{-3} [32]. We used reported biomass allometry for *C. pentandra* in plantation at 4 years of age [44] and used the height of 300 cm which was the average height of surviving trees at approximately that age (299 cm ± 182). Leaf and wood biomass distribution was based on the reported aboveground biomass [43]. Wood N concentration was reported [45], as well as leaf N concentration [38].

Manioc stems were planted at the beginning of the second rotation in July 2014, and biomass was measured in July 2015 by placing a 1 × 1 m PVC frame at four randomly selected points in each plot. All biomass from within the vertical plane of the frame was collected, aboveground and belowground. Manioc biomass was divided into three compartments: leaf, stem, and root. Biomass compartments were dried in a forced-air oven at 65 °C and weighed.

2.6. Statistical Analyses

All statistical analyses were performed using SAS University Edition (SAS, Cary, NC, USA). Data distributions were evaluated prior to analysis and transformed as needed. Analyses were performed using Proc MIXED. In the factorial model, the main-plot treatment was with or without P+K fertilization, and the sub-plot treatment was with or without the N-fixing tree species *I. edulis*. Fert and Nfix were treated as fixed effects, and block as a random effect. Repeated measures analysis was used to account for covariance among dates, with plot as the subject. Contrast statements were used to test differences in growth rates between specific years.

Multivariate regressions (Proc REG) were used to evaluate the role of soil variables (mineral soil and O horizon) on tree and manioc growth. Sampling dates used in the multivariate regressions were: July 2013, May 2014 during Rotation 1, and July 2015 during Rotation 2. Macronutrients N, P, K, Ca, and Mg were used as regressors for both mineral soil and O horizon, as well the C/N ratio. Additionally, O horizon (mulch layer) mass and macronutrient content were used as regressors.

3. Results

3.1. Survival

At the end of the first rotation in May 2014, the survival of trees ranged from 0% for *C. odorata* in PK+I+ to 76% for *P. multijuga* in PK−I+ (Table 1). Both Fert and NFix treatments affected the survival of the planted trees, although P+K fertilization more consistently decreased the survival than did the presence of *I. edulis* (Table 1). In the first two years after the establishment of Rotation 2, P+K fertilization continued to have a significant effect on the survival of four of the five planted species (Table S2). The survival for all species was higher 20 months after planting in the second rotation in 2014, than 24 months after planting in 2005 (Table 1). *C. odorata* had the most dramatic increase in survival in the full-factorial treatment (PK+I+), with 66.7% survival after 20 months in Rotation 2 vs. 0% after 24 months in Rotation 1. However, *C. odorata* survival decreased without fertilization in Rotation 2 relative to Rotation 1. The only other species with reduced survival in Rotation 2 relative to Rotation 1 without fertilization was *I. edulis* (84 vs. 92%, respectively). The presence of *I. edulis* (I+) did not significantly affect survival for any species 20 months after planting in Rotation 2 relative to 24 months after planting in Rotation 1 (Table S2).

3.2. Planted Trees

With all species grouped together (henceforth, All-Species), planted tree growth was enhanced by P+K fertilization (Figure 3) during Rotation 1, with all measured attributes responding to fertilization (Table 2). However, by Year 9, the differences in growth rate ceased to be significant in most instances for individual species (Table S3). During Rotation 2, the presence of the N fixer *I. edulis* increased GLD ($p = 0.02$) and height ($p = 0.01$) in the All-Species grouping and DBH for *C. pentandra* ($p = 0.03$), but there was no Fert–Nfix interaction in planted tree growth during the second rotation. During Year 2 of Rotation 2, only GLD differed for All-Species with P+K fertilization ($p = 0.06$). Fertilization increased height for *C. odorata*, *I. edulis*, and *S. amazonicum* during the Year 2 Rotation 2 (Table S3; Figure 4).

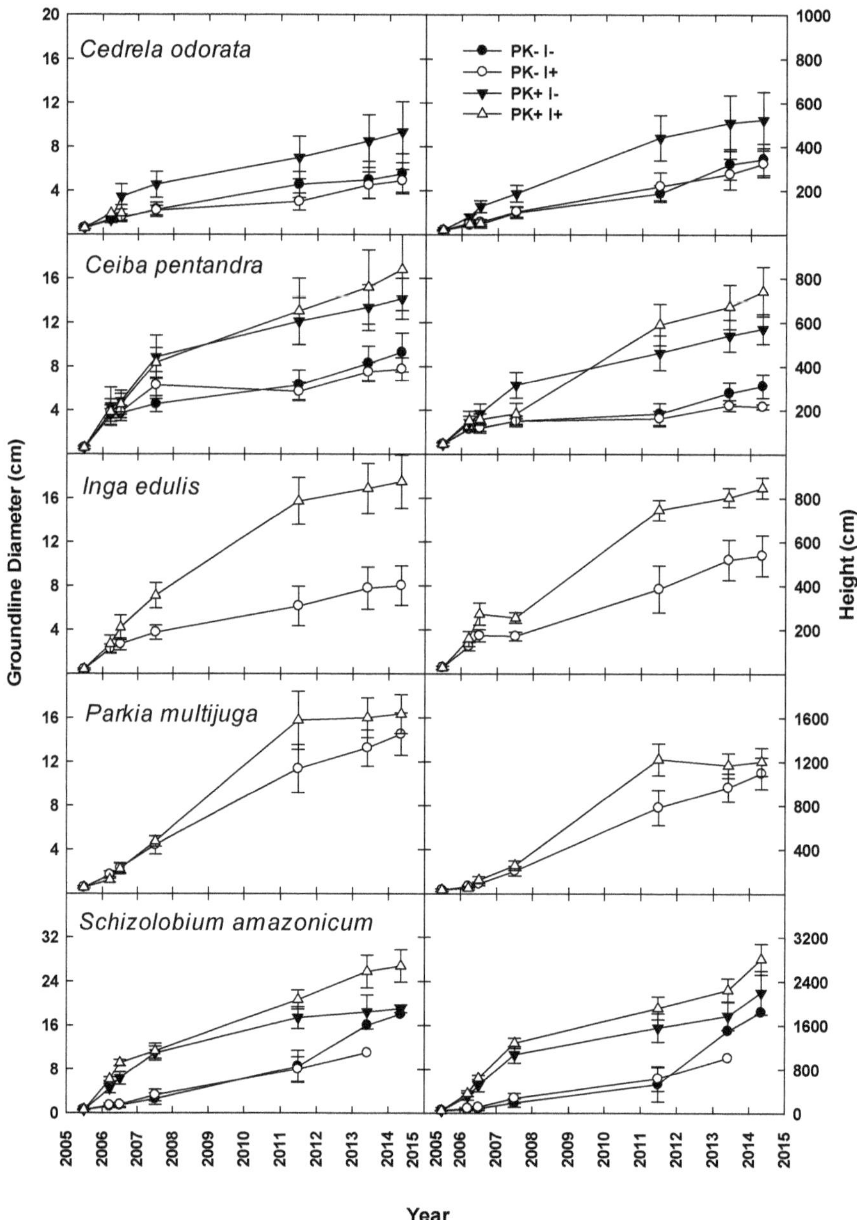

Figure 3. Groundline diameter and height of five native tree species grown in mixed culture during the first crop–fallow cycle of an improved-fallow slash-and-mulch agroforestry system in Eastern Amazonia of Brazil

Table 1. Survival rates of five native tree species, mean (SE), planted in an improved-fallow, slash-and-mulch agroforestry system in Pará State, Brazil. Capital letters indicate significant differences in the main-plot (P+K fertilization, Fert), and lower-case letters indicate differences in the sub-plot (*I. edulis*, Nfix) treatments.

Species	Treatment [1]	2006	2007	Rotation 1 2011	2013	May 2014	Rotation 2 2015	2016
					% Survival (SE)			
Cedrela odorata	PK− I−	85 (18)	73 (22)	70 (23)	53 (25)	51 (25) A	89 (16)	64 (24)
	PK− I+	82 (20)	68 (24)	68 (24)	55 (26)	55 (26) A	89 (12)	52 (26)
	PK+ I−	14 (17)	11 (16)	9 (15)	9 (15)	9 (15) B	81 (20)	42 (25)
	PK+ I+	10 (15)	—	—	—	— B	95 (11)	67 (24)
Ceiba pentandra	PK− I−	100 (0)	85 (18)	50 (25)	33 (24)	26 (22) a	84 (18)	84 (18)
	PK− I+	100 (0)	83 (19)	54 (25)	25 (22)	17 (19) b	93 (13)	93 (13)
	PK+ I−	44 (25)	44 (25)	46 (25)	43 (25)	41 (25) a	80 (19))	81 (20)
	PK+ I+	28 (23)	24 (22)	20 (20)	20 (20)	20 (20) b	85 (19)	85 (19)
Inga edulis	PK− I−							
	PK− I+	99 (5)	92 (14)	72 (23)	50 (25)	46 (22) A	93 (13)	84 (19)
	PK+ I−							
	PK+ I+	51 (25)	47 (25)	41 (25)	34 (24)	27 (22) B	89 (16)	86 (18)
Parkia multijuga	PK− I−						90 (15)	87 (17) b
	PK− I+	91 (15)	81 (20)	82 (20)	76 (22)	76 (22) A	95 (12)	95 (12) a
	PK+ I−						85 (18)	85 (18) b
	PK+ I+	19 (20)	19 (20)	19 (20)	76 (22)	14 (18) B	90 (15)	90 (15) a
Schizolobium amazonicum	PK− I−	62 (25)	23 (21)	7 (13)	2 (7)	2 (7) B	62 (25)	44 (25)
	PK− I+	71 (23)	38 (25)	19 (20)	5 (11)	— B	87 (17)	78 (21)
	PK+ I−	46 (25)	42 (25)	42 (25)	35 (24)	35 (24) A	83 (19)	81 (20)
	PK+ I+	33 (24)	33 (24)	33 (24)	28 (23)	28 (23) A	85 (18)	78 (21)

[1] Treatments are indicated by: P+K fertilization (PK+) or without (PK−) in the main-plot, and the presence of Inga edulis and Parkia multijuga in the planting mix (I+) or without (I−).

Table 2. Statistical output of ground line diameter (GLD), diameter at breast height (DBH), and height data for five species of native trees grown during two crop–fallow rotations in a 12-year-old slash-and-mulch improved-fallow agroforestry system in Eastern Amazonia of Brazil.

	Variable	All Species			Cedrela odorate			Ceiba pentandra			Inga edulis			Parkia multijuga			Schizolobium amazonicum		
		GLD	DBH	Height	GLD	DBH	Height	GLD	DBH	Height	GLD	DBH	Height	GLD	DBH	Height	GLD	DBH	Height
		Pr > F			Pr > F			Pr > F			Pr > F			Pr > F			Pr > F		
Rotation 1	Fert [1]	0.004 *	0.003 *	0.001 *	0.25	.	0.6	0.08 +	0.03 *	0.02 *	0.01 *	0.003 *	0.008 *	0.55	0.3	0.4	0.007 *	0.01 *	0.004 *
	Nfix [2]	0.03 *	0.4	0.005 *	0.94	.	0.17	0.6	0.18	0.8							0.3	0.9	0.18
	Date	<0.001 *	<0.001 *	<0.001 *	<0.001 *	.	<0.001 *	<0.001 *	<0.001 *	<0.001 *	<0.001 *	<0.001 *	<0.001 *	<0.001 *	<0.001 *	<0.001 *	<0.001 *	<0.001 *	<0.001 *
	Fert*Nfix	0.004 *	0.001 *	0.001 *	0.7	.	0.080 +	>0.9	0.10 +	0.18	.	.	.	.	.	.	0.5	0.4	1.0
	Fert*Date	<0.001 *	<0.001 *	<0.001 *	0.8	.	0.5	<0.001 *	0.001 *	<0.001 *	<0.001 *	<0.001 *	<0.001 *	0.2	0.5	0.3	<0.001 *	0.03 *	<0.001 *
	Nfix*Date	<0.001 *	<0.001 *	<.001 *	0.3	.	0.7	0.8	0.3	0.5							0.3	0.2	0.3
	Fert*Nfix*Date	0.02 *	0.4	0.3	0.01 *	.	0.11	0.2	0.5	0.04 *							0.03 *	0.4	0.4
Rotation 2	Fert	0.07 +	0.06 +	0.02 *	0.09 +	.	0.19	0.07 +	0.02 *	0.03 *	0.14	0.4	0.2	0.08 +	0.4	0.07 +	0.04 *	0.2	0.02 *
	Nfix	0.02 *	0.9	0.01 *	0.9	.	0.4	0.6	0.03 *	0.7	.	.	.	0.9	0.7	1.0	0.8	1.0	0.7
	Date	<0.001 *	0.3	<0.001 *	<0.001 *	.	<0.001 *	<0.001 *	0.02 *	<0.001 *	<0.001 *	0.83	<0.001 *	<0.001 *	0.002 *	<0.001 *	<0.001 *	<0.001 *	<0.001 *
	Fert*Nfix	0.7	0.8	0.13	0.19	.	0.14	0.2	0.01 *	0.4	.	.	.	0.2	0.07 +	0.14	0.7	0.6	0.3
	Fert*Date	<0.001 *	0.6	<.001 *	0.3	.	0.05 *	0.01 *	0.05 *	0.02 *	0.11	.	0.004 *	0.002 *	0.02 *	<0.001 *	<0.001 *	0.05 *	<0.001 *
	Nfix*Date	0.02 *	0.4	0.6	0.8	.	1.0	0.4	0.002 *	0.3	.	.	.	0.4	0.12	0.6	0.9	0.7	1.0
	Fert*Nfix*Date	0.9	0.4	0.009 *	0.5	.	0.9	0.6	0.004 *	0.9	.	.	.	1.0	0.06 +	0.3	0.9	0.4	0.08 +

[1] Fert is the main-plot treatment of P+K fertilization, or without. [2] Nfix is the sub-plot treatment inclusion of the N-fixing tree species Inga edulis in the planting mix, or not; * indicates significant difference at the $p < 0.05$ level; + indicates significant difference at the $p < 0.10$ level.

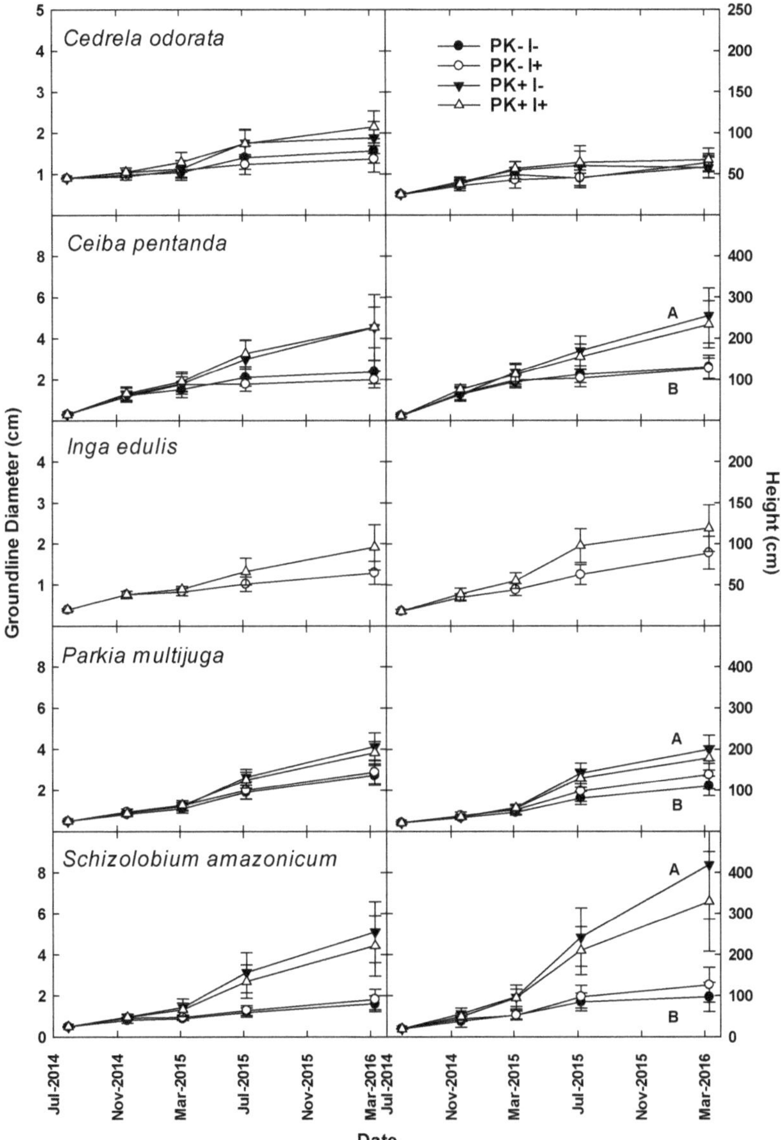

Figure 4. GLD and height of five native tree species grown in mixed culture during the second crop–fallow cycle of an improved-fallow slash-and-mulch agroforestry system in Eastern Amazonia of Brazil. Capital letters indicate significant differences for the indicated species by main-plot P + K fertilization treatment.

By the end of Rotation 1, the estimated N content of the trees ranged from 100 to 600 kg N ha^{-1} in PK−I− and PK+I−, respectively (Figure 5). The estimated N content of planted trees decreased after mulching and tree harvest for Rotation 2 by 70–90%, despite the retention of "leave trees" that were not harvested or felled and mulched (Table S4).

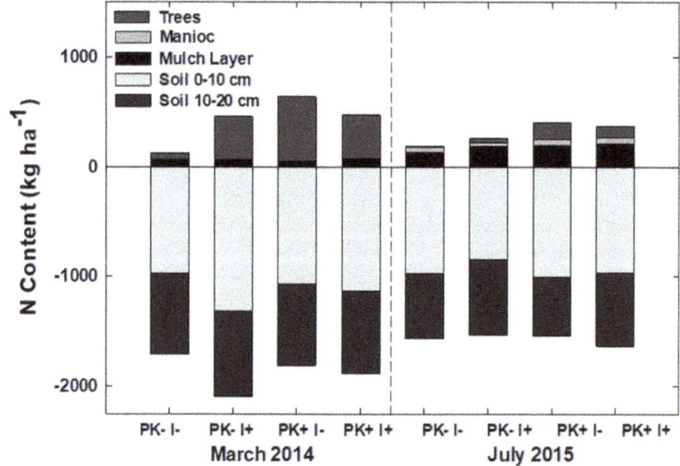

Figure 5. Agroecosystem N content (kg ha^{-1}) during the final year of Rotation 1 and the first year of Rotation 2 of an improved-fallow slash-and-mulch agroforestry system in Eastern Amazonia of Brazil.

3.3. Manioc

Ten years after P+K fertilization at the initiation of Rotation 1, both manioc biomass ($p = 0.07$) and N content ($p < 0.10$) during Rotation 2 were greater in fertilization plots (Figure 6; Table S5). In contrast, the presence of the N-fixing tree species *I. edulis* did not affect the N-content of manioc in Rotation 2.

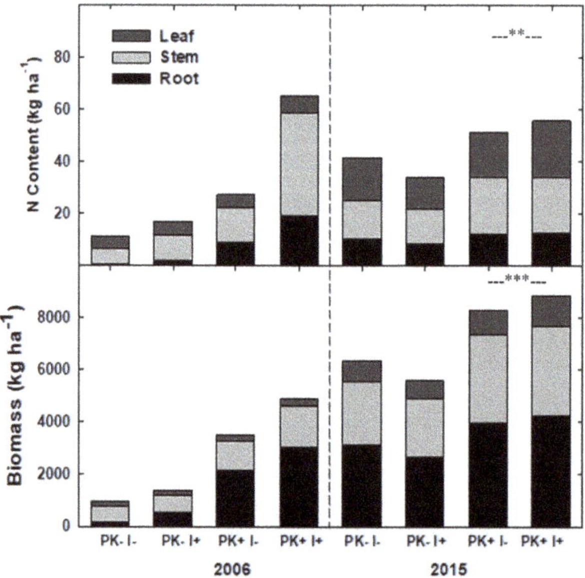

Figure 6. Nitrogen content and biomass (kg ha^{-1}) of *Manihot esculenta* by root, stem, and leaf compartments measured in Year 1 after the establishment of Rotation 1 (2006) and Rotation 2 (2015) in the crop–fallow system in Eastern Amazonia of Brazil; ** indicates $p = 0.07$, and *** indicates $p < 0.10$.

3.4. Mulch Layer

The mulch layer resulting from site preparation at the beginning of Rotation 1 had a mass of 54 Mg ha^{-1} (Figure 7). By Year 6 after site establishment, the mulch layer achieved a steady mass that remained between 10 and 15 Mg ha^{-1} until Year 9. Mulching of secondary forest biomass in Year 9 in 2014 generated between 100 and 140 Mg ha^{-1} of mulch layer biomass. After mulching the site for Rotation 2, the presence of the N-fixing *I. edulis* contributed to the increase in mulch layer biomass and C content (p = 0.04 and 0.03, respectively) but did not increase N content (p = 0.2; Table S6). Concentrations of C (26–41%) and N (30–58%) in the mulch did vary with date (p = 0.001 for both C and N) but did not differ with treatment. From establishment to Year 1 of Rotation 2, the mulch material decreased by an average of 57%, from ~120 to 55 Mg ha^{-1}, and the mulch layer N content also declined during this period, from ~370 to 160 kg ha^{-1} (Figure 7).

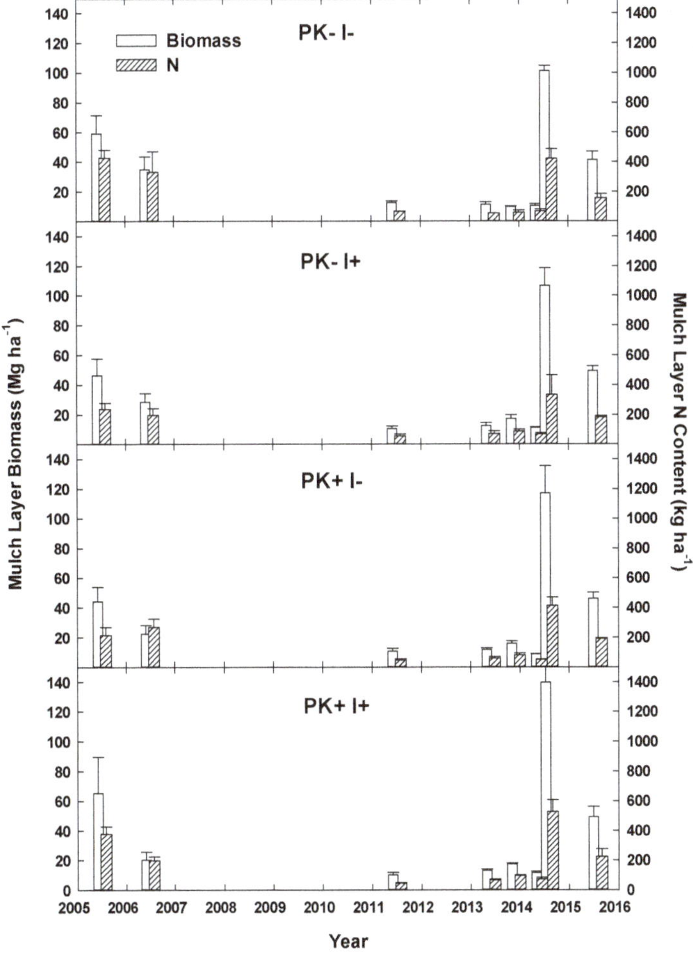

Figure 7. Mulch layer mass and N content of an improved-fallow slash-and-mulch agroforestry system in Eastern Amazonia of Brazil. Site establishment via mulching tractor took place in June 2005 and was abandoned to secondary succession in 2007 before the second rotation was established via mulching tractor in July 2014.

3.5. Soil

Soil N concentration at a 0–10 cm depth increased slightly during secondary succession from 2013 to 2014 from 0.09 to 0.10% ($p = 0.14$) but decreased to 0.08% ($p = 0.002$) one year after mulching for Rotation 2. There were no significant changes in soil C at 0–10 cm but soil C/N decreased during secondary succession from 2013 to 2014 (Tables S7 and S8) from 16.1 to 14.2 ($p = 0.02$) and increased to 18.5 ($p < 0.0001$) one year after mulching for Rotation 2.

Soil C and N concentrations differed at 0–10 and 10–20 cm depths during the pre- and post-mulching sampling dates, and trends of higher soil C and lower soil N were observed during Rotation 2 sampling in July 2015 with the inclusion of *I. edulis* in the planting mix (I+; Table 3 and Table S9). The sum of soil C content across both depths was not different between May 2014 and July 2015 ($p = 0.3$), or with treatment on either date. The sum of soil N content was 20% greater with *Inga* (I+) than without (I−) in May 2014 ($p = 0.02$) but not in July 2015 ($p = 0.9$).

Table 3. Soil C and N concentrations by depth increments in an improved-fallow slash-and-mulch agroforestry system in Eastern Amazonia, Brazil. Site establishment via mulching tractor took place in June 2005 and was abandoned to secondary succession in 2007 before the second rotation was established via mulching tractor in July 2014.

Date	Fert	Nfix	Depth	C	N
			cm	\% (SE)	
	N	N	0–10	1.20 (0.07)	0.08 (0.00)
			10–20	0.80 (0.07)	0.06 (0.00)
	N	Y	0–10	1.08 (0.25)	0.11 (0.02)
			10–20	0.89 (0.08)	0.07 (0.00)
May 2014	Y	N	0–10	1.39 (0.03)	0.09 (0.01)
			10–20	0.81 (0.05)	0.06 (0.00)
	Y	Y	0–10	1.31 (0.07)	0.09 (0.01)
			10–20	0.84 (0.03)	0.06 (0.00)
	N	N	0–10	1.50 (0.08)	0.08 (0.00)
			10–20	0.77 (0.01)	0.04 (0.00)
	N	Y	0–10	1.34 (0.14)	0.07 (0.01)
			10–20	0.72 (0.03)	0.07 (0.03)
July 2015	Y	N	0–10	1.56 (0.19)	0.08 (0.01)
			10–20	0.73 (0.03)	0.04 (0.00)
	Y	Y	0–10	1.44 (0.07)	0.08 (0.00)
			10–20	0.86 (0.03)	0.05 (0.00)

Soil P concentrations were greater at the 0–10 cm depth in 2013 with P+K fertilization ($p = 0.02$), but not in 2014 or 2015, while soil K concentrations were not different with fertilization on any measurement date. Soil P concentrations were greater with *I. edulis* ($p = 0.09$) in 2013 and again in 2015 ($p = 0.08$), but not in 2014, while soil K concentration did not differ with *I. edulis* on any measurement date (Table 4 and Table S10).

Soil P concentration was a significant predictor of All-Species height, total biomass, and total N across the three sampling dates explaining 57%, 31%, and 30% of the variance (Table 5). Soil K or N were never included in the multivariate regressions for all species, while soil Ca and Al were included for height prediction (Table 5). On an individual species bases, soil P was the best predictor of biomass for *C. pentandra*, *P. multijuga*, and *S. amazonicum*, explaining 32%, 80%, and 20% of the variance, respectively (Table 6). For *I. edulis* soil Mg was the best predictor, and soil K was the best predictor for *C. odorata*, followed by soil C (Table 6).

Table 4. Soil P and K concentrations at the 0–10 cm horizon in a slash-and-mulch agroforestry system in Eastern Amazonia of Brazil; ** indicates statistical significance in the main-plot treatment within the year, and [+] indicates statistical significance in the sub-plot N-fixer treatment within the year.

Year	Treatment	P	K
		mg kg^{-1} (SE)	
2013	PK–I–	1.01 (0.02)	5.91 (0.48)
	PK–I+	1.04 (0.06) [+]	6.38 (0.84)
	PK+I–	1.25 (0.11) **	6.62 (1.01)
	PK+I+	1.53 (0.08) ** [+]	7.15 (0.37)
2014	PK–I–	0.82 (0.10)	12.47 (1.36)
	PK–I+	0.72 (0.06)	12.29 (1.93)
	PK+I–	1.06 (0.06)	13.31 (1.98)
	PK+I+	1.25 (0.05)	14.28 (2.41)
2015	PK–I–	0.99 (0.04)	5.86 (0.68)
	PK–I+	1.11 (0.07) [+]	6.94 (0.83)
	PK+I–	1.31 (0.16)	6.75 (0.75)
	PK+I+	1.41 (0.06) [+]	5.82 (0.37)

Table 5. Predictors of height, biomass and N content after the first rotation for five native tree species grown in a mixed-culture, crop-fallow agroforestry system in eastern Amazonia of Brazil. Trees and crops were initially planted in 2005, abandoned to secondary succession in 2007 until 2014, when the second rotation of the crop–fallow system was planted and subsequently abandoned to secondary succession in 2016.

Dependent variable	Independent Variable	Sample Date	Partial R^2	Model R^2	C(p)	F Value	Pr > F
Height	Soil P	5/2014 [1]	0.481	0.481	−1.673	13.0	0.003 *
	Soil P	7/2015 [2]	0.591	0.591	−0.791	20.2	0.0005 *
	Soil Al		0.059	0.650	−0.405	2.2	0.163
	Soil P	By Date [3]	0.566	0.566	15.881	18.3	0.0008 *
	Soil Ca		0.225	0.791	3.428	14.0	0.003 *
	Soil Al		0.057	0.848	1.775	4.5	0.056 †
Sum Biomass	Soil P	5/2014	0.238	0.238	−2.925	14.4	0.0004 *
	Soil P	7/2015	0.128	0.128	−2.579	10.3	0.002 *
	Soil P		0.312	0.312	−2.572	21.4	<0.0001 *
Sum N	Soil P	5/2014	0.209	0.209	−3.564	12.2	0.001 *
	Soil P	7/2015	0.114	0.114	−2.986	9.0	0.004 *
	Soil P	By Date	0.302	0.302	−1.967	20.3	<0.0001 *

[1] Sampling Date 5/2014 was the final sampling event of Rotation 1, 9 Years after site establishment; [2] Sampling Date 7/2015 was Year 1 sampling event of Rotation 2, 10 years after site establishment; [3] By Date represents regression analyses of 3 sampling events in 7/2013, 5/2014, and 7/2015. * indicates significant difference at the $p < 0.05$ level; † indicates significant difference at the $p < 0.10$ level.

Table 6. Multivariate regression analysis of biomass and N Content of five species of native trees grown in a mixed-culture, crop-fallow agroforestry system in eastern Amazonia of Brazil. Trees and crops were initially planted in 2005, abandoned to secondary succession in 2007 until 2014, when the second phase of the crop–fallow system was planted and subsequently abandoned to secondary succession in 2016.

Dependent Variable	Species	Independent Variable	Model R^2	C(p)	F Value	Pr > F
Sum Biomass	*Cedrela odorata*	Soil K	0.533	5.3	11.4	0.007
		Soil C	0.604	5.2	1.6	0.2
	Ceiba pentandra	Soil P	0.323	15.3	5.7	0.03
		Soil K	0.496	10.9	3.8	0.08
	Inga edulis	Soil Mg	0.596	9.1	2.5	0.15
		Soil P	0.921	.	70.4	<0.001
		Soil N	0.955	.	3.7	0.11
	Parkia multijuga	Soil P	0.802	.	16.2	0.02
		Soil N	0.972	.	18.2	0.02
		Soil K	0.999	.	71.3	0.01
		Soil Al	1.000	.	44.0	0.095
	Schizolobium amazonicum	Soil P	0.198	8.2	1.7	0.2
Sum Biomass N	*Cedrela odorata*	Soil K	0.455	7.2	8.3	0.02
		Soil C	0.555	6.4	2.0	0.2
	Ceiba pentandra	Soil P	0.323	15.4	5.7	0.03
		Soil K	0.496	10.9	3.8	0.08
		Soil Mg	0.596	9.1	2.5	0.15
	Inga edulis	Soil P	0.919	.	68.4	<0.001
		Soil N	0.948	.	2.8	0.16
	Parkia multijuga	Soil P	0.675	.	8.3	0.05
		Soil N	0.935	.	12.0	0.04
		Soil K	0.975	.	3.2	0.2
	Schizolobium amazonicum	Soil P	0.198	8.2	1.7	0.2

Multivariate regression analyses of soil and litter macronutrient concentrations showed that P, K, and N concentrations and C/N ratio were significant predictors of GLD, DBH, and height (Table 7). The most frequent significant regressor for growth measurements for individual tree species was soil P concentration (10 times), followed by C/N ratio (6), and soil K concentration (3).

Table 7. Multivariate regression analysis of soil and mulch layer nutrient concentrations and contents against planted tree GLD, DBH, and height in an improved-fallow slash-and-mulch agroforestry systems in Eastern Amazonia, Brazil. Rotation 1 extended from site preparation and planting in 2005 until 2014, and Rotation 2 measurements extended from clearing of secondary forest and re-planting the site in 2014 until the final measurement in 2015.

Species	Variable	Rotation		GLD Parameter Estimate	GLD Partial R^2	GLD F Value	GLD Pr > F	DBH Parameter Estimate	DBH Model R^2	DBH F Value	DBH Pr > F	Height Parameter Estimate	Height Model R^2	Height F Value	Height Pr > F
Cedrela odorata	Soil	1	C:N	0.101	0.377	6.7	0.03								
		1	P (%)					1.410	0.349	5.4	0.04				
		1	Ca (%)					1.375	0.192	3.8	0.08				
		2	K (%)	0.290	0.312	6.3	0.02								
		2	C:N					0.743	0.556	10.0	0.01				
	Mulch	1 2	C:N	0.004	0.261	5.0	0.04					0.018	0.436	10.8	0.005
Ceiba pentandra	Soil	1	P (%)	1.361	0.605	19.9	0.0006	3.308	0.712	32.2	<.0001	2.340	0.702	30.7	<.0001
		2	P (%)	1.774	0.312	6.4	0.02	2.095	0.244	4.5	0.05	1.871	0.309	6.3	0.03
	Mulch	1	P (%)	1.496	0.492	12.6	0.004	2.999	0.394	8.5	0.01	2.060	0.566	17.0	0.001
		2	C:N									0.013	0.098	3.5	0.09
Inga edulis	Soil	1	P (%)	1.969	0.795	23.2	0.003	1.683	0.353	30.8	0.003	1.109	0.721	15.5	0.008
		1	K (%)	−1.861	0.106	5.3	0.07								
		2	C:N	0.036	0.706	14.4	0.009	0.219	0.571	0.6	0.05	−0.249	0.805	24.8	0.003
		2	Ca (%)					−0.971	0.246	0.8	0.08				
	Mulch	1	P (%)					2.275	0.452	18.4	0.008				
		1	K (%)					−1.858	0.425	4.4	0.08				
		2	C:N	0.002	0.787	22.1	0.003					−0.016	0.832	29.6	0.002
Parkia multijuga	Soil	1	P (%)					0.318	0.147	6.5	0.06				
		1	Mg (%)					0.298	0.074	14.0	0.03				
		1	Al (%)					−0.842	0.764	16.1	0.01				
		2	Al (%)	−1.483	0.345	7.4	0.02	−3.364	0.301	6.0	0.03	−2.029	0.286	5.6	0.03
		2	K (%)					2.369	0.283	8.9	0.01	1.178	0.164	3.9	0.07
Schizolobium amazonicum	Soil	1	P (%)	0.591	0.770	43.4	<.0001	3.018	0.636	22.7	0.0004	4.749	0.737	36.5	<0.0001
		1	C:N	0.109	0.057	4.0	0.07					0.258	0.061	3.6	0.08
		2	P (%)	0.291	0.589	20.1	0.0005	0.935	0.155	4.4	0.06	1.824	0.579	19.2	0.0006
		2	C (%)	0.339	0.167	8.9	0.01								
		2	N (%)					1.405	0.419	9.4	0.009	0.997	0.113	4.7	0.05
	Mulch	1	P (%)	3.433	0.398	8.6	0.01					4.057	0.416	9.3	0.009
		2	C:N	−0.017	0.329	6.9	0.02					−0.022	0.379	8.6	0.01

4. Discussion

4.1. Survival

Fertilization reduced the survival when pooled across the five planted tree species and reduced the survival of *C. odorata* and *P. multijuga* by the end of Rotation 1. Competing vegetation from secondary forest succession can suppress the survival of planted trees, and stimulated growth by P+K fertilization can enhance this competitive interaction [45]. For example, *C. odorata*, which suffered nearly 50% reduction of survival with P+K fertilization, was likely unable to escape light competition due to increased competing vegetation growth with fertilization. On the other hand, *S. amazonicum* survival was enhanced by over 30% with P+K fertilization, indicating that it was able to outgrow competitors with fertilization.

The presence of *I. edulis* reduced the survival of *S. amazonicum* to 0% in PK−I+ and that of *C. pentandra* to 0% in PK+I+ by the end of Rotation 1. Despite potential N benefits from N fixation, *Inga* is also a competitor with neighboring trees for light, water, and nutrient resources. For example, *Inga* suppressed secondary vegetation in a Costa Rican AFS [46], and in tropical secondary forests, evidence indicates that N-fixing species survive better than non-fixers in young (i.e., 10–11-year-old) forests [47]. Annualized mortality rates reported here ranged from 6% to 9% over Rotation 1, which were similar to the 7% mortality rate of non-fixing trees described [47]. Evidence presented here indicates that the presence of the N-fixing *I. edulis* can affect individual species survival.

The survival of planted tree species was greater 20 months after planting in the second rotation, relative to survival 24 months after site establishment of Rotation 1. During the second rotation, nearly all *C. odorata* individuals were attacked by *Hypsipyla grandella* insects at the meristems, causing deformation of the main stem and open wounds, both of which likely increased mortality. Other, unknown insects attacked *C. pentandra*, causing wounds on the stem, which sometimes led to meristem death and, in some cases, tree death.

4.2. Planted Tree Growth

Fertilization with P+K increased planted tree growth rates and size through Year 6 [22]. During the final year of Rotation 1 (Year 9), however, although trees were still larger in the presence of P+K fertilization, there was no longer a significant effect on growth rates in most cases. This is consistent with other research from Eastern Amazonia that showed that P and/or K fertilization increased secondary forest growth [45,48]. The presence of the N-fixing *I. edulis* was not associated with increased growth, survival, or volume of any other planted tree species. As such, no companion tree benefit can be ascribed to planting of this N-fixer. We had hypothesized that root and foliar turnover, particularly over the course of 9 years, would increase soil N and accelerate plant growth, but this was not supported by our data, though it may also indicate that P or K are more limiting nutrients in this system [45]. One premise that supports the use of many forms of AFS is that the use of N fixers increases N availability in an agroecosystem, so these results are somewhat surprising. However, other recent research has similarly indicated that N-fixing trees may neither benefit nor inhibit neighboring trees [18] and may not confer a competitive or facilitative effect on secondary growth [19]. It may be necessary to convert N fixers into green manure to capture N benefits.

Fertilization in Rotation 1 had a residual effect, as shown by increased seedling growth during the first 20 months of Rotation 2. Competing vegetation was severely reduced with mulching, and reduced light and rooting competition during the initial months of Rotation 2 may have improved conditions for initial seedling growth, as was the case during the first two years of Rotation 1, when the fertilization effect was greater. Although competition control is likely to have benefitted planted tree growth performance during the first 20 months of the second rotation in all treatments, P+K fertilization was shown to have legacy effects on planted tree growth for up to 20 years [26].

Higher soil P enhanced tree growth as a whole and increased the N content of planted trees. Fertilizer P can also enhance N-fixation in symbiotic N-fixing legumes [49], although the concentrations of wood and leaf N in *I. edulis* reported here were lower than in other studies [50]. At the end of Rotation 1, *S. amazonicum* was the dominant tree when present, accounting for 50–70% of planted tree N.

4.3. Plant and Soil N Stocks

Agroecosystem N stock in trees, manioc, mulch, and 0–20 cm soil compartments contained as much as 2500 kg N ha^{-1} at the end of the first Rotation (Figure 7). The N stock decreased from the last sampling event prior to site conversion via mulching tractor in March 2014 to the Year 1 sampling of Rotation 2 (Figure 7), with 4–30% (PK−I− and PK+I+, respectively) more N at the end of Rotation 1. Between 3% and 19% of the mean difference (422 kg ha^{-1}) was contained in the planted trees, 0.2% of which was exported with *S. amazonicum* stems in the PK−I− treatment and 2.8% in the PK+I− treatment, and between 0.06% and 0.12% was lost as N$_2$O gas [51]. Additional 0.9 and 18 kg ha^{-1} N (PK−I− and PK+I+, respectively) were removed in the first manioc harvest. The 136–215 kg N ha^{-1} accumulated by the improved-fallow vegetation during Rotation 1, as compared to the secondary vegetation prior to site preparation, was more than adequate to compensate for N lost to *S. amazonicum* timber export and manioc export.

Soil N content in the 0–10 cm depth ranged from 850 to 1300 kg ha^{-1}, which was similar to the 1000–1250 kg ha^{-1} of soil N contents previously reported from soils at this same research station [23]. Declines of N content in the 0–10 cm soil horizon from pre- to post-mulching, although not statistically significant, were about equal to the gain of N in the mulch horizon during that time. This could indicate that N was immobilized by the high C/N ratio woody mulch material added to the soil surface. However, leaching losses of mobile NO$_3^-$ beyond the surface rooting zone after forest disturbance are also possible [52], although any inputs from the surface horizon were not recovered in the 10–20 cm horizon as higher reported soil N.

4.4. Mulch Layer and Secondary Succession Biomass

The 7-year-old secondary forest that was converted to AFS in this research project produced 53.8 ± 5.1 Mg ha^{-1} of mulched biomass or 7.7 ± 0.7 Mg ha^{-1} yr^{-1} of secondary forest biomass accumulation prior to initial site conversion. This mass is comparable to the mean secondary vegetation of an Amazon basin-wide sampling study (53 Mg ha^{-1}; [53]) and slightly higher than the mean annual accumulation determined in a secondary forest study from the same municipality (6.6 Mg ha^{-1} yr^{-1}; [54]). Implementation of the improved-fallow slash-and-mulch AFS appeared to enhance secondary forest biomass growth by 116% relative to unimproved fallow preceding it. The improved fallow enhanced fallow vegetation growth to 12.9 ± 1.0 Mg ha^{-1} yr^{-1}, an annual rate increase of 60%.

Increases in improved-fallow biomass relative to secondary forest prior to site preparation resulted in mulched fallow biomass that was 13% greater with *I. edulis* (I +) than without it (I −) at site conversion for Rotation 2. This result is consistent with short-fallow AFS in Central Amazonia [55], although the increase was not as large as that reported for *I. edulis* in improved-fallow plantings in Eastern Amazonia (46%; [47]). Increases in fallow biomass were not necessarily a response to increased soil N, as planted trees did not grow larger in the presence of *I. edulis*, but *I. edulis* itself was a good grower and biomass accumulator. Increased planted tree biomass was related to available soil P, as both the height and biomass of planted trees increased with soil P, which ranged only from 4 to 6 µg dm^{-3}. In a secondary forest chronosequence study, [56] the suggested tighter N cycling in <10-year-old forests indicated greater N limitation to growth. In our study, however, soil N was a significant predictor of growth only for *I. edulis*, but here it is difficult to know if greater soil N with larger *I. edulis* is responsible for growth or rather a response to it. Fertilization studies in 6- [44] and 26-year-old secondary forests [57], also in the state of Pará, similarly demonstrated no tree growth response to P fertilization alone. In our study, P was always applied with K, and soil K was as significant predictor of growth for *C. odorata*,

C. pentandra, and *P. multijuga*. This study suggests tree growth was improved with P fertilization and in the presence of higher soil P.

4.5. Manioc

Manioc growth was also improved with higher site P. In fact, total manioc biomass and N content were higher in the second rotation than in the first one across all treatments, and P+K fertilization increased manioc biomass and N content in all biomass compartments. These increases indicate that improved-fallow slash-and-mulch system was able to efficiently accumulate nutrients to sustain crop growth. This conclusion is supported by the increase in mulch layer biomass, which reflects secondary vegetation biomass accumulation between mulching events 1 and 2. The associated increase in mulch layer nutrients, even in the control treatment, exceeded and thus could sustain, nutrient export for manioc root and *S. amazonicum* stem harvests. Fertilization may not have stimulated manioc growth in Year 10 as much as in Year 1, but its effect on manioc growth was still evident 10 years after application.

5. Conclusions

Results from this research indicate that the simultaneous planting of the staple crop manioc with native tree species, in conjunction with slash-and-mulch site preparation, can yield two crop harvests and a merchantable timber harvest through the course of one crop–fallow rotation with P + K fertilization, while leaving other tree species for harvest in future crop–fallow cycles. This combination of site preparation techniques may give producers who utilize agroforestry systems multiple points of income (e.g., during the first two crop harvests at Years 1 and 2 and at the conclusion of the fallow phase at Year 9, in this case), as opposed to only generating income with crop harvests, which is typical without simultaneous improved-fallow plantings.

The improved-fallow design increased the annualized fallow biomass accumulation during Rotation 1, and the increased biomass provided enough nutrients to replace those exported via crop and timber export. The slash-and-mulch site preparation, when combined with improved-fallow planting, can sustain, and even increase, nutrient stocks as compared to nutrient losses prevalent in slash-and-burn site preparation, even with crop export at 12 and 20 months and timber export at Year 9. The inclusion of rapid-growing pioneer species such as *S. amazonicum* and *I. edulis* in an improved-fallow design with P + K fertilization will allow producers to harvest manioc and merchantable timber after one crop–fallow rotation, while returning enough nutrients to the agroecosystem via green manure to sustain crop harvest.

After the end of a full rotation and the initiation of a second, this research justifies the use of the leguminous species *S. amazonicum* and *Parkia multijuga* in a mixed-species planting design, with the inclusion of *I. edulis* for use as green manure. Harvest of P+K fertilized *S. amazonicum* at the end of Rotation 1, mulching, and then re-planting *I. edulis* and leaving *P. multijuga* for harvest at the end of Rotation 2 are supported by this study. Due to widespread pest attacks which caused deformation of the stems, and therefore loss of timber value, the results of this study cannot justify the inclusion of the slower growing, although higher value, timber species *C. odorata* or *C. pentandra*.

Supplementary Materials: The following are available online at http://www.mdpi.com/1999-4907/10/12/1125/s1, Figure S1: Mulch layer C content (kg ha^{-1}) of an improved-fallow slash-and-mulch agroforestry system in Eastern Amazonia of Brazil., Figure S2: Soil-C content by depth increments 0–10 and 10–20 cm in an improved-fallow slash-and-mulch agroforestry system in Eastern Amazonia, Brazil., Table S1: Soil particle-size distribution at different depths in an improved-fallow slash-and-mulch agroforestry system in Eastern Amazonia, Brazil., Table S2: Differences in survival of five species of native trees after 24 months after planting of Rotation 1 and newly planted trees at 20 months of Rotation 2 of a crop-fallow agroforestry system in eastern Amazonia of Brazil., Table S3: Statistical contrasts of Height, DBH, and GLD of five species of native trees grown in mixed-culture, crop-fallow agroforestry system in eastern Amazonia of Brazil., Table S4: Sum of estimated total planted-tree aboveground biomass (kg) by plot and total biomass-N of planted-trees (g), and Percent (%) by Species of Total Biomass and Total N, at four different dates., Table S5: Statistical output of *Manihot esculenta* biomass and N content (kg ha^{-1}) by root, stem, leaf compartments, and sum of all compartments, measured at Year 1 after establishment of Rotation 2 (2015) of crop-fallow system in Eastern Amazonia of Brazil., Table S6: Statistical output of Mulch layer Mass, N and C concentration, and N and C content of an improved-fallow slash-and-mulch agroforestry system in Eastern Amazonia of Brazil, Table S7: Statistical output of significant differences for plot-level soil sampling at 0 – 10 cm depth of an improved-fallow slash-and-mulch agroforestry system in Eastern Amazonia, Brazil., Table S8: Soil N and C concentrations (%) in the 0–10 and 10–20 cm depths at three different years during the first rotation of a slash-and-mulch agroforestry system in eastern Amazonia of Brazil., Table S9: Statistical output of significant differences for plot-level soil sampling at 0–10 vs. 10–20 cm depths of an improved-fallow slash-and-mulch agroforestry system in Eastern Amazonia, Brazil., Table S10: Statistical significance of P and K concentrations and content of Manioc by compartment, Mulch layer, and 0–10 cm Soil horizon.

Author Contributions: Conceptualization, A.H.J., O.R.K., L.M., and D.M.; Data curation, A.H.J.; Formal analysis, A.H.J., and D.M.; Funding acquisition, O.R.K., L.M., and D.M.; Investigation, A.H.J., L.M., and D.M.; Methodology, A.H.J., O.R.K., L.M., and D.M.; Project administration, A.H.J., S.S.V., F.d.A.O., O.R.K., L.M., and D.M.; Resources, S.S.V., F.d.A.O., O.R.K., L.M., and D.M.; Supervision, F.d.A.O., O.R.K., L.M., and D.M.; Writing—Review & editing, A.H.J., L.M., and D.M.

Funding: This research was funded by: Warnell School of Forestry & Natural Resources, University of Georgia, Athens, GA, USA; Tinker Foundation, Graduate Student Summer Research Travel Awards; and The Climate Food and Farming (CLIFF) Research Network, CGIAR—Climate Change, Agriculture and Food Security (CCAFS), PhD Research Grants for Climate Change Mitigation in Smallholding Agriculture.

Acknowledgments: I would like to thank Lori Sutter for her help in the lab, as well as my fellow graduate students Uma Nagendra, Ross Pringle, Rachel Ryland, Maryam Foroughi, Diego Barcellos, Karla McGee, and Peter Baas for their encouragement, inspiration, stimulation, and assistance. I want to thank meu amigo Livy Madeira for our spirited exchanges about science and knowledge in general, and him and his band, Alpha Doze, for letting me make my international rock n' roll debut. I would also like to thank Vermelha, the FEIGA farm dog, for her company, carinhos, and for always being super excited to see me when I would come to visit.

Conflicts of Interest: The authors declare no conflict of interest. Additionally, the funders had no role in the design of the study; in the collection, analyses, or interpretation of data; in the writing of the manuscript, or in the decision to publish the results.

References

1. World Agroforestry Centre. *Annual Report 2008–2009: Agroforestry—A Global Land Use*; World Agroforestry Centre: Nairobi, Kenya, 2009.

2. Leakey, R.R.B. *Living with the Trees of Life: Towards the Transformation of Tropical Agriculture*; CABI: Wallingford, UK, 2012; p. 200.

3. Brondizio, E.S.; Cak, A.; Caldas, M.M.; Mena, C.; Bilsborrow, R.; Futemma, C.T.; Ludewigs, T.; Moran, E.F.; Batistella, M. Small farmers and deforestation in Amazonia. In *Amazonia and Global Change*; Keller, M., Bustamante, M., Gash, J., Dias, P.S., Eds.; American Geophysical Union: Washington, DC, USA, 2009; pp. 117–143. Available online: https://www.alice.cnptia.embrapa.br/handle/doc/663208 (accessed on 12 November 2019).

4. Denich, M.; Vielhauer, K.; Kato, M.S.A.; Block, A.; Kato, O.R.; Sá, T.D.A.; Lücke, W.; Vlek, P.L.G. Mechanized land preparation in forest-based fallow systems: The experience from Eastern Amazonia. *Agrofor. Syst.* **2004**, *61*, 91–106.

5. Metzger, J.P. Landscape dynamics and equilibrium in areas of slash-and-burn agriculture with short and long fallow period (Bragantina region, NE Brazilian Amazon). *Landsc. Ecol.* **2002**, *17*, 419–431. [CrossRef]

6. Buschbacher, R.; Uhl, C.; Serrão, E.A.S. Abandoned pastures in Eastern Amazonia II: Nutrient stocks in the soil and vegetation. *J. Ecol.* **1998**, *76*, 682–699. [CrossRef]

7. Sanchez, P.A. Improved fallows come of age in the tropics. *Agrofor. Syst.* **1999**, *47*, 3–12. [CrossRef]

8. Szott, L.T.; Palm, C.A. Nutrient stocks in managed and natural humid tropical fallows. *Plant Soil* **1996**, *186*, 293–309. [CrossRef]

9. Barrios, E.; Cobo, J.G. Plant growth, biomass production and nutrient accumulation by slash/mulch agroforestry systems in tropical hillsides of Colombia. *Agrofor. Syst.* **2004**, *60*, 255–265. [CrossRef]

10. Basamba, T.A.; Barrios, E.; Singh, B.R.; Rao, I.M. Impact of planted fallows and a crop rotation on nitrogen mineralization and phosphorus and organic matter fractions on a Colombian volcanic-ash soil. *Nutr. Cycl. Agroecosystems* **2007**, *7*, 127–141. [CrossRef]

11. Pinedo-Vasquez, M.; Zarin, D.J.; Coffey, K.; Padoch, C.; Rabelo, F. Post-boom logging in Amazonia. *Hum. Ecol.* **2001**, *29*, 219–239. [CrossRef]

12. Joslin, A.; Markewitz, D.; Morris, L.; Kato, O.R.; Figueiredo, R.; Oliveira, F.A. Crescimento de Cinco Espécies Nativas em Successão Natural na Amazônia Oriental. In Proceedings of the VIII Congresso Brasileiro de

Sistemas Agroflorestais, Belém, Brasil, 21–25 November 2011; Available online: http://www.alice.cnptia.embrapa.br/alice/handle/doc/910528 (accessed on 12 Nov., 2019).

13. Brienza, S., Jr. Biomass dynamics of fallow vegetation enriched with leguminous trees in the Eastern Amazon of Brazil. PhD. Thesis, University of Göttingen, Göttingen, Germany, 1999.

14. Batterman, S.A.; Hedin, L.O.; van Breugel, M.; Ransijn, J.; Craven, D.J.; Hall, J.S. Key role of symbiotic dinitrogen fixation in tropical forest secondary succession. *Nature* **2013**, *502*, 224–227. [CrossRef]

15. Bauhus, J.; Khanna, P.K.; Menden, N. Aboveground and belowground interactions in mixed plantations of *Eucalyptus globulus* and *Acacia mearnsii*. *Can. J. For. Res.* **2000**, *30*, 1886–1894. [CrossRef]

16. Iglesias, L.; Salas, E.; Leblanc, H.A.; Nygren, P. Response of *Theobroma cacao* and *Inga edulis* seedlings to cross-inoculated populations of arbuscular mycorrhizal fungi. *Agrofor. Syst.* **2011**, *83*, 63–73. [CrossRef]

17. Jalonen, R.; Nygren, P.; Sierra, J. Transfer of nitrogen from a tropical legume to an associated fodder grass via root exudation and common mycelial networks. *Plantcell Environ.* **2009**, *32*, 1366–1376. [CrossRef] [PubMed]

18. Taylor, B.N.; Chazdon, R.L.; Bachelot, B.; Menge, D.N.L. Nitrogen-fixing trees inhibit growth of regenerating Costa Rican rainforests. *Proc. Natl. Acad. Sci. USA* **2017**, *114*, 8817–8822. [CrossRef] [PubMed]

19. Lai, H.R.; Hall, J.S.; Batterman, S.A.; Turner, B.L.; van Breugel, M. Nitrogen fixer abundance has no effect on biomass recovery during tropical secondary forest succession. *J. Ecol.* **2018**, *106*, 1415–1427. [CrossRef]

20. Kato, M.S.A.; Kato, O.R.; Denich, M.; Vlek, P.L.G. Fire-free alternatives to slash-and-burn for shifting cultivation in the eastern Amazon region: The role of fertilizers. *Field Crop. Res.* **1999**, *62*, 225–237. [CrossRef]

21. Joslin, A.H.; Markewitz, D.; Morris, L.A.; Oliveira, F.A.; Figueiredo, R.O.; Kato, O.R. Five native tree species and manioc under slash-and-mulch agroforestry in the eastern Amazon of Brazil: Plant growth and soil responses. *Agrofor. Syst.* **2011**, *81*, 1–14. [CrossRef]

22. Joslin, A.; Markewitz, D.; Morris, L.A.; Assis, F.A.; Kato, O.R. Improved fallow: Growth and nitrogen accumulation of five native tree species in Brazil. *Nutr. Cycl. Agroecosyst.* **2016**, *106*, 1–15. [CrossRef]

23. Hölscher, D.; Ludwig, B.; Müller, R.F.; Fölster, H. Dynamic of soil chemical parameters in shifting agriculture in the Eastern Amazon. *Agric. Ecosyst. Environ.* **1997**, *66*, 153–163. [CrossRef]

24. Fassbender, H.W.; Beer, J.; Heuveldop, J.; Imbach, A.; Enriquez, G.; Bonnemann, A. Ten year balances of organic matter and nutrients. *For. Ecol. Manag.* **1991**, *45*, 173–183. [CrossRef]

25. Joslin, A.H.; Markewitz, D.; Morris, L.A.; Oliveira, F.A.; Figueiredo, R.O.; Kato, O.R. Soil and plant N-budget one year after planting of a slash-and-mulch agroforestry system in the Eastern Amazon of Brazil. *Agrofor. Syst.* **2013**, *87*, 1339–1349. [CrossRef]

26. Pritchett, W.L.; Comerford, N.B. Long-term Response to Phosphorus Fertilization on Selected Southeastern Coastal Plain Soils. *Soil Sci. Soc. Am. J.* **1982**, *46*, 640–644. [CrossRef]

27. Rego, R.S.; Silvam, B.N.R.; Raimundo, S.O. Detailed soil survey in an area in the municipality of Igarapé Açu, Pará. In *Summaries of Lectures and Posters, Presented at the 1st SHIFT-Workshop, Belém, Brazil, 8–13 March 1993*; Junk, W.J., Bianchi, H., Eds.; Bundesministerium für Forschung und Technologie (BMFT): Geesthact, Germany, 1993.

28. Gee, G.W.; Bauder, J.W. Particle Size Analysis. In *Methods of Soil Analysis*, 2nd ed.; Klute, A., Ed.; Soil Science Society of America: Madison, WI, USA, 1986; pp. 383–411.

29. Kato, O.R.; Kato, M.S.A.; de Carvalho, C.R.; Figueiredo, R.O.; Sá, T.D.A.; Vielhauer, K.; Denich, M. Manejo de Vegetação Secundária na Amazônia Visando ao Aumento da Sustentabilidade do uso Agrícola do Solo. In Proceedings of the XXX Congresso Brasileiro de Ciência do Solo, Recife, Brazil, 12–22 July 2005.

30. United States Environmental Protection Agency (U.S. EPA). *Method 3050B: Acid Digestion of Sediments, Sludges, and Soils*, 2nd, ed.; United States Environmental Protection Agency: Washington, DC, USA, 1996.

31. Mehlich, A. *Determination of P, Ca, Mg, K, Na, NH_4*; Soil Testing Division Publication No. 1–53; North Carolina Department of Agriculture, Agronomic Division: Raleigh, NC, USA, 1953.

32. Fearnside, P. Wood density for estimating forest biomass in Brazilian Amazonia. *For. Ecol. Manag.* **1997**, *90*, 59–87. [CrossRef]

33. Keefe, K.; Schulze, M.D.; Pinheiro, C.; Zweede, J.C.; Zarin, D. Enrichment planting as a silvicultural option in the eastern Amazon: Case study of Fazenda Cauaxi. *For. Ecol. Manag.* **2009**, *258*, 1950–1959. [CrossRef]

34. Silva, A.K.L.; Vasconcelos, S.S.; de Carvalho, C.J.R.; Cordeiro, I.M.C.C. Litter dynamics and fine root production in Schizolobium parahyba var. amazonicum plantations and regrowth forest in Eastern Amazon. *Plant Soil* **2011**, *347*, 377–386. [CrossRef]

35. Poorter, L. The Relationships of Wood-, Gas- and Water Fractions of Tree Stems to Performance and Life History Variation in Tropical Trees. *Ann. Bot.* **2008**, *102*, 367–375. [CrossRef]

36. Cole, T.G.; Ewel, J.J. Allometric equations for four valuable tropical tree species. *For. Ecol. Manag.* **2006**, *229*, 351–360. [CrossRef]

37. Menalled, F.D.; Kelty, M.J. Crown structure and biomass allocation strategies of three juvenile tropical tree species. *Plant Ecol.* **2001**, *152*, 1–11. [CrossRef]

38. Dreschel, P.; Zech, W. Foliar nutrient levels of broad-leaved tropical trees: A tabular review. *Plant Soil* **1991**, *131*, 29–46.

39. Heitz, P.; Dünisch, O.; Wanek, W. Long-Term Trends in Nitrogen Isotope Composition and Nitrogen Concentration in Brazilian Rainforest Trees Suggest Changes in Nitrogen Cycle. *Environ. Sci. Technol.* **2010**, *44*, 1191–1196. [CrossRef]

40. Bayala, J.; Mando, A.; Teklehaimano, Z.; Ouedraogo, S.J. Nutrient release from decomposing leaf mulches of karité (*Vitellaria paradoxa*) and néré (*Parkia biglobosa*) under semI–arid conditions in Burkina Faso, West Africa. *Soil Biol. Biochem.* **2005**, *37*, 533–539. [CrossRef]

41. Ometo, J.P.H.B.; Ehleringer, J.R.; Domingues, T.F.; Berry, J.A.; Ishida, F.Y.; Mazzi, E.; Higuchi, N.; Flanagan, L.B.; Nardoto, G.B.; Martinelli, L.A. The stable carbon and nitrogen isotopic composition of vegetation in tropical forests of the Amazon Basin, Brazil. *Biogeochemistry* **2006**, *79*, 251–274. [CrossRef]

42. Kern, J.; Kreibich, H.; Koschorreck, M.; Darwich, A. Nitrogen Balance of a Floodplain Forest of the Amazon River: The Role of Nitrogen Fixation. In *Amazonian Floodplain Forests. Ecological Studies (Analysis and Synthesis)*; Junk, W., Piedade, M., Wittmann, F., Schöngart, J., Parolin, P., Eds.; Springer: Dordrecht, The Netherlands, 2010; Volume 210.

43. Da Costa, K.C.P.; Ferraz, J.B.S.; Bastos, R.P.; Ferreira, M.; Trinidade, A.S.; Piotto, G. Biomass and Nutrients in three species of Parkia plantings on degraded area in Central Amazon. *Florestas* **2014**, *44*, 637–646.

44. Yeboah, D.; Burton, A.J.; Astorer, A.J.; OpunI–Frimpong, E. Variation in wood density and carbon content of tropical plantation tree species from Ghana. *New For.* **2014**, *45*, 35–52. [CrossRef]

45. Davidson, E.A.; Carvalho, C.J.R.; Vieira, I.C.G.; Figueiredo, R.O.; Moutinho, P.; Ishida, F.Y.; dos Santos, M.T.P.; Guerrero, J.B.; Kalif, K.; Sabá, R.T. Nitrogen and phosphorus limitation of biomass growth in a tropical secondary forest. *Ecol. Appl.* **2004**, *14*, S150–S163. [CrossRef]

46. Kettler, J.S. Fallow enrichment of a traditional slash/mulch system in southern Costa Rica: Comparisons of biomass production and crop yield. *Agrofor. Syst.* **1996**, *2*, 165–176. [CrossRef]

47. Menge, D.N.L.; Chazdon, R.L. Higher survival drives the success of nitrogen-fixing trees through succession in Costa Rican forests. *New Phytol.* **2016**, *209*, 965–977. [CrossRef] [PubMed]

48. Rangel-Vasconcelos, L.G.T.; Kato, O.R.; Vasconcelos, S.S.; Oliveira, F.A. Phosphorus fertilization increases biomass and nutrient accumulation under improved fallow management in a slash-and-mulch system in Eastern Amazonia, Brazil. *Revista Brasileira de Ciência do Solo* **2017**, *41*, 1806–9657. [CrossRef]

49. Binkley, D.; Sencok, R.; Cromack, K., Jr. Phosphorus limitation on nitrogen fixation by *Falcataria* seedlings. *For. Ecol. Manag.* **2003**, *186*, 171–176. [CrossRef]

50. Lojka, B.; Preininger, D.; Lojkova, J.; Banout, J.; Polesny, Z. Biomass growth and farmer knowledge of *Inga edulis* in Peruvian Amazon. *Agric. Trop. Et Subtrop.* **2005**, *38*, 44.

51. Joslin, A.; Vasconcelos, S.S.; Oliveira, F.A.; Kato, O.R.; Morris, L.A.; Markewitz, D. Do Nitrogen-fixing Trees in an Improved-Fallow Slash-and-Mulch Agroforestry System affect Nitrous Oxide and Methane Efflux? *Glob. Chang. Biol.*. (under review).

52. Vitousek, P.M.; Gosz, J.R.; Grier, C.C.; Melillo, J.M.; Reiners, W.A. A Comparative Analysis of Potential Nitrification and Nitrate Mobility in Forest Ecosystems. *Ecol. Monogr.* **1982**, *52*, 155–177. [CrossRef]

53. Saatchi, S.S.; Houghton, R.A.; dos Santos Alvalá, R.C.; Soares, J.; Yu, Y. Distribution of aboveground live biomass in the Amazon basin. *Glob. Chang. Biol.* **2007**, *13*, 816–837. [CrossRef]

54. Davidson, E.A.; Sá, T.D.A.; Carvalho, C.J.R.; Figueiredo, R.O.; Kato, M.S.A.; Kato, O.R.; Ishida, F.Y. An integrated greenhouse gas assessment of an alternative to slash-and-burn agriculture in eastern Amazonia. *Glob. Chang. Biol.* **2008**, *14*, 998–1007. [CrossRef]

55. Staver, C. Shortened bush fallow rotations with relay-cropped *Inga edulis* and *Desmodium ovalifolium* in wet central Amazonian Peru. *Agrofor. Syst.* **1989**, *8*, 173–196. [CrossRef]

56. Davidson, E.A.; Carvalho, C.J.R.; Figueira, A.M.; Ishida, F.Y.; Ometto, J.P.H.B.; Nardoto, G.B.; Sabá, R.T.; Hayashi, S.N.; Leal, E.C.; Vieira, I.C.G.; et al. Recuperation of nitrogen cycling in Amazonian forests following agricultural abandonment. *Nature* **2007**, *447*, 995–998. [CrossRef] [PubMed]

57. Markewitz, D.; Figueiredo, R.O.; Carvalho, C.J.R.; Davidson, E.A. Soil and tree response to P fertilization in a secondary tropical forest supported by an Oxisol. *Biol. Fertil. Soils* **2012**, *48*, 665–678. [CrossRef]

MDPI

St. Alban-Anlage 66

4052 Basel

Switzerland

Tel. +41 61 683 77 34

Fax +41 61 302 89 18

www.mdpi.com

Forests Editorial Office

E-mail: forests@mdpi.com

www.mdpi.com/journal/forests

www.ingramcontent.com/pod-product-compliance
Lightning Source LLC
LaVergne TN
LVHW071518180726
843512LV00014B/1114